Modern Drying Technology

Edited by
Evangelos Tsotsas and Arun S. Mujumdar

1807–2007 Knowledge for Generations

Each generation has its unique needs and aspirations. When Charles Wiley first opened his small printing shop in lower Manhattan in 1807, it was a generation of boundless potential searching for an identity. And we were there, helping to define a new American literary tradition. Over half a century later, in the midst of the Second Industrial Revolution, it was a generation focused on building the future. Once again, we were there, supplying the critical scientific, technical, and engineering knowledge that helped frame the world. Throughout the 20th Century, and into the new millennium, nations began to reach out beyond their own borders and a new international community was born. Wiley was there, expanding its operations around the world to enable a global exchange of ideas, opinions, and know-how.

For 200 years, Wiley has been an integral part of each generation's journey, enabling the flow of information and understanding necessary to meet their needs and fulfill their aspirations. Today, bold new technologies are changing the way we live and learn. Wiley will be there, providing you the must-have knowledge you need to imagine new worlds, new possibilities, and new opportunities.

Generations come and go, but you can always count on Wiley to provide you the knowledge you need, when and where you need it!

William J. Pesce
President and Chief Executive Officer

Peter Booth Wiley
Chairman of the Board

Modern Drying Technology

Volume 1: Computational Tools at Different Scales

Edited by
Evangelos Tsotsas and Arun S. Mujumdar

WILEY-VCH Verlag GmbH & Co. KGaA

The Editors

Prof. Evangelos Tsotsas
Otto-von-Guericke-University
Thermal Process Engineering
Universitätsplatz 2
39106 Magdeburg
Germany

Prof. Arun S. Mujumdar
National University of Singapore
Mechanical Engineering/Block EA,07-0
9 Engineering Drive 1
Singapore 117576
Singapore

All books published by Wiley-VCH are carefully produced. Nevertheless, authors, editors, and publisher do not warrant the information contained in these books, including this book, to be free of errors. Readers are advised to keep in mind that statements, data, illustrations, procedural details or other items may inadvertently be inaccurate.

Library of Congress Card No.:
applied for

British Library Cataloguing-in-Publication Data
A catalogue record for this book is available from the British Library.

Bibliographic information published by the Deutsche Nationalbibliothek
The Deutsche Nationalbibliothek lists this publication in the Deutsche Nationalbibliografie; detailed bibliographic data are available on the Internet at http://dnb.d-nb.de.

© 2007 WILEY-VCH Verlag GmbH & Co. KGaA, Weinheim

All rights reserved (including those of translation into other languages). No part of this book may be reproduced in any form – by photoprinting, microfilm, or any other means – nor transmitted or translated into a machine language without written permission from the publishers. Registered names, trademarks, etc used in this book, even when not specifically marked as such, are not to be considered unprotected by law.

Typesetting Thomson Digital, India
Printing betz-druck GmbH, Darmstadt
Binding Litges & Dopf GmbH, Heppenheim
Wiley Bicentennial Logo Richard J. Pacifico

Printed in the Federal Republic of Germany
Printed on acid-free paper
ISBN 978-3-527-31556-7

Contents

Series Preface XI
Preface of Volume 1 XV
List of Contributors XXI
Recommended Notation XXV
EFCE Working Party on Drying: Address List XXXI

1   **Comprehensive Drying Models based on Volume Averaging: Background, Application and Perspective** *1*
     P. Perré, R. Rémond, I.W. Turner
1.1   Microscopic Foundations of the Macroscopic Formulation *1*
1.2   The Macroscopic Set of Equations *6*
1.3   Physical Phenomena Embedded in the Equations *7*
1.3.1   Low-temperature Convective Drying *7*
1.3.1.1   The Constant Drying Rate Period *8*
1.3.1.2   The Decreasing Drying Rate Period *9*
1.3.2   Drying at High Temperature: The Effect of Internal Pressure on Mass Transfer *10*
1.4   Computational Strategy to Solve the Comprehensive Set of Macroscopic Equations *11*
1.4.1   The Control-volume Finite-element (CV-FE) Discretization Procedure *13*
1.4.2   Evaluation of the Tensor Terms at the CV Face *14*
1.4.3   Solution of the Nonlinear System *15*
1.4.3.1   Outer (Nonlinear) Iterations *16*
1.4.3.2   Construction of the Jacobian *17*
1.4.3.3   Inner (Linearized System) Iterations *17*
1.5   Possibilities Offered by this Modeling Approach: Convective Drying *19*
1.5.1   High-temperature Convective Drying of Light Concrete *19*
1.5.1.1   Test 1: Superheated Steam *20*
1.5.1.2   Tests 2 and 3: Moist Air, Soft and Severe Conditions *22*

1.5.2	Typical Drying Behavior of Softwood: Difference Between Sapwood and Heartwood	25
1.6	Possibilities Offered by this Modeling Approach: Less-common Drying Configurations	29
1.6.1	Drying with Volumetric Heating	29
1.6.2	The Concept of Identity Drying Card (IDC)	32
1.6.3	Drying of Highly Deformable Materials	34
1.7	Homogenization as a Way to Supply the Code with Physical Parameters	37
1.8	The Multiscale Approach	42
1.8.1	Limitations of the Macroscopic Formulation	42
1.8.2	The Stack Model: An Example of Multiscale Model	43
1.8.2.1	Global Scale	46
1.8.2.2	Local Scale	46
1.8.2.3	Coupling Approach	46
1.8.2.4	Samples Simulations	47
1.8.2.5	Accounting for Wood Variability	49
1.8.2.6	Accounting for Drying Quality	50
	Conclusion	52

2 Pore-network Models: A Powerful Tool to Study Drying at the Pore Level and Understand the Influence of Structure on Drying Kinetics *57*

T. Metzger, E. Tsotsas, M. Prat

2.1	Introduction	57
2.2	Isothermal Drying Model	58
2.2.1	Model Description	58
2.2.1.1	Network Geometry and Corresponding Data Structures	59
2.2.1.2	Boundary-layer Modeling	60
2.2.1.3	Saturation of Pores and Throats	62
2.2.1.4	Vapor Transfer	63
2.2.1.5	Capillary Pumping of Liquid	64
2.2.1.6	Cluster Labeling	65
2.2.1.7	Drying Algorithm	66
2.2.2	Simulation Results and Experimental Validation	68
2.2.3	Gravity and Liquid Viscosity – Stabilized Drying Front	71
2.2.3.1	Modeling Gravity	71
2.2.3.2	Modeling Liquid Viscosity	72
2.2.3.3	Dimensionless Numbers and Length Scales	75
2.2.3.4	Phase Distributions and Drying Curves	77
2.2.4	Film Flow	79
2.2.5	Wettability Effects	83
2.2.6	First Drying Period	85
2.3	Model Extensions	87

2.3.1	Heat Transfer 87
2.3.2	Multicomponent Liquid 92
2.4	Influence of Pore Structure 92
2.4.1	Pore Shapes 92
2.4.2	Coordination Number 94
2.4.3	Bimodal Pore-size Distributions 95
2.4.4	Outlook 100
2.5	Towards an Assessment of Continuous Models 100

3   Continuous Thermomechanical Models using Volume-averaging Theory   103
F. Couture, P. Bernada, M. A. Roques

3.1	Introduction 103
3.2	Modeling 105
3.2.1	Nature of Product Class 106
3.2.2	Averaged Internal Equations 107
3.2.2.1	State Equations and Volume Conservation 108
3.2.2.2	Mass-conservation Equations 109
3.2.2.3	Momentum-conservation Equations 109
3.2.2.4	Energy-conservation Equations 112
3.2.3	Boundary Conditions for Convective Drying 113
3.3	Simulation 114
3.3.1	Numerical Resolution Technique 114
3.3.2	Comparison between Real Viscoelastic and Assumed Elastic Behavior 115
3.4	Liquid Pressure as Driving Force 120
3.5	Conclusions 122

4   Continuous Thermohydromechanical Model using the Theory of Mixtures   125
S. J. Kowalski

4.1	Preliminaries 125
4.2	Global Balance Equations 126
4.3	Constitutive Equations in the Skeletal Frame of Reference 130
4.4	Rate Equations for Heat and Mass Transfer 132
4.5	Differential Equations for Heat and Mass Transfer 134
4.5.1	Differential Equation for Heat Transfer 134
4.5.2	Determination of the Microwave Heat Source $\Re$ 135
4.5.3	Differential Equation for Mass Transfer 139
4.6	Thermomechanical Equations for a Drying Body 141
4.6.1	Physical Relations 141
4.6.2	Differential Equations for Body Deformation 143
4.7	Drying of a Cylindrical Sample made of Kaolin 144
4.7.1	Convective Drying of a Kaolin Cylinder 144
4.7.2	Microwave Drying of a Kaolin Cylinder 150

4.8	Final Remarks	*152*
	Acknowledgments	*152*
	Additional Notation used in Chapter 4	*153*

5 CFD in Drying Technology – Spray-Dryer Simulation *155*
S. Blei, M. Sommerfeld

5.1	Introduction	*155*
5.1.1	Introduction to CFD	*155*
5.1.2	Introduction to Multiphase Flow Modeling	*158*
5.1.3	State-of-the-art in Spray-dryer Computations	*160*
5.2	The Euler–Lagrange Approach: an Extended Model for Spray-dryer Calculations	*162*
5.2.1	Fluid-phase Modeling	*163*
5.2.2	Fundamentals of Lagrangian Particle Tracking	*166*
5.2.2.1	Drag Force	*167*
5.2.2.2	Virtual Mass Force	*168*
5.2.2.3	Basset History Force	*168*
5.2.2.4	Forces Caused by Pressure Gradients in the Fluid	*168*
5.2.2.5	Magnus Force	*168*
5.2.2.6	Saffman Force	*169*
5.2.2.7	Gravitational Force	*169*
5.2.3	Particle Tracking	*169*
5.2.4	Particle Turbulent Dispersion Modeling	*171*
5.2.5	Two-way Coupling Procedure	*173*
5.3	Droplet-drying Models	*173*
5.3.1	Introduction	*173*
5.3.2	Review of Droplet-drying Models	*175*
5.3.3	Exemplary Drying Model for Whey-based Milk Products	*176*
5.3.4	Numerical Implementation	*178*
5.4	Collisions of Particles	*181*
5.4.1	Introduction	*181*
5.4.2	Extended Stochastic Collision Model	*182*
5.4.3	Modeling of Particle Collisions: Coalescence and Agglomeration	*187*
5.4.3.1	Surface-tension Dominated Droplets (STD Droplets)	*187*
5.4.3.2	Droplets Dominated by Viscous Forces (VD Droplets)	*188*
5.4.3.3	Dry Particles	*189*
5.4.4	Collisions of Surface-tension Dominated Droplets (STD–STD)	*190*
5.4.5	Collisions of Viscous Droplets	*190*
5.4.6	Collisions of Dry Particles	*191*
5.5	Example of a Spray-dryer Calculation	*192*
5.5.1	Geometry and Spatial Discretization of the Spray Dryer	*192*
5.5.2	Results for the Fluid Phase	*193*
5.5.3	Results of the Dispersed Phase	*195*
5.6	Prediction of Product Properties	*200*
5.6.1	Particle-size Distribution	*200*

5.6.2	Heat Damage	*201*
5.6.3	Particle Morphology	*201*
5.7	Summary	*203*
	Additional Notation used in Chapter 5	*204*

6 Numerical Methods on Population Balances *209*
J. Kumar, M. Peglow, G. Warnecke, S. Heinrich, E. Tsotsas, L. Mörl, M. Hounslow, G. Reynolds

6.1	Introduction	*209*
6.2	Pure Breakage	*214*
6.2.1	Population-balance Equation	*214*
6.2.2	Numerical Methods	*214*
6.2.2.1	The Cell-average Technique	*216*
6.2.2.2	The Finite-volume Scheme	*222*
6.3	Pure Aggregation	*225*
6.3.1	Population-balance Equation	*225*
6.3.2	Numerical Methods	*226*
6.3.2.1	The Fixed-pivot Technique	*226*
6.3.2.2	The Cell-average Technique	*227*
6.3.2.3	The Finite-volume Scheme	*231*
6.4	Pure Growth	*233*
6.4.1	Population balance Equation	*233*
6.4.2	Numerical Methods	*233*
6.5	Combined Aggregation and Breakage	*239*
6.6	Combined Aggregation and Nucleation	*242*
6.7	Combined Growth and Aggregation	*244*
6.8	Combined Growth and Nucleation	*245*
6.9	Multidimensional Population Balances	*247*
6.9.1	Reduced Model	*247*
6.9.2	Complete Model	*250*
	Additional Notation used in Chapter 6	*256*

7 Process-systems Simulation Tools *261*
I. C. Kemp

7.1	Introduction	*261*
7.1.1	Summary of Contents	*261*
7.1.2	The Solids Processing Challenge	*262*
7.1.3	Types of Software for Dryers	*263*
7.2	Numerical Calculation Procedures	*263*
7.2.1	Categorization of Dryer Models	*264*
7.2.2	Equipment and Material Model	*265*
7.2.3	Parametric Models	*266*
7.3	Heat and Mass Balances	*268*
7.4	Scoping Design Methods	*269*
7.4.1	Continuous Convective Dryers	*269*

7.4.2	Continuous-contact Dryers	*270*
7.4.3	Batch Dryers	*270*
7.4.4	Simple Allowance for Falling-rate Drying	*271*
7.5	Scaling Methods	*272*
7.5.1	Basic Scale-up Principles	*273*
7.5.2	Integral Model	*274*
7.5.3	Application to Fluidized-bed Dryers	*274*
7.6	Detailed Design Models	*276*
7.6.1	Incremental Model	*277*
7.6.2	Application to Pneumatic Conveying, Rotary and Band Dryers	*278*
7.6.2.1	Pneumatic Conveying Dryers	*278*
7.6.2.2	Cascading Rotary Dryers	*281*
7.6.3	Advanced Methods – Computational Fluid Dynamics (CFD)	*281*
7.7	Ancillary Calculations	*283*
7.7.1	Processing Experimental Data	*283*
7.7.2	Humidity and Psychrometry	*284*
7.7.2.1	British Standard BS1339 for Humidity Calculations	*284*
7.7.2.2	Plotting Psychrometric Charts	*286*
7.7.3	Physical-properties Databanks	*286*
7.8	Process Simulators	*287*
7.8.1	Current Simulators and their Limitations	*287*
7.8.2	Potential Developments	*288*
7.9	Expert Systems and Decision-making Tools	*289*
7.9.1	Dryer Selection	*289*
7.9.1.1	Tree-search Algorithms	*289*
7.9.1.2	Matrix-type Rule-based Algorithms	*289*
7.9.1.3	Qualitative Information	*292*
7.9.1.4	Alternative Tree-search Approach	*292*
7.9.2	Troubleshooting and Problem Solving in Dryers	*294*
7.10	Knowledge Bases and Qualitative Information	*295*
7.10.1	Internet Websites	*295*
7.10.2	The Process Manual Knowledge Base	*295*
7.11	Commercialization of Drying Software	*296*
7.11.1	Barriers to Drying-software Development	*297*
7.11.1.1	Complexity of the Calculations	*297*
7.11.1.2	Difficulties in Modeling Solids	*297*
7.11.1.3	Limited Market and Lack of Replicability	*298*
7.11.1.4	Changes in Operating-system Software	*298*
7.11.2	The Future: Possible Ways Forward	*300*
7.12	Conclusions	*301*
7.12.1	Range of Application of Software in Drying	*301*
7.12.2	Overall Conclusion	*302*
	Additional Notation used in Chapter 7	*303*

Index *307*

Series Preface

The present series is dedicated to drying, i.e. to the process of removing moisture from solids. Drying has been conducted empirically since the dawn of the human race. In traditional scientific terms it is a unit operation in chemical engineering. The reason for the continuing interest in drying and, hence, the motivation for the series concerns the challenges and opportunities. A permanent challenge is connected to the sheer amount and value of products that must be dried – either to attain their functionalities, or because moisture would damage the material during subsequent processing and storage, or simply because customers are not willing to pay for water. This comprises almost every material used in solid form, from foods to pharmaceuticals, from minerals to detergents, from polymers to paper. Raw materials and commodities with a low price per kilogram, but with extremely high production rates, and also highly formulated, rather rare but very expensive specialties have to be dried.

This permanent demand is accompanied by the challenge of sustainable development providing welfare, or at least a decent living standard, to a still-growing humanity. On the other hand, opportunities emerge for drying, as well as for any other aspect of science or living, from either the incremental or disruptive development of available tools. This duality is reflected in the structure of the book series, which is planned for five volumes in total, namely:

Volume 1: Computational tools at different scales
Volume 2: Experimental techniques
Volume 3: Product quality and formulation
Volume 4: Energy savings
Volume 5: Process intensification

As the titles indicate, we start with the opportunities in terms of modern computational and experimental tools in Volumes 1 and 2, respectively. How these opportunities can be used in fulfilling the challenges, in creating better and new products, in reducing the consumption of energy, in significantly improving existing or introducing new processes will be discussed in Volumes 3, 4 and 5. In this sense, the first two volumes of the series will be driven by science; the last three will try to show how engineering science and technology can be translated into progress.

Modern Drying Technology. Edited by Evangelos Tsotsas and Arun S. Mujumdar
Copyright © 2007 WILEY-VCH Verlag GmbH & Co. KGaA. All rights reserved
ISBN: 978-3-527-31556-7

In total, the series is designed to have both common aspects with and essential differences from an extended textbook or a handbook. Textbooks and handbooks usually refer to well-established knowledge, prepared and organized either for learning or for application in practice, respectively. On the contrary, the ambition of the present series is to move at the frontier of "modern drying technology", describing things that have recently emerged, mapping things that are about to emerge, and also anticipating some things that may or should emerge in the near future. Consequently, the series is much closer to research than textbooks or handbooks can be. On the other hand, it was never intended as an anthology of research papers or keynotes – this segment being well covered by periodicals and conference proceedings. Therefore, our continuing effort will be to stay as close as possible to a textbook in terms of understandable presentation and as close as possible to a handbook in terms of applicability.

Another feature in common with an extended textbook or a handbook is the rather complete coverage of the topic by the entire series. Certainly, not every volume or chapter will be equally interesting for every reader, but we do hope that several chapters and volumes will be of value for graduate students, for researchers who are young in age or thinking, and for practitioners from industries that are manufacturing or using drying equipment. We also hope that the readers and owners of the entire series will have a comprehensive access not to all, but to many significant recent advances in drying science and technology. Such readers will quickly realize that modern drying technology is quite interdisciplinary, profiting greatly from other branches of engineering and science. In the opposite direction, not only chemical engineers, but also people from food, mechanical, environmental or medical engineering, material science, applied chemistry or physics, computing and mathematics may find one or the other interesting and useful results or ideas in the series.

The mentioned interdisciplinary approach implies that drying experts are keen to abandon the traditional chemical engineering concept of unit operations for the sake of a less rigid and more creative canon. However, they have difficulties of identification with just one of the two new major trends in chemical engineering, namely process-systems engineering or product engineering. Efficient drying can be completely valueless in a process system that is not efficiently tuned as a whole, while efficient processing is certainly valueless if it does not fulfil the demands of the market (the customer) regarding the properties of the product. There are few topics more appropriate in order to demonstrate the necessity of simultaneous treatment of product and process quality than drying. The series will try to work out chances that emerge from this crossroads position.

One further objective is to motivate readers in putting together modules (chapters from different volumes) relevant to their interests, creating in this manner individual, task-oriented threads trough the series. An example of one such thematic thread set by the editors refers to simultaneous particle formation and drying, with a focus on spray fluidized beds. From the point of view of process-systems engineering, this is process integration – several "unit operations" take place in the same equipment. On the other hand, it is product engineering, creating structures – in many cases nanostructures – that correlate with the desired application properties. Such properties are distributed over the ensemble (population) of particles, so that it is necessary to discuss mathematical methods (population balances) and numerical tools able to resolve the respective distributions in one chapter of Volume 1. Measuring techniques providing access to properties and states of the particle system will be treated in one chapter of Volume 2. In Volume 3, we will attempt to combine the previously introduced theoretical and experimental tools with the goal of product design. Finally, important issues of energy consumption and process intensification will appear in chapters of Volumes 4 and 5. Our hope is that some thematic combinations we have not even thought about in our choice of contents will arise in a similar way.

As the present series is a series of edited books, it can not be as uniform in either writing style or notation as good textbooks are. In the case of notation, a list of symbols has been developed and will be printed in the beginning of every volume. This list is not rigid but foresees options, at least partially accounting for the habits in different parts of the world. It has been recently adopted as a recommendation by the Working Party on Drying of the European Federation of Chemical Engineering (EFCE). However, the opportunity of placing short lists of additional or deviant symbols at the end of every chapter has been given to all authors. The symbols used are also explained in the text of every chapter, so that we do not expect any serious difficulties in reading and understanding.

The above indicates that the clear priority in the edited series was not in uniformity of style, but in the quality of contents that are very close to current international research from academia and, where possible, also from industry. Not every potentially interesting topic is included in the series, and not every excellent researcher working on drying contributes to it. However, we are very confident about the excellence of all research groups that we were able to gather together, and we are very grateful for the good cooperation with all chapter authors. The quality of the series as a whole is set mainly by them; the success of the series will primarily be theirs. We would also like to express our acknowledgements to the team of Wiley-VCH who have done a great job in supporting the series from the first idea to realization. Furthermore, our thanks go to Mrs Nicolle Degen for her additional work, and to our families for their tolerance and continuing support.

Last but not least, we are grateful to the members of the Working Party on Drying of the EFCE for various reasons. First, the idea about the series came up during the annual technical and business meeting of the working party 2005 in Paris. Secondly, many chapter authors could be recruited among its members. Finally, the Working Party continues to serve as a panel for discussion, checking and readjustment of our conceptions about the series. The list of the members of the working party with their affiliations is included in every volume of the series in the sense of acknowledgement, but also in order to promote networking and to provide access to national working parties, groups and individuals. The present edited books are complementary to the regular activities of the EFCE Working Party on Drying, as they are also complementary to various other regular activities of the international drying community, including well-known periodicals, handbooks, and the International Drying Symposia.

June 2007

Evangelos Tsotsas
Arun S. Mujumdar

Preface of Volume 1

As indicated in the general preface, Volume 1 of the "Modern Drying Technology" series is dedicated to "Computational Tools at Different Scales". It contains seven chapters, namely:

Chapter 1: Comprehensive drying models based on volume averaging: Background, application and perspective
Chapter 2: Pore-network models: A powerful tool to study drying at the pore level and understand the influence of structure on drying kinetics
Chapter 3: Continuous thermomechanical models using volume-averaging theory
Chapter 4: Continuous thermohydromechanical model using the theory of mixtures
Chapter 5: CFD in drying technology: Spray-dryer simulation
Chapter 6: Numerical methods on population balances
Chapter 7: Process-systems simulation tools

The choice of starting the series with the fundamentals reflects our opinion that optimal answers to the various challenges of modern industrial drying may require the use of one or more of the computational tools currently available, so that the nature, potential, restrictions and perspectives of such tools must be known and critically understood from the beginning. The severity of the problems to be solved by modeling and simulation is indicated in the title of Volume 1 by reference to the different scales of phenomena relevant to drying, which are:

- The molecular scale, where moisture molecules interact with each other, with further species in the liquid or in the gas, and with the surface of the solid;
- The pore scale, as the smallest topological entity for expressing the transport of momentum, mass and heat in the interior of drying particles or single bodies;
- The particle scale, which smears away local phenomena, but still lets us identify the single drying body on an individual basis;
- The particle-system scale, where equipment has to be designed and properly operated by not necessarily focusing on individuals, but still understanding how particles interact with each other, the gas flow, and the apparatus.
- The process-systems scale, which must work well as a whole in order to satisfy the demands of present and future markets.

Modern Drying Technology. Edited by Evangelos Tsotsas and Arun S. Mujumdar
Copyright © 2007 WILEY-VCH Verlag GmbH & Co. KGaA. All rights reserved
ISBN: 978-3-527-31556-7

Chapter 1 of this volume describes the transition from the pore scale to the particle scale by volume averaging. This transition starts with the continuous formulation of the conservation laws for mass, heat and momentum in microscopic, single-phase (solid, liquid, gas) domains. The subsequent volume averaging leads to, again, continuous equations that apply to the particle. The price of the transition is that the influence of microstructure is concealed and has to be *a posteriori* reconstructed by fitting effective parameters to experimental results. Nevertheless, the resulting continuous particle-scale model is much better founded and more comprehensive than any model written on the basis of experience or intuition. Consequently, it can be used to reliably describe both the evolution of moisture and temperature profiles within the drying body, and the overall drying kinetics. Additionally, it helps to predict the overpressure developing within the product in the course of drying. Such overpressure can lead to mechanical dewatering and is important when processing products like wood at high temperatures, provided either by hot air or by electromagnetic heating, as various examples in Chapter 1 demonstrate.

Furthermore, with the choice of wood as the main considered product, the effects of anisotropy are discussed in Chapter 1. This can be done by retrofitting different permeability values in longitudinal and transverse directions. A more fundamental alternative would be to calculate for realistically depicted microstructures, replacing volume averaging by genuine homogenization. This is a still-developing approach, which tries to regain a good part of the information that is lost during volume averaging with the goal of reducing or even removing the need to derive effective properties by fitting at the particle scale. The transition from the particle scale to the particle-system scale is also addressed in Chapter 1 with the specific example of static and rather large "particles", namely a stack of wood boards. Even for this simple particle system, iteration is necessary since the drying kinetics of every individual board depend on the local air conditions, and vice versa. Additionally, drying kinetics also depend on the properties of the considered board, which can not be expected to be the same throughout the stack – a feature that is modeled by randomization and a Monte-Carlo procedure. Because large deformations or cracks are not exactly what the buyers of dried wood boards expect, hints about the coupling between the transport (drying) and the mechanical problem are also given in Chapter 1.

A completely different approach is described in Chapter 2. Its essential feature is that it does not rely on volume averaging or homogenization in order to achieve the transition from the pore to the particle scale, but tries to generate the entire drying body by a combination of discrete, pore-scale elements. Respective pore networks can be used for systematic studies of the influence of the structure of a porous medium on drying kinetics – an influence that is at least partially flared out by continuous particle-scale models. Various examples in Chapter 2 illustrate the potential of discrete modeling to serve as a virtual laboratory with the ultimate objective of better understanding how structures correlate with properties and, thus, how superior functional products could be developed. Its capacity to treat percolation problems, applicability to processes involving liquid migration without drying, and a straightforward explanation for the appearance of a constant rate period during the drying of many materials are further advantages of the pore-network model. Challenges of programming, definition of clusters, incorporation of all significant transport phenomena, consideration of stochastic variability and experimental validation are discussed in Chapter 2 step by step.

While such difficulties appear treatable, the ultimate challenge of any discrete approach is one of size. On the one hand, it is more logical to combine discrete elements instead of deriving continuous models expressed by differential equations that can not be solved, except by discretization. On the other hand, the natural discrete elements can be much smaller and, thus, much more numerous than the numerical ones. As a consequence, pore-network representations of relatively large but nanostructured objects are extremely large and, computationally, extremely expensive. A possible outcome indicated at the end of Chapter 2 could involve the solution of relatively small discrete problems, derivation of effective properties from such solutions, and subsequent continuous modeling for particles or single bodies. The success of this and of other intelligent strategies for addressing the problem of network size will decide how intensively we will be able to use the potential of pore-network models in the near future.

The problem of deformations and stresses during drying, which has been introduced in Chapter 1, is treated in much more detail in Chapter 3. The theoretical background is the same, namely volume averaging. However, Chapter 3 focuses on fully saturated, highly deformable two-phase media like colloids (particulate or macromolecular gels). The shrinkage of such materials during drying is composed of a linear (ideal) and a nonlinear (viscoelastic) constituent, whereby the viscoelastic influence depends on the temporal trajectory of the process. Both this time dependence and the necessity of a full, bilateral coupling between the transport part and the rheological part of the problem make numerical solutions quite demanding. Higher accuracy in the prediction of stresses is the return on such numerical investment, as demonstrated in Chapter 3 by examples and by comparison with calculations that assume elastic behavior. The effect of stress-profile inversion (migration of maximal stress from the surface to the interior of the body in the course of drying) is properly predicted. Different ways of combining the equations for liquid momentum, solid momentum and total momentum with each other and with empirical expressions for, e.g., the relationship between liquid pressure and liquid content are critically discussed – revealing some intricate aspects of closure when the mechanical part of the problem is also considered, and pointing out perspectives for further improvement.

In Chapter 4 a combination of thermodynamics and continuum mechanics is applied to drying. The use of thermodynamics implies a treatment of the multiphase system that is analogous to the treatment of multicomponent molecular mixtures, with the advantage of a clear derivation within a stringent and well-established semantic frame. Such clarity is only possible by assigning physical phenomena to compositions (and not to structures) throughout the derivation, transferring the task of identifying transport parameters in dependence of state variables and phase topology to the macroscopic level. Unequivocally, the respective model is even more clearly a particle-scale model than in case of volume averaging. Because more empiricism is involved, some intricacies of closure that have played a role in Chapter 3 are easier to avoid, and the model can more easily be extended to unsaturated (three-phase) drying media. Results are illustrated on a kaolin cylinder subjected to deformation and stress during either convective or microwave drying. As in Chapter 3, results for elastic and viscoelastic behavior are compared with each other. The discussion of microwave heating goes into more detail than in Chapter 1. On the contrary, the discussion of overpressure in the interior of the body is much shorter and, from the modeling point of view, more empirical.

Chapter 5 refers to a particle system that is much more complex than the stack of wood boards treated in Chapter 1, namely droplets on their way of transformation to particles in a spray dryer. The droplets are very strongly affected in their movement by the turbulent gas flow, which, therefore, must be computed as accurately as possible. However, droplets and particles influence in their turn the flow of the continuous phase, creating a coupling that has to be resolved by application of so-called Euler–Lagrange methods. Moreover, particles interact with each other and may or may not agglomerate. If they do so, this is going to have an influence on particle movement and on gas flow. On the other hand, conditions for agglomeration depend on the flow, and they also depend on the state of drying of the droplets as it results from the flow and from mass and heat transfer at the particle scale. A still more intricate interconnection of the involved phenomena is obtained, which is denoted by "four-way coupling" in Chapter 5.

It is evident that, apart from particle-scale drying models, models for mechanical particle–particle interactions need to be implemented. Such local models may, in perspective, go down to the molecular scale. It is also obvious that, since not every collision can be computed, some sort of stochastic sampling has to be done from the particle system. All steps of this procedure are discussed in Chapter 5, along with a general introduction in computational fluid dynamics that will be valuable for many applications other than spray drying. The outcome expected from the very significant, but nowadays manageable, computational effort is process intensification, but also a better access to properties of the product like particle-size distribution, particle morphology and the, hopefully low, extent of thermal damage.

While sampling or Monte-Carlo techniques are an essential part of discrete approaches, the distribution of properties in particle systems can also be described in a continuous manner by population balances. The respective formalism and the necessary mathematics are explained in Chapter 6. Herein, and because population balances can be relatively easily written but are difficult to solve, the mathematics are, again, mathematics of discretization. A new numerical method, called the "cell-average method", is presented in detail and compared systematically with established alternatives. This comparison uses the few existing analytical solutions as a benchmark and shows advantages of the cell-average technique in terms of accuracy, convergence and stability. Another advantage is the flexibility in successfully treating aggregation, growth, nucleation and breakage, as well as their combinations. Such phenomena are very common in processes that combine particle formulation with drying, e.g. in spray fluidized beds. And, they are similar to the phenomena governing processes like crystallization, bubble-column absorption, biocatalysis or polymerization, providing a wide applicability of the discussed methods and principles.

Though only one distributed property is usually considered in population balances, namely the volume (size) of particles, extensions to more "internal coordinates" (for example: particle size and particle moisture) is possible. Such extensions are also presented in Chapter 6. The result of calculations by population dynamics can, however, never be better than the kinetics implemented in the equations for agglomeration or breakage, for nucleation or growth. To this end, we can either fit the so-called kernels to property distributions measured for samples taken out of large particle ensembles, or we must return by experiment and/or modeling to the particle or even to the molecular scale for a more fundamental consideration. Both strategies will be illustrated in further volumes of the series, completing the background that is necessary for application of population dynamics as discussed in Chapter 6.

From the process-systems perspective, drying is just one, though often crucial, step on the way to attractive and competitive products. As in landscape photography, it may be reasonable to sacrifice some resolution in the details for the sake of a good overview, rather looking at the forest than at the trees. The appropriate (or even necessary) amount of sacrifice depends on the goals, circumstances and resources, and changes usually in the course of an industrial project. This is illustrated in Chapter 7 by many examples on balance, scoping, scaling and detailed calculations for convective or contact dryers in continuous or batch operation that may significantly contribute to initial process development, design, commissioning or debottlenecking. It is pointed out that expert systems, decision-making tools and knowledge bases may essentially support models, especially in tasks like dryer selection, troubleshooting and problem solving. It is explained why process simulators that are successful with liquids and gases run into major difficulties when solids are involved. The state-of-the-art and potential developments in solids-processing simulators are reviewed. Finally, opportunities for, but also barriers of, either scientific or economic nature to the commercialization of drying software are critically discussed.

The transition from the molecular scale to the pore scale is perhaps the least pronounced element in the present volume, implying the applicability of classical thermodynamics, which is not a bad assumption for uniform interfaces and small molecules. However, it becomes less satisfactory as functionalized interfaces and large, structured molecules (e.g., biomolecules) are considered, so that it has to be supported or replaced by molecular dynamics or quantum mechanics. Hints about such approaches will be given in subsequent volumes. Their implementation is relatively straightforward, especially in the framework of pore-network models. Further discrete models for particle systems (especially discrete-element methods for fluidized or mechanically agitated beds) will also be discussed in considerable depth in the rest of the series. In general, computational tools at different scales or for scale transitions will continue to play an important role in subsequent volumes, though more from the specific point of view of, e.g., product quality or process intensification.

Although not all scales and not all transitions could be and have been treated with the same intensity in Volume 1, almost all of them are addressed both directly as well as by numerous citations of the primary literature in every chapter. We hope that so many pieces of the puzzle indicate that a big picture does exist, leading from molecules to marketplace products and production plants. The hierarchical nature of this picture shows that we (fortunately) need not treat simultaneously every scale with the resolution and accuracy necessary for the smallest one. On the other hand, we must continue to invest in research, especially in research concerning the scale transitions. From the practical point of view it shows that the most complex approach is not always the best one – simple solutions can be good if they fit the requirements of the task, and complex solutions can be bad, if they do not.

From the links to other chemical engineering topics and other disciplines of engineering and science mentioned in the series introduction, several are especially pronounced in Volume 1, due to its thematic content. Within these are interconnections to numerical mathematics, computing, mechanics, fluid dynamics, heat and mass transfer, heterogeneous catalysis, but also to systems biology, hydrology and geology. We have learnt from the respective communities and hope that they will also find interesting results and transferable ideas in the topics of Volume 1.

As to the acknowledgments, they are for Volume 1 identical to those in the series preface. We would like to stress them by reference, but not repeat them here.

June 2007

Evangelos Tsotsas
Arun S. Mujumdar

List of Contributors

Editors

Prof. Evangelos Tsotsas
Lehrstuhl für Thermische
Verfahrenstechnik
Otto-von-Guericke-Universität
Magdeburg
P.O. Box 4120
39016 Magedeburg
Tel.: 0391 67 18784
Fax: 0391 67 11160
E-mail: Evangelos.Tsotsas@vst.uni-magdeburg.de

Prof. Arun S. Mujumdar
University of Singapore
Dept. of Mechanical Engineering
9 Engineering Drive 1
Singapore 117576
Singapore
Tel.: 65-6778 6033
E-mail: mpeasm@nus.edu.sg

Authors

Dr. Philippe Bernada
Université de Pau et des Pays de
l'Adour
Laboratoire de Thermique Energétique
et Procédés – E.N.S.G.T.I
5 Rue Jules Ferry
64075 Pau Cedex
France
E-mail: philippe.bernada@univ-pau.fr

Dr. Stefan Blei
BASF Aktiengesellschaft
GCT/T – L540
67056 Ludwigshafen
Germany

Dr. Frédéric Couture
Université de Pau et des Pays
de l'Adour
Laboratoire de Thermique Energétique
et Procédés – E.N.S.G.T.I
5 Rue Jules Ferry
64075 Pau Cedex
France

Jun. Prof. Stefan Heinrich
Otto-von-Guericke-University
Magdeburg
Department of Process and Systems
Engineering
39016 Magdeburg
Germany

Prof. Mike Hounslow
Department of Chemical and Process
Engineering
University of Sheffield
UK

Modern Drying Technology. Edited by Evangelos Tsotsas and Arun S. Mujumdar
Copyright © 2007 WILEY-VCH Verlag GmbH & Co. KGaA. All rights reserved
ISBN: 978-3-527-31556-7

Ir. Ian C. Kemp
Glaxo SmithKline plc
42 Peachcroft Road
Abingdon
Oxfordshire OX14 2NA
UK
E-mail: ianandsue.kemp@ukgateway.net

Prof. Stefan J. Kowalski
Poznań University of Technology
Institute of Technology and Chemical Engineering
Department of Process Engineering
Pl. Marii Sklodowskiej Curie 2
60-965 Poznan
Poland
E-mail: stefan.j.kowalski@put.poznan.pl

Dr. Jitendra Kumar
Otto-von-Guericke University Magdeburg
Department of Mathematics
39016 Magdeburg
Germany

Dr. Thomas Metzger
Otto-von-Guericke-Universität Magdeburg
Lehrstuhl für Thermische Verfahrenstechnik
P.O. Box 4120
39016 Magdeburg
Germany
E-mail: Thomas.Metzger@VST.Uni-Magedeburg.de

Prof. Lothar Mörl
Otto-von-Guericke University Magdeburg
Department of Process and Systems Engineering
39016 Magdeburg
Germany

Dr. Mirko Peglow
Otto-von-Guericke-Universität Magdeburg
Lehrstuhl für Thermische Verfahrenstechnik
P.O. Box 4120
39016 Magdeburg
Germany
E-mail: Mirko.Peglow@VST.Uni-Magdeburg.de

Prof. Patrick Perré
LERMAB (Integrated Wood Research Unit)
UMR 1093
INRA/ENGREF/University H. Poincaré Nancy I
14 Rue Girardet
54042 Nancy
France
E-mail: perre@engref.fr

Dr. Marc Prat
Institut de Mécanique des Fluides de Toulouse
Allée du Professeur Camille Soula
31400 Toulouse
France
E-mail: Marc@Prat@im ft.fr

Dr. Romain Rémond
LERMAB (Integrated Wood Research Unit)
UMR 1093
INRA/ENGREF/University H. Poincaré Nancy I
14 Rue Girardet
54042 Nancy
France

Dr. Gavin Reynolds
Department of Chemical and
Process Engineering
University of Sheffield
UK

Prof. Michel Roques
Université de Pau et des Pays de
l'Adour
Laboratoire de Thermique Energétique
et Procédés – E.N.S.G.T.I
5 Rue Jules Ferry
64075 Pau Cedex
France

Prof. Martin Sommerfeld
Zentrum für Ingenieurwissenschaften,
Mechanische Verfahrenstechnik
Martin-Luther-Universität
Halle-Wittenberg
Geusaer Strafie
06217 Merseburg or: 06099 Halle (Saale)
Germany
E-mail: martin.sommerfeld@iw.
uni-halle.de

Prof. Evangelos Tsotsas
Lehrstuhl für Thermische
Verfahrenstechnik
Otto-von-Guericke-Universität
Magdeburg
P.O. Box 4120
39016 Magedeburg
Tel.: 0391 67 18784
Fax: 0391 67 11160
E-mail: Evangelos.Tsotsas@vst.uni-magdeburg.de

Prof. Ian W. Turner
School of Mathematical Sciences
Queensland University of
Technology
GPO Box 2434
Brisbane Q4001
Australia

Prof. Gerald Warnecke
Otto-von-Guericke University
Magdeburg
Department of Mathematics
39016 Magdeburg
Germany

Recommended Notation

- Alternative symbols are given in brackets
- Vectors are denoted by bold symbols, a single bar, an arrow or an index (e.g., index: i)
- Tensors are denoted by bold symbols, a double bar or a double index (e.g., index: i, j)
- Multiple subscripts should be separated by colon (e.g., $\rho_{p:dry}$: density of dry particle)

A	surface area	m^2
a_w	water activity	-
B	nucleation rate	$kg^{-1} m^{-1} s^{-1}$
b	breakage function	m^{-3}
$C\ (K)$	constant or coefficient	various
c	specific heat capacity	$J\ kg^{-1} K^{-1}$
D	equipment diameter	m
$D\ (\delta)$	diffusion coefficient	$m^2 s^{-1}$
d	diameter or size of solids	m
E	energy	J
F	mass flux function	-
$F(\dot{V})$	volumetric flow rate	$m^3 s^{-1}$
f	relative (normalized) drying rate	-
f	multidimensional number density	-
G	shear function or modulus	Pa
G	growth rate	$kg\ s^{-1}$
g	acceleration due to gravity	$m\ s^{-2}$
H	height	m
H	enthalpy	J
H	Heaviside step function	-
h	specific enthalpy (dry basis)	$J\ kg^{-1}$
$h(\alpha)$	heat-transfer coefficient	$W\ m^{-2} K^{-1}$
$\tilde{h}(h_N)$	molar enthalpy	$J\ mol^{-1}$
Δh_v	specific enthalpy of evaporation	$J\ kg^{-1}$
I	total number of intervals	-

Modern Drying Technology. Edited by Evangelos Tsotsas and Arun S. Mujumdar
Copyright © 2007 WILEY-VCH Verlag GmbH & Co. KGaA. All rights reserved
ISBN: 978-3-527-31556-7

J	numerical flux function	-
J	Jacobian matrix	various
$j(\dot{m}, J)$	mass flux, drying rate	$\text{kg m}^{-2}\text{s}^{-1}$
K	dilatation function or bulk modulus	Pa
$k(\beta)$	mass transfer coefficient	m s^{-1}
L	length	m
$M(m)$	mass	kg
$\tilde{M}(M, M_N)$	molecular mass	kg kmol^{-1}
$\dot{M}(W)$	mass flow rate	kg s^{-1}
$\dot{m}(J, j)$	mass flux, drying rate	$\text{kg m}^{-2}\text{s}^{-1}$
$\dot{m}$	volumetric rate of evaporation	$\text{kg m}^{-3}\text{s}^{-1}$
N	number	-
N	molar amount	mol
$\dot{N}(W_N)$	molar flow rate	mol s^{-1}
n	molar density, molar concentration	mol m^{-3}
n	number density	m^{-3}
n	outward normal unit vector	
$\dot{n}(J_N)$	molar flux	$\text{mol m}^{-2}\text{s}^{-1}$
P	power	W
P	total pressure	kg m s^{-2}
p	partial pressure/vapor pressure of component	kg m s^{-2}
$\dot{Q}(Q)$	heat flow rate	W
$\dot{q}(q)$	heat flux	W m^{-2}
R	equipment radius	m
R	individual gas constant	$\text{J kg}^{-1}\text{K}^{-1}$
$\tilde{R}(R_N)$	universal gas constant	$\text{J kmol}^{-1}\text{K}^{-1}$
r	radial coordinate	m
r	pore (throat) radius	m
S	saturation	-
S	selection function	s^{-1}
s	boundary-layer thickness	m
T	temperature	K, °C
t	time	s
u	velocity, usually in z-direction	m s^{-1}
u	displacement	m
V	volume, averaging volume	m^3
$\dot{V}(F)$	volumetric flow rate	m^3s^{-1}
v	specific volume	m^3kg^{-1}
v	general velocity, velocity in x-direction	m s^{-1}
W	weight force	N
$W(\dot{M})$	mass flow rate	kg s^{-1}
w	velocity, usually in y-direction	m s^{-1}
X	solids moisture content (dry basis)	-

x	mass fraction in liquid phase	-
x	particle volume in population balances	m³
x	general Eulerian coordinate, coordinate (usually lateral)	m
x_0	general Lagrangian coordinate	m
$\tilde{x}(x_N)$	molar fraction in liquid phase	-
Y	gas moisture content (dry basis)	-
y	spatial coordinate (usually lateral)	m
$y\,(\omega)$	mass fraction in gas phase	-
$\tilde{y}(y_N)$	molar fraction in gas phase	-
z	spatial coordinate (usually axial)	m

Operators

∇	gradient operator	
$\nabla\cdot$	divergence operator	
Δ	difference operator	

Greek letters

$\alpha(h)$	heat-transfer coefficient	W m⁻² K⁻¹
$\beta(k)$	mass-transfer coefficient	m s⁻¹
β	aggregation kernel	s⁻¹
δ	Dirac-delta distribution	
$\delta(D)$	diffusion coefficient	m² s⁻¹
ε	voidage	-
ε	emissivity	-
ε	small-scale parameter for periodic media	-
ε	strain	-
η	efficiency	-
θ	angle, angular coordinate	rad
κ	thermal diffusivity	m² s⁻¹
λ	thermal conductivity	W m⁻¹ K⁻¹
μ	dynamic viscosity	kg m⁻¹ s⁻¹
μ	moment of the particle-size distribution	various
ν	kinematic viscosity	m² s⁻¹
π	circular constant	-
ρ	density, mass concentration	kg m⁻³
Σ	summation operator	
σ	surface tension	N m⁻¹
σ	Stefan–Boltzmann constant for radiative heat transfer	W m⁻² K⁻⁴
σ	standard deviation (of pore-size distribution)	m
σ	stress	Pa
τ	dimensionless time	-

Φ	characteristic moisture content	-
φ	relative humidity	-
φ	phase potential	Pa
ω	angular velocity	rad
ω (y)	mass fraction in gas phase	-

Subscripts

a	at ambient conditions
as	at adiabatic saturation conditions
b	bound water
bed	bed
c	cross section
c	capillary
cr	at critical moisture content
D	drag
dry	dry
dp	at dewpoint
eff	effective
eq	equilibrium (moisture content)
f	friction
g	gas (dry)
H	wet (humid) gas
i	inner
i,1,2,...	component index, particle index
i,j,k	coordinate index, $i,j,k = 1$ to 3
in	inlet value
l	liquid (alternative: as a superscript)
m	mean value
max	maximum
mf	at minimum fluidization
min	minimum
N	molar quantity
o	outer
out	outlet value
P	at constant pressure
p	particle
pbe	population balance equation
ph	at the interface
r	radiation
rel	relative velocity
s	solid (compact solid phase), alternative: as a superscript
S	at saturation conditions
surf	surface
V	based on volume

v	vapor, evaporation
w	water
w	wall
wb	at wet-bulb conditions
wet	wet
∞	at large distance from interface

Superscripts, special symbols

v	volumetric strain
$*$	rheological strain
$*$	at saturation conditions
$^{-}$ or $\langle \rangle$	average, phase average
$^{-a}$ or $\langle \rangle^{\alpha}$	intrinsic phase average
$\sim$	spatial deviation variable

EFCE Working Party on Drying: Address List

Dr. Odilio Alves-Filho (guest)
Grupo de Análisis y Simulación de
Procesos Agroalimentarios
Departamento de Tecnologia de
Alimentos
Escuela Técnica Superior de Ingenieros
Agrónomos
Universidad Politécnica de Valencia
Camino de Vera s/n
46022 Valencia
Spain
Tel.: +34 96 387 73 68
Fax: +34 96 387 98 39
E-mail: odialfil@tal.upv.es

Prof. Julien Andrieu (delegate)
UCB Lyon I/ESCPE
LAGEP UMR CNRS 5007
batiment 308 G
43 boulevard du 11 novembre 1918
69622 Villeurbanne cedex
France
Tel.: +33 4 72 43 18 43
Fax: +33 4 72 43 16 82
E-mail: andrieu@lagep.univ-lyon1.fr

Dr. Paul Avontuur (guest industry)
Glaxo Smith Kline
New Frontiers Science Park H89
Harlow CM19 5AW
United Kingdom
Tel.: +44 1279 64 3797
E-mail: Paul.Avontuur@gsk.com

Dr. Christopher G. J. Baker (guest)
Drying Associates
Harwell International Business Centre
404/13 Harwell Didcot
Oxfordshire OX11 ORA
United Kingdom
Tel.: +44 1235 432245
Fax: +44 1235 435405
E-mail: baker@kuc01.kuniv.edu.kw

Prof. Antonello Barresi (delegate)
Dip. Scienza dei Materiali e Ingegneria
Chimica
Politecnico di Torino
Corso Duca degli Abruzzi 24
10129 Torino
Italy
Tel.: +39 011 5644658
Fax: +39 011 5644699
E-mail: antonello.barresi@polito.it

Dr. Rainer Bellinghausen (delegate industry)
Bayer Technology Services GmbH
BTS-PT-PT-PDSP
Building E 41
51368 Leverkusen
Germany
Tel.: +49 214 30 61867
Fax: +49 214 30 9661867
E-mail: rainer.bellinghausen@bayer-technology.com

Modern Drying Technology. Edited by Evangelos Tsotsas and Arun S. Mujumdar
Copyright © 2007 WILEY-VCH Verlag GmbH & Co. KGaA. All rights reserved
ISBN: 978-3-527-31556-7

Dr. Carl-Gustav Berg (guest)
Abo Akademi
Process Design Laboratory
Biskopsgatan 8
20500 Abo
Finland
Tel.: +358 40 7792 396
Fax: +358 50 7830 2247
E-mail: cberg@abo.fi

Prof. Jean-Jacques Bimbenet (honorary guest)
ENSIA
1 Avenue des Olympiades
91744 Massy cedex
France
Tel.: +33 169 935069
Fax: +33 169 935185
E-mail: bimbenet@ensia.fr

Dr. Catherine Bonazzi (delegate)
ENSIA – INRA
JRU for Food Process Engineering
1 Avenue des Olympiades
91744 Massy cedex
France
Tel.: +33 1 69 93 50 69
Fax: +33 1 69 93 51 85
E-mail: bonazzi@ensia.fr

Mr. Pascale Bridou-Buffet (EFCE office)
EFCE-Paris Office
Societe de Chimie Industrielle
28 Rue Saint-Dominique
75007 Paris
France
Tel.: +33 1 53 59 02 18
Fax: +33 1 45 55 40 33
E-mail: SCI.fr@wanadoo.fr

Dr. Bojan Cermak (delegate industry)
Korunni 60
12000 Praha 2
Czech Republic
Tel.: +42 222 516 499
Fax: +42 222 516 499
E-mail: bojan.cermak@email.cz

Paul Deckers M.Sc. (delegate industry)
Bodec
Process Optimization and Development
Industrial Area 't Zand
Bedrijfsweg 1
5683 CM Best
The Netherlands
Tel.: +31 499 335888
Fax: +31 499 335889
E-mail: deckers@bodec.nl

Prof. Stephan Ditchev (guest)
University of Food Technology
26 Maritza Blvd.
4002 Plovdiv
Bulgaria
Tel.: +359 32 64 28 41
Fax: +359 32 64 28 41
E-mail: sditchev@mail.bg

Prof. Anatoly A. Dolinsky (delegate)
Institute of Engineering Thermophysics
2a Zhelyabov St.
252057 Kiev
Ukraine
Tel.: +7 44 44 69 053
Fax: +7 44 44 66 091

Dr. German I. Efremov (guest)
Pavla Korchagina 22
129278 Moscow
Russia
Tel.: +7 (095) 282 2053
Fax: +7 (095) 952 1744
E-mail: efremov_german@mail.ru

Prof. Trygve Eikevik (guest)
Norwegian University of Science and Technology
Dep. of Energy and Process Engineering
Kolbjørn Hejes vei 1B
7491 Trondheim
Norway
Tel.: +47 73 593921
Fax: +47 73 593950
E-mail: Trygve.M.Eikevik@ntnu.no

Dr.-Ing. Ioannis Evripidis (guest industry)
Dow Deutschland GmbH & Co. OHG
P.O. Box 1120
21677 Stade
Germany
Tel.: + 49 4146 913517
Fax: + 49 4146 912326
E-mail: evripidis@dow.com

Prof. Dr. Istvan Farkas (delegate)
Dep. of Physics and Process Control
Szent Istvan University
Pater K. u. 1
2103 Godollo
Hungary
Tel.: +36 28 522055
Fax: +36 28 410804
E-mail: Farkas.Istvan@gek.szie.hu

Andrew Furlong (EFCE office)
Head of External Relations
IChemE
Davis Building
Rugby CV21 3HQ
United Kingdom
Tel.: +44 1788 534 484
Fax: +44 1788 560 833
E-mail: AFurlong@icheme.org

Dr.-Ing. Dietrich Gehrmann (guest)
Wilhelm-Hastrich-Str. 12
51381 Leverkusen
Germany
Tel.: +49 2171 31431
Fax: +49 2171 33981
E-mail: Dietrich.Gehrmann@t-online.de

Dr. Adrian-Gabriel Ghiaus (delegate)
Thermal Engineering Department
Technical University of Civil Engineering
Bd. P. Protopopescu 66
021414 Bucharest
Romania
Tel.: +40 21 2524280
Fax: +40 21 2526880
E-mail: ghiaus@mech.upatras.gr

Ms. Ines Honndorf (EFCE office)
EFCE c/o Dechema
P.O. Box 150104
60061 Frankfurt/M
Germany
Tel.: +49 69 7564209
Fax: +49 69 7564201
E-mail: honndorf@dechema.de

Prof. Dr. Ing. Gheorghita Jinescu (guest)
Department of Chemical Engineering
Faculty of Industrial Chemistry
University "Politehnica" din Bucuresti
1 Polizu street
Building F, Room F210
78126 Bucharest
Romania
Tel.: +40 1 650 3289 ext. -268, -291
Fax: +40 1 410 0285
E-mail: g_jinescu@chim.upb.ro

Prof. Dr. Gligor Kanevce (guest)
St. Kliment Ohridski University
Faculty of Technical Sciences
ul. Ivo Ribar Lola b.b.
Bitola
Macedonia
Tel.: +38 996 263 256
Fax: +38 996 263 256
E-mail: kanevce@osi.net.mk

Prof. Dr. Markku Karlsson (delegate)
UPM-Kymmene Corporation
P.O. Box 380
00101 Helsinki
Finland
Tel.: +358 204 15 0228
Fax: +358 204 15 0343
E-mail: markku.karlsson@upm-kymmene.com

Ir. Ian C. Kemp (delegate, immediate past chairman)
GMS
GSK
Priory Street
Ware SG12 0XA
United Kingdom
Tel.: +44 1920 862271
E-mail: iankemp@btinternet.com

Prof. Dr. Ir. P. J. A. M. Kerkhof (guest)
Eindhoven University of Technology
Dept. of Chemical Engineering
P. O. Box 513
5600 MB Eindhoven
The Netherlands
Tel.: +31 40 2472970
Fax: +31 40 2439303
E-mail: p.j.a.m.kerkhof@tue.nl

Prof. Matthias Kind (guest)
Institut für Thermische Verfahrenstechnik
Universität Karlsruhe (TH)
Kaiserstr. 12
76128 Karlsruhe
Germany
Tel.: +49 721 608 2390
Fax: +49 721 608 3490
E-mail: matthias.kind@ciw.uni-karlsruhe.de

Prof. Eli Korin (guest)
Chemical Engineering Department
Ben-Gurion University of the Negev
Beer-Sheva 84105
Israel
Tel.: +972 8 6461820
Fax: +972 8 6477656
E-mail: ekorin@bgumail.bgu.ac.il

Emer. Prof. Ram Lavie (guest)
Department of Chemical Engineering
Technion – Israel Institute of Technolgy
Technion City
Haifa 32000
Israel
Tel.: +972 4 8292934
Fax: +972 4 8230476
E-mail: lavie@tx.technion.ac.il

Dr. Ir. Angélique Léonard (delegate)
Laboratoire de Génie Chimique
Département de Chimie Appliquée
Université de Liège
Bâtiment B6c – Sart-Tilman
4000 Liège
Belgium
Tel.: +32 4 366 47 22
Fax: +32 4 366 28 18
E-mail: A.Leonard@ulg.ac.be

Prof. Natalia Menshutina (guest)
Mendeleyev University of Chemical Technology of Russia (MUCTR)
Department of Cybernetics of Chemical Technological Processes
125047 Muisskaya sq.9
Moscow
Russia
Tel.: +7 (095) 9787417
Fax: +7 (095) 9787417
E-mail: chemcom@muctr.edu.ru

Dr. Thomas Metzger (secretary)
Thermal Process Engineering
Otto-von-Guericke University
P. O. Box 4120
39016 Magdeburg
Germany
Tel.: +49 391 6711362
Fax: +49 391 6711160
E-mail: thomas.metzger@vst.uni-magdeburg.de

Prof. Antonio Mulet Pons (delegate)
Universitat Politecnica de Valencia
Departament de Tecnologia d'Aliments
Cami de Vera s/n
46071 Valencia
Spain
Tel.: +34 96 3877368
Fax: +34 96 3877369
E-mail: amulet@tal.upv.es

Prof. Zdzislaw Pakowski (delegate)
Faculty of Process and Environmental Engineering
Technical University of Lodz
ul. Wolczanska 213
93-005 Lodz
Poland
Tel.: +48 42 6313731
Fax: +48 42 6365663
E-mail: pakowski@wipos.p.lodz.pl

Prof. Patrick Perré (guest)
LERMAB – ENGREF
14 Rue Girardet
54042 Nancy
France
Tel.: +33 383 396890
Fax: +33 383 396847
E-mail: perre@engref.fr

Prof. Michel Roques (guest)
Universite de Pau et des Pays de l'Adour
ENSGTI
5 Rue Jules-Ferry
64000 Pau
France
Tel.: +33 559 407832
Fax: +33 559 407801
E-mail: michel.roques@univ-pau.fr

Dr. Carmen Rosselló (delegate)
University of Illes Baleares
Dep. Quimica
Ctra. Valldemossa km 7.5
07122 Palme Mallorca
Spain
Tel.: +34 71 173239
Fax: +34 71 173426
E-mail: Carmen.rossello@uib.es

Emer. Prof. G. D. Saravacos (delegate)
Nea Tiryntha
21100 Nauplion
Greece
Tel.: +30 75236491
Fax: +30 75236491
E-mail: gsaravac@otenet.gr

Prof. Dr.-Ing. Ernst-Ulrich Schluender (honorary guest)
Lindenweg 10
76275 Ettlingen
Germany
E-mail: euschluender@t-online.de

Dr. Michael Schönherr (guest industry)
Research Manager Drying
Process Engineering
BASF Aktiengesellschaft
GCT/T – L 540
67056 Ludwigshafen
Germany
Tel.: +49 621 60-55108
Fax. +49 621 60-74795
E-mail: michael.schoenherr@basf.com

Dr. Alberto M. Sereno (delegate)
University of Porto
Dept. of Chemical Engineering
Rua Dr. Roberto Frias
4200-465 Porto
Portugal
Tel.: +351 22 508 1655
Fax: +351 22 508 1449
E-mail: sereno@fe.up.pt

Dr. Milan Stakic (guest)
Thermal Process Engineering
Otto-von-Guericke University
P.O. Box 4120
39016 Magdeburg
Germany
Tel.: +49 391 6712455
Fax: +49 391 6711160
E-mail: milan.stakic@vst.uni-magdeburg.de
stakicm@yahoo.com

Prof. Stig Stenstrom (delegate)
Lund University
Institute of Technology
Dept. of Chemical Engineering
P.O. Box 124
22100 Lund
Sweden
Tel.: +46 46 108298
Fax: +46 46 104526
E-mail: stig.stenstrom@chemeng.lth.se

Prof. Ingvald Strommen (delegate)
Dept. of Energy and Process Engineering
Norwegian University of Science and Technology
Kolbjørn Hejes vei 1b
7491 Trondheim
Norway
Tel.: +47 73 59 37 42
Fax: +47 73 59 35 80
E-mail: ingvald.strommen@ntnu.no

Prof. Czeslaw Strumillo (delegate)
Technical University of Lodz
Faculty of Process and Environmental Engineering
Lodz Technical University
ul. Wolczanska 213
93-005 Lodz
Poland
Tel.: +48 42 6313735
Fax: +48 42 6365663
E-mail: cstrumil@wipos.p.lodz.pl

Prof. Radivoje Topic (delegate)
Faculty of Mechanical Engineering
University of Belgrade
27 Marta 80
11000 Beograd
Serbia
Fax: +381 11 337 03 64
E-mail: r.topic@eunet.yu

Prof. Dr.-Ing. Evangelos Tsotsas (delegate, chairman of WP)
Thermal Process Engineering
Otto-von-Guericke University
P.O. Box 4120
39016 Magdeburg
Germany
Tel.: +49 391 6718784
Fax: +49 391 6711160
E-mail: evangelos.tsotsas@vst.uni-magdeburg.de

Dr. Henk C. van Deventer (delegate)
TNO Quality of Life
P.O. Box 342
7300 AH Apeldoorn
The Netherlands
Tel.: +31 55 549 3805
Fax: +31 55 549 3386
E-mail: henk.vandeventer@tno.nl

Michael Wahlberg M.Sc. (guest)
Niro
Gladsaxevej 305
2860 Soeborg
Denmark
Tel.: +45 3954 5454
Fax: +45 3954 5107
E-mail: mw@niro.dk

Prof. Roland Wimmerstedt (honorary guest)
Lund University
Institute of Technology
Dept. of Chemical Engineering
P.O. Box 124
22100 Lund
Sweden
Tel.: +46 46 2228298
Fax: +46 46 2224526
E-mail: Roland.Wimmerstedt@chemeng.lth.se

Prof. Ireneusz Zbicinski (guest)
Faculty of Process and Environmental Engineering
Lodz Technical University
ul. Wolczanska 213
93-005 Lodz
Poland
Tel.: +48 42 6313773
Fax: +48 42 6364923
E-mail: zbicinsk@mail.p.lodz.pl

1
Comprehensive Drying Models based on Volume Averaging: Background, Application and Perspective

Patrick Perré, Romain Rémond, Ian W. Turner

1.1
Microscopic Foundations of the Macroscopic Formulation

The drying of a wet porous medium is a process that involves coupled and simultaneous heat, mass and momentum transfer. Modeling this complex process requires the development of transport equations derived from the standard conservation laws (see for example Bird et al., 1960). The challenge, however, is to overcome the problems associated with structural dependencies and the complex geometries evident in the internal pore network within the medium. Typically, transport phenomena are represented according to macroscopic equations valid at the relevant level of description. Selecting a representative elementary volume, or averaging volume, containing many pores and assuming that the porous material can be represented as a fictitious continuum (Bear and Corapcioglu, 1987) is certainly one way to achieve this description. Another possibility is to rigorously derive the macroscopic equations from microscopic balance equations by means of volume averaging (Gray, 1975; Whitaker, 1977, 1998; Marle, 1982). In this section a brief exposition of the volume-averaging strategy is presented for the liquid-phase and gas-phase water-vapor transport equations and we refer the interested reader to Whitaker (1998) for the complete derivation of the macroscopic drying equations.

The technique of volume averaging is described in Slattery (1967, 1972) and has been used by numerous authors over the last three decades to model transport in porous media (Carbonell and Whitaker, 1983, 1984; Glatzmaier and Ramirez, 1988; Quintard and Whitaker, 1993, 1994, 1995). The underlying idea is to average the dependent variable (for example liquid ρ_w or the gas-phase water-vapor density ρ_{gv}) over some representative localized volume, as depicted in Fig. 1.1. The averaging volume V comprises the individual phase volumes

$$V = V_w(\mathbf{x}, t) + V_g(\mathbf{x}, t) + V_s(\mathbf{x})$$

Modern Drying Technology. Edited by Evangelos Tsotsas and Arun S. Mujumdar
Copyright © 2007 WILEY-VCH Verlag GmbH & Co. KGaA. All rights reserved.
ISBN: 978-3-527-31556-7

1 Comprehensive Drying Models based on Volume Averaging

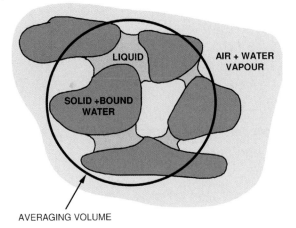

Fig. 1.1 Schematic view of averaging volume in a porous medium.

each of which can vary with space, as well as time for the liquid and gas phases. Averages are then defined in terms of these volumes and are said to be associated with the centroid of the averaging volume V, which assumes the existence of a representative volume that is large enough for the averaged quantities to be defined and small enough to avoid variations due to macroscopic gradients and nonequilibrium configurations at the microscopic level (Fig. 1.2).

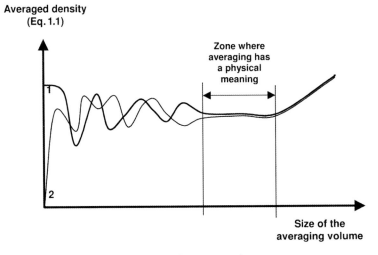

Fig. 1.2 The averaging method assumes the existence of a representative elementary volume, large enough for the pore effect to be smoothed and small enough for macroscopic variations and nonequilibrium effects to be avoided. Two curves are presented for this simple example on density: point 1 is situated in the solid phase and point 2 in a pore.

1.1 Microscopic Foundations of the Macroscopic Formulation

The development of the volume-averaged transport equations requires the introduction of what are called superficial and intrinsic averages. For example, the superficial average of the density of the liquid phase is given by

$$\langle \rho_w \rangle = \frac{1}{V} \int_{V_w} \rho_w \, dV \tag{1.1}$$

and the intrinsic average by

$$\langle \rho_w \rangle^w = \frac{1}{V_w} \int_{V_w} \rho_w \, dV \tag{1.2}$$

where V_w is the volume of the liquid phase contained in V. One also notes the relationship $\langle \rho_w \rangle = \varepsilon_w \langle \rho_w \rangle^w$ in which $\varepsilon_w = \frac{V_w}{V}$ is the volume fraction of the liquid phase. The latter average is claimed to be the best representation in the sense that if ρ_w were a constant given by ρ_w^0 say, then the intrinsic average gives $\langle \rho_w \rangle^w = \rho_w^0$, whereas the superficial average gives $\langle \rho_w \rangle = \frac{V_w}{V} \rho_w^0$.

Consider now the liquid phase continuity equation

$$\frac{\partial \rho_w}{\partial t} + \nabla \cdot (\rho_w \mathbf{v}_w) = 0 \tag{1.3}$$

and then form the superficial average of Eq. 1.3 to arrive at

$$\left\langle \frac{\partial \rho_w}{\partial t} \right\rangle + \langle \nabla \cdot (\rho_w \mathbf{v}_w) \rangle = 0 \tag{1.4}$$

It is clear from Eq. 1.4 that it is necessary to deal with the volume averages of both the time derivative and the divergence term, which requires the utilization of the spatial averaging theorem and the general transport theorem (Whitaker, 1998). The averaging theorem enables the average of the divergence term to be expressed as

$$\langle \nabla \cdot (\rho_w \mathbf{v}_w) \rangle = \nabla \cdot \langle \rho_w \mathbf{v}_w \rangle + \frac{1}{V} \iint_{A_{wg}} \rho_w \mathbf{v}_w \cdot \mathbf{n}_{wg} \, d\sigma + \frac{1}{V} \iint_{A_{ws}} \rho_w \mathbf{v}_w \cdot \mathbf{n}_{ws} \, d\sigma \tag{1.5}$$

where A_{wg}, A_{ws} represent, respectively, the area of the liquid/gas and liquid/solid interfaces contained within the averaging volume V and $\mathbf{n}_{wg}$, $\mathbf{n}_{ws}$ represent the unit normal vectors directed from the liquid phase towards the gas and solid phases respectively. One then notes that due to the liquid/solid interface being impermeable, $\mathbf{v}_w \cdot \mathbf{n}_{ws} = 0$ and the last term in Eq. 1.5 vanishes.

The general transport theorem enables the superficial average of the time derivative (or accumulation term) to be expressed as

$$\left\langle \frac{\partial \rho_w}{\partial t} \right\rangle = \frac{\partial \langle \rho_w \rangle}{\partial t} - \frac{1}{V} \iint_{A_{wg}} \rho_w \mathbf{w} \cdot \mathbf{n}_{wg} \, d\sigma - \frac{1}{V} \iint_{A_{ws}} \rho_w \mathbf{w} \cdot \mathbf{n}_{ws} \, d\sigma \tag{1.6}$$

in which $\mathbf{w} \cdot \mathbf{n}_{wg}$ and $\mathbf{w} \cdot \mathbf{n}_{ws}$ represent respectively the speed of displacement of the liquid/gas and liquid/solid interfaces. For a rigid porous medium the latter can be taken as zero so that the last term in Eq. 1.6 vanishes. Combining the results in Eq. 1.5 and Eq. 1.6 and substituting into Eq. 1.4 we obtain

$$\frac{\partial \langle \rho_w \rangle}{\partial t} + \nabla \cdot \langle \rho_w \mathbf{v}_w \rangle + \frac{1}{V} \iint_{A_{wg}} \rho_w (\mathbf{v}_w - \mathbf{w}) \cdot \mathbf{n}_{wg} \, d\sigma = 0 \tag{1.7}$$

Since the liquid-phase density is treated as constant, the superficial average density can be written as $\langle \rho_w \rangle = \varepsilon_w \langle \rho_w \rangle^w = \varepsilon_w \rho_w$ and by defining the mass rate of evaporation as

$$\langle \dot{m} \rangle = \frac{1}{V} \iint_{A_{wg}} \rho_w (\mathbf{v}_w - \mathbf{w}) \cdot \mathbf{n}_{wg} \, d\sigma$$

we obtain the final form of the volume-averaged continuity equation for the liquid phase as

$$\frac{\partial}{\partial t} (\rho_w \varepsilon_w) + \nabla \cdot (\rho_w \langle \mathbf{v}_w \rangle) + \langle \dot{m} \rangle = 0 \tag{1.8}$$

The entire process described above can now be repeated for the gas-phase water-vapor continuity equation

$$\frac{\partial \rho_{gv}}{\partial t} + \nabla \cdot (\rho_{gv} \mathbf{v}_{gv}) = 0 \tag{1.9}$$

to obtain

$$\frac{\partial}{\partial t} (\varepsilon_g \langle \rho_{gv} \rangle^g) + \nabla \cdot \langle \rho_{gv} \mathbf{v}_{gv} \rangle + \frac{1}{V} \iint_{A_{gw}} \rho_{gv} (\mathbf{v}_{gv} - \mathbf{w}) \cdot \mathbf{n}_{gw} \, d\sigma = 0 \tag{1.10}$$

where the relation $\langle \rho_{gv} \rangle = \varepsilon_g \langle \rho_{gv} \rangle^g$ has been used. Assuming that there is no excess surface mass of either species at the gas/liquid interface leads to the jump condition

$$\rho_{gv} (\mathbf{v}_{gv} - \mathbf{w}) \cdot \mathbf{n}_{gw} = \rho_w (\mathbf{v}_w - \mathbf{w}) \cdot \mathbf{n}_{wg} \quad \text{at } A_{gw} \tag{1.11}$$

Substitution of Eq. 1.11 in Eq. 1.10 gives

$$\frac{\partial}{\partial t} \left(\varepsilon_g \langle \rho_{gv} \rangle^g \right) + \nabla \cdot \langle \rho_{gv} \mathbf{v}_{gv} \rangle - \langle \dot{m} \rangle = 0 \tag{1.12}$$

Next we decompose the species velocity by assuming that the gas-phase mass-average velocity is determined by Darcy's law and then utilize Fick's law to write

$$\rho_{gv} \mathbf{v}_{gv} = \rho_{gv} \mathbf{v}_g - \rho_g D_v \nabla \omega_v \tag{1.13}$$

where D_v is the vapor phase diffusivity and ω_v is the mass fraction of water vapor. Substituting Eq. 1.13 into Eq. 1.12 we obtain

$$\frac{\partial}{\partial t}\left(\varepsilon_g \langle \rho_{gv} \rangle^g\right) + \nabla \cdot \langle \rho_{gv} \mathbf{v}_g \rangle - \langle \dot{m} \rangle = \nabla \cdot \langle \rho_g D_v \nabla \omega_v \rangle \tag{1.14}$$

Manipulating the second (advection) term in Eq. 1.14 is more demanding than for the liquid phase and requires the introduction of the concept of spatial deviation variables (Whitaker, 1998), where the point quantities ρ_{gv}, $\mathbf{v}_g$ are expressed in terms of the average quantity and the spatial deviation variable, namely:

$$\rho_{gv} = \langle \rho_{gv} \rangle^g + \tilde{\rho}_{gv} \quad \text{and} \quad \mathbf{v}_g = \langle \mathbf{v}_g \rangle^g + \tilde{\mathbf{v}}_g$$

This process represents a decomposition of length scales, where for example $\langle \rho_{gv} \rangle^g$ undergoes significant changes over the large length scale, whereas the characteristic length associated with the deviation variable $\tilde{\rho}_{gv}$ is the small length scale. Carbonell and Whitaker (1983) then use some rather detailed and elegant analysis to obtain the result

$$\frac{\partial}{\partial t}\left(\varepsilon_g \langle \rho_{gv} \rangle^g\right) + \nabla \cdot \left(\langle \rho_{gv} \rangle^g \langle \mathbf{v}_g \rangle\right) - \langle \dot{m} \rangle = \nabla \cdot \langle \rho_g D_v \nabla \omega_v \rangle - \nabla \cdot (\tilde{\rho}_{gv} \tilde{\mathbf{v}}_g) \tag{1.15}$$

The quantity $\nabla \cdot (\tilde{\rho}_{gv} \tilde{\mathbf{v}}_g)$ in Eq. 1.15 represents dispersive transport, while the term $\nabla \cdot \langle \rho_g D_v \nabla \omega_v \rangle$ represents diffusive transport. We now focus on the treatment of the diffusive transport term. Whitaker (1998) employs the averaging theorem, together with the decomposition $\omega_v = \langle \omega_v \rangle^g + \tilde{\omega}_v$ and the relation $\langle \omega_v \rangle = \varepsilon_g \langle \omega_v \rangle^g$ to express

$$\langle \nabla \omega_v \rangle = \varepsilon_g \nabla \langle \omega_v \rangle^g + \frac{1}{V}\iint_{A_{gw}} \mathbf{n}_{gw} \tilde{\omega}_v d\sigma + \frac{1}{V}\iint_{A_{gs}} \mathbf{n}_{gs} \tilde{\omega}_v d\sigma$$

Based on length-scale constraints and the relative magnitudes of the spatial deviations, Whitaker then arrives at the following volume-averaged mass diffusive flux, which is represented in terms of intrinsic averaged quantities determined by the volume-averaged transport equations and spatial deviation quantities that must be determined by means of closure problems:

$$\langle \rho_g D_v \nabla \omega_v \rangle = \langle \rho_g \rangle^g D_v \left\{ \varepsilon_g \nabla \langle \omega_v \rangle^g + \frac{1}{V}\iint_{A_{gw}} \mathbf{n}_{gw} \tilde{\omega}_v d\sigma + \frac{1}{V}\iint_{A_{gs}} \mathbf{n}_{gs} \tilde{\omega}_v d\sigma \right\} \tag{1.16}$$

Substitution of Eq. 1.16 into Eq. 1.15 leads to the rather complicated volume-averaged gas-phase water-vapor transport equation

$$\frac{\partial}{\partial t}\left(\varepsilon_g \langle \rho_{gv} \rangle^g\right) + \nabla \cdot \left(\langle \rho_{gv} \rangle^g \langle \mathbf{v}_g \rangle\right) - \langle \dot{m} \rangle$$

$$= \nabla \cdot \left[\langle \rho_g \rangle^g D_v \left\{ \varepsilon_g \nabla \langle \omega_v \rangle^g + \frac{1}{V}\iint_{A_{gw}} \mathbf{n}_{gw} \tilde{\omega}_v d\sigma + \frac{1}{V}\iint_{A_{gs}} \mathbf{n}_{gs} \tilde{\omega}_v d\sigma \right\}\right] - \nabla \cdot (\tilde{\rho}_{gv} \tilde{\mathbf{v}}_g) \tag{1.17}$$

Further analysis (Whitaker,-1998) highlights the fact that the first term on the right-hand side of Eq. 1.17 leads to the classical diffusive flux expressed in terms of an *effective diffusivity tensor*. To achieve this, it is assumed that the surface-integral terms containing the fluctuation components of the mass fraction ω_v are proportional to its macroscopic gradient $\nabla \langle \omega_v \rangle^g$.

Whitaker (1998) also derives the volume-averaged closed forms of the two momentum equations as

$$\langle \mathbf{v}_w \rangle = -\frac{\mathbf{K}_w}{\mu_w} \left(\nabla \langle \rho_w \rangle^w - \rho_w \mathbf{g} \right) + \mathbf{K}_{wg} \langle \mathbf{v}_g \rangle$$

$$\langle \mathbf{v}_g \rangle = -\frac{\mathbf{K}_g}{\mu_g} \left(\nabla \langle \rho_g \rangle^g - \rho_g \mathbf{g} \right) + \mathbf{K}_{gw} \langle \mathbf{v}_w \rangle$$

in which $\mathbf{K}_w$ and $\mathbf{K}_g$ are the liquid and gas permeability tensors and $\mathbf{K}_{wg}$ and $\mathbf{K}_{gw}$ are the viscous drag tensors. Arguments by Dullien and Dong (1996) suggest that for modeling the drying process the viscous coupling terms can be omitted, which leads to the traditional form of the volume-averaged momentum equations.

1.2
The Macroscopic Set of Equations

In the literature, several variants of the macroscopic equation set have been proposed for simulating the drying process. In this section, we focus on the most comprehensive set of equations used at the macroscopic level, which considers three independent state variables. This formulation, as proposed below, originates for the most part from Whitaker's work (Whitaker, 1977) with minor changes required to account for bound water diffusion and drying with internal overpressure (Perré and Degiovanni, 1990).

As a reminder (see Section 1.1), the reader must be aware that all variables are averaged over the REV (representative elementary volume), hence the expression "macroscopic". In all of these equations the subscript *eff* denotes the "effective" property that has to be determined either experimentally or by using a predictive scaling approach (see Section 1.6). In order to simplify the notation, the averaged values of, for example, variable ρ_b as defined in Eq. 1.1 and Eq. 1.2 are indicated with a bar as $\bar{\rho}_b$. Quantities involving the double bar, for example $\bar{\bar{\mathbf{D}}}_{eff}$, indicate a tensor.

Water conservation

$$\frac{\partial}{\partial t}(\varepsilon_w \rho_w + \varepsilon_g \rho_v + \bar{\rho}_b) + \nabla \cdot (\rho_w \bar{\mathbf{v}}_w + \rho_v \bar{\mathbf{v}}_g + \overline{\rho_b \mathbf{v}_b}) = \nabla \cdot (\rho_g \bar{\bar{\mathbf{D}}}_{eff} \nabla \omega_v) \quad (1.18)$$

Air conservation

$$\frac{\partial}{\partial t}(\varepsilon_g \rho_a) + \nabla \cdot (\rho_a \bar{\mathbf{v}}_g) = \nabla \cdot (\rho_g \bar{\bar{\mathbf{D}}}_{eff} \nabla \omega_a) \quad (1.19)$$

Energy conservation

$$\frac{\partial}{\partial t}\left(\varepsilon_w \rho_w h_w + \varepsilon_g(\rho_v h_v + \rho_a h_a) + \overline{\rho_b} \overline{h_b} + \rho_s h_s - \varepsilon_g P_g\right)$$
$$+ \nabla \cdot \left(\rho_w h_w \overline{\mathbf{v}}_w + (\rho_v h_v + \rho_a h_a)\overline{\mathbf{v}}_g + h_b \overline{\rho_b \mathbf{v}_b}\right) \quad (1.20)$$
$$= \nabla \cdot \left(\rho_g \overline{\overline{\mathbf{D}}}_{\text{eff}}(h_v \nabla \omega_v + h_a \nabla \omega_a) + \overline{\overline{\lambda}}_{\text{eff}} \nabla T\right) + \Phi$$

where the gas and liquid phase velocities are given by the *generalized Darcy's law*.

$$\overline{\mathbf{v}}_l = -\frac{\overline{\overline{\mathbf{K}}}_l \overline{\overline{\mathbf{k}}}_l}{\mu_l} \nabla \varphi_l, \nabla \varphi_l = \nabla P_l - \rho_l g \nabla \chi \quad \text{with } l = w, g \quad (1.21)$$

and the bound-water flux by a simple diffusion expression

$$\overline{\rho_b \mathbf{v}_b} = -\overline{\overline{\mathbf{D}}}_b \nabla \overline{\rho}_b \quad (1.22)$$

The quantities φ are known as the phase potentials and χ is the depth scalar.

Boundary conditions

For the external drying surfaces of the sample, the boundary conditions are assumed to be

$$\mathbf{J}_w|_{x=0^+} \cdot \mathbf{n} = h_m c M_v \ln\left(\frac{1 - x_\infty}{1 - x_v|_{x=0}}\right)$$
$$P_g|_{x=0^+} = P_{\text{atm}} \quad (1.23)$$
$$\mathbf{J}_e|_{x=0^+} \cdot \mathbf{n} = h(T|_{x=0} - T_\infty)$$

where $\mathbf{J}_w$ and $\mathbf{J}_e$ represent the fluxes of total moisture and total enthalpy at the boundary, respectively; x denotes the position from the boundary along the external unit normal. The mass and heat transfer coefficients are denoted by h_m and h, respectively, and x_v and x_∞ are the molar fractions of vapor at the exchange surface and in the airflow.

1.3 Physical Phenomena Embedded in the Equations

1.3.1 Low-temperature Convective Drying

When the role of internal gaseous pressure is almost negligible in a convective drying process, this process can be denoted as low-temperature convective drying. These conditions prevail in many industrial dryers operated with heated air. Usually, two periods of drying may be distinguished: a constant rate and a decreasing rate period.

1.3.1.1 The Constant Drying Rate Period

While liquid water is present at the exchange surface, the mass flux (mass loss per unit of time and area) is constant and depends only on the external conditions (temperature, relative humidity, velocity and flow configuration). This stage is commonly referred to in literature as the constant-rate stage or first drying period. The existence of a constant drying rate period denotes an efficient internal mass transfer through capillary forces.

The constant rate stage is observed for porous media such as aerated concrete and sapwood dried at moderate conditions (Perré and Martin, 1994). In the case of food, conflicting results have been reported. For example, constant-rate stages have been reported for air drying of tropical marine fish (Kannan and Bandyopadhyand, 1995); cod muscle (Jason, 1958); potatoes, carrots, onions, garlic, apples, pears, peaches and grapes (Saravacos and Charm, 1962) and apples (Jung et al., 1986). The absence of constant-rate stages has been indicated for carrot (Sereno and Medeiros, 1990); corn, potato, prune, apple (Bimbenet et al., 1985); potato slices (Magee and Wilkinson, 1992; Zhao and Poulsen, 1988). Recent works prove that shrinkage and shape changes during drying are responsible for these contradictory results (May and Perré, 2002, Perré and May, 2007). In such cases, the constant-rate stage is in fact a constant drying flux period (drying rate per actual exchange area).

In the case of consolidated porous media with negligible shrinkage (e.g. most building materials and natural mineral products like fragmented rocks), the relationship between gaseous porosity and moisture content is obvious; as the drying process progresses, liquid water is simply replaced by gas. However, for highly deformable materials such as food, a balance exists between volume change and porosity. It becomes necessary to know whether the loss in moisture content turns into volume reduction or into an increase of porosity (Perré and May, 2001).

Coupled heat and vapor transfer occurs across the boundary layer depicted in Fig. 1.3. The heat flux supplied by the airflow is used solely for transforming the liquid water into vapor. During this stage, the temperature at the surface is equal to the wet-bulb temperature. Moreover, because no energy transfer occurs within the medium during this period, the product temperature remains at the wet-bulb temperature throughout the thickness.

The exposed surface is supplied with liquid water from the inside of the product by capillary action: the liquid migrates from regions with high moisture content (liquid/gas interfaces within large pores) towards regions with low moisture content (liquid/gas interfaces within small pores). This liquid flow is expressed by Darcy's law (permeability × gradient of capillary pressure).

The constant drying rate period lasts as long as the surface is supplied with liquid. Its duration depends strongly on the drying conditions (magnitude of the external flux) and on the medium properties.

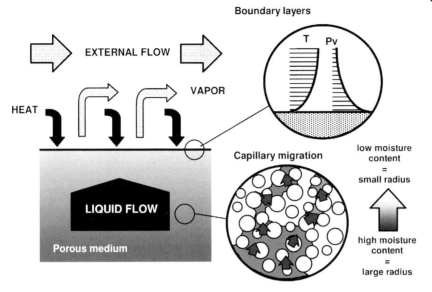

Fig. 1.3 Constant drying rate period: the moisture migrates inside the medium mostly by capillary forces, evaporation occurs at the exchange surface with a dynamic equilibrium within the boundary layer between the heat and the vapor flows [after Perré, 1996].

1.3.1.2 The Decreasing Drying Rate Period

Once the surface attains the hygroscopic range, the vapor pressure becomes smaller than the saturated vapor pressure (Fig. 1.4). Consequently, the external vapor flux is reduced and the heat flux supplied to the medium is temporarily greater than what is necessary for liquid evaporation. The excess energy is used to heat the product, the surface at first, followed by the inner part by conduction. A new, more subtle, dynamic equilibrium takes place. The surface vapor pressure, and hence the external vapor flow, depends on both temperature and moisture content. To maintain the energy balance, the surface temperature increases as the surface moisture content decreases. This leads to a decreasing drying rate where the heat supplied by the airflow becomes progressively smaller and smaller.

A two-zone process develops inside the porous medium: an inner zone where liquid migration prevails and a surface zone, where both bound-water and water-vapor diffusion take place. During this period, a conductive heat flux must exist inside the medium to increase the temperature and to evaporate the liquid driven by gaseous diffusion. The region of liquid migration naturally reduces as the drying progresses and finally disappears. The process is finished when the temperature and the moisture content attain, respectively, the outside air temperature and the equilibrium moisture content.

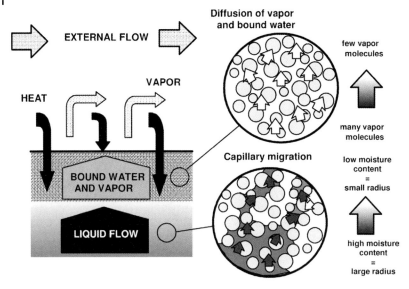

Fig. 1.4 Second drying period: A region in the hygroscopic range develops from the exposed surface. In that region, both vapor diffusion and bound water diffusion act. Evaporation takes place partly inside the medium. Consequently, a heat flux has to be driven towards the inner part of the material by conduction [after Perré, 1996].

1.3.2
Drying at High Temperature: The Effect of Internal Pressure on Mass Transfer

When the total gaseous pressure, which is the sum of the partial pressures of air and vapor, exceeds the external pressure, a pressure gradient drives the moisture (liquid and/or vapor) towards the exchange surfaces (Lowery, 1979; Kamke and Casey, 1988; Perré, 1995). This is the definition of high-temperature conditions, which are a common way to reduce the drying time.

In the presence of liquid water within the material, the mentioned condition is inevitably fulfilled when the product temperature is above the boiling point of water (100 °C at atmospheric pressure). This is the aim of convective drying at high temperature (moist air or superheated steam) and a possible aim of contact drying or drying with an electromagnetic field (microwave or radio-frequency).

However, as shown in Fig. 1.5, it is possible to reduce the boiling point of water by decreasing the external pressure and, consequently, to obtain a high-temperature effect with relatively moderate drying conditions. This is the principle of vacuum drying, which is particularly useful for products that would be damaged by high temperature levels.

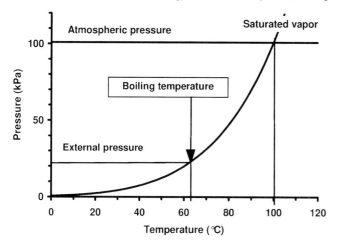

Fig. 1.5 Vacuum drying seeks to reduce the boiling point of water in order to obtain a high-temperature effect with moderate drying conditions.

Due to the very anisotropic behavior of wood, in terms of permeability, this process is especially efficient for lumber drying. Whenever an overpressure exists inside a board, the large anisotropy ratios produce intricate transfer mechanisms. Heat is often supplied in the direction of the thickness, while, in spite of the length, the effect of the pressure gradient on gaseous (important for low moisture content) or liquid migration (important for high moisture content) takes place in the longitudinal direction (Fig. 1.6). This is a result of the anatomical features of wood. In the case of very intensive internal transfer, the endpiece can be fully saturated, and sometimes, moisture can leave the sample in the liquid state. This phenomenon is clearly observable also during microwave heating.

1.4
Computational Strategy to Solve the Comprehensive Set of Macroscopic Equations

TransPore is a finite-volume-based computational model (Patankar, 1980; Ferguson and Turner, 1996; Jayantha and Turner, 2003; Turner and Perré, 1996). The finite-volume method was chosen as the preferred spatial discretization strategy due to its conservative nature, which ensures conservation at any discrete level, across individual control volumes (CVs) or a group of CVs, within the mesh (Patankar, 1980). The flexibility of the implementation permits an arbitrarily shaped mesh to be employed for the computational domain and any meshing software can be used to construct this mesh. For the applications reported throughout this chapter, where wood growth ring structures are captured within the computational domain, we typically tessellate the domain with triangles using the Easymesh software (Bojan Niceno) and construct control volumes around the triangular element vertices in what is known as a

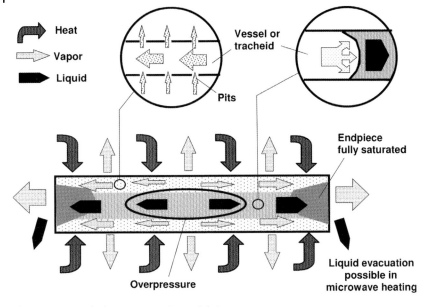

Fig. 1.6 Drying at high temperature (second drying period): a high-temperature regime means that an overpressure develops inside the medium. Depending on the moisture content this overpressure induces liquid and/or gaseous flow; In wood, which is strongly anisotropic, the majority of the flow occurs in the longitudinal direction (magnified views).

vertex-centred sense (Perré and Turner, 2000, 2002; Jayantha and Turner, 2003). Each triangle is given specific material characteristics, such as density for example, that are used throughout the simulations. This discretization approach, which is often referred to as the control volume finite element (CV-FE) method, is the ideal numerical technique for drying because mesh elements have partial subcontrol volumes (SCVs) defined within them that comprise part of a CV, and fluxes that need to be approximated at SCV faces lie entirely within a given element that has a unique set of material properties associated with it (Perré and Turner, 2002). In this way, the CV-FE method allows the heterogeneous nature of the porous medium to be dealt with in a seamless manner. Rapid changes in material properties that arise at any given SCV face no longer cause difficulty to flux conservation using this approach. Several options exist for the construction of the CVs, provided of course that when assembled together they span the computational domain and do not overlap with one another. Figure 1.7 depicts the strategy we have adopted for constructing the SCVs within the mesh elements. For any given element vertex there are two SCV faces where fluxes must be evaluated. The SCV face is defined by the line joining the element centroid to the midpoint of an element edge. Each element then contains a total of three SCV faces, each of which contributes one part to the control volume surrounding each vertex.

1.4 Computational Strategy to Solve the Comprehensive Set of Macroscopic Equations | 13

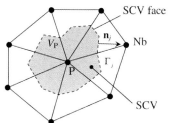

Fig. 1.7 A single control volume and notation.

1.4.1
The Control-volume Finite-element (CV-FE) Discretization Procedure

The conservation laws (Eq. 1.18, Eq. 1.19 and Eq. 1.20) given above can be cast into the following conservative form

$$\frac{\partial \psi_l}{\partial t} + \nabla \cdot \mathbf{J}_l = 0, \quad l = \text{water, air, energy} \tag{1.24}$$

Throughout this section we choose to illustrate the CV-FE discretization process only for the water-conservation law where $\psi_w = \varepsilon_w \rho_w + \varepsilon_g \rho_v + \rho_b$ and

$$\mathbf{J}_w = -\rho_w \frac{\overline{\overline{\mathbf{K}}}_w \overline{\overline{\mathbf{k}}}_w}{\mu_w} \nabla \varphi_w - \rho_v \frac{\overline{\overline{\mathbf{K}}}_g \overline{\overline{\mathbf{k}}}_g}{\mu_g} \nabla \varphi_g - \overline{\overline{\mathbf{D}}}_{\text{eff}} \rho_g \nabla \omega_v - \overline{\overline{\mathbf{D}}}_b \nabla \overline{\rho}_b$$

(Eq. 1.18, Eq. 1.21 and Eq. 1.22).

The procedure is identical for the remaining two conservation laws and a similar exposition results. The discretized form of the partial differential Eq. 1.24 is obtained by integrating over the CV denoted V_p in Fig. 1.7 and applying the Gauss divergence theorem to obtain

$$\frac{d\overline{\psi}_w}{dt} + \frac{1}{\Delta V_p} \sum_{j=1}^{Nf_p} \int_{\Gamma_j} \mathbf{J}_w \cdot \mathbf{n} \, d\sigma = 0 \tag{1.25}$$

where $\overline{\psi}_w = \frac{1}{\Delta V_p} \int_{V_p} \psi_w dV$ is the cell averaged value of ψ_w over the CV, ΔV_p is the volume (or more correctly area in two dimensions) of the CV and the line integral over the control-volume boundary has been expressed as the sum of the integrals over each SCV face Γ_j, $j = 1, \ldots, Nf_p$ that has unit normal $\mathbf{n}_j$. Nf_p is the set of faces that constitute the CV surrounding the vertex p. We then follow the classical finite-volume theory (Patankar, 1980; Jayantha and Turner, 2003) and employ a single midpoint quadrature rule, which yields second-order spatial accuracy when the flux $(\mathbf{J}_w \cdot \mathbf{n})_j$ is known exactly at the midpoint of the SCV face assumed here to have length A_j

$$\frac{d\overline{\psi}_w}{dt} + \frac{1}{\Delta V_p} \sum_{j=1}^{Nf_p} (\mathbf{J}_w \cdot \mathbf{n})_j A_j = 0 \tag{1.26}$$

Note, however, that other integration rules based on Gauss quadrature may be employed for Eq. 1.25 to increase the spatial accuracy of the approximation (Moroney and Turner, 2006). Assuming now that $\overline{\psi}_p \approx \psi_p$ the value at the representative node for the CV (which is in this case an element vertex), and considering the time integral from $n\delta t$ to $(n + 1)\delta t$, the discrete analogue of Eq. (1.24) becomes

$$\psi_p^{n+1} - \psi_p^n + \frac{\delta t}{\Delta V_p} \sum_{j=1}^{Nf_p} (\mathbf{J}_w \cdot \mathbf{n})_j^{n+1} A_j = 0 \tag{1.27}$$

where δt is the discrete time step, and the superscripts $n, n + 1$ denote the current and next time levels, respectively. The accuracy of this approximation greatly depends on the evaluation of the flux terms at the SCV faces. The particular flux estimates utilized for *TransPore* are summarized throughout the following subsections.

The flux function $\mathbf{J}_w$ comprises advective and diffusive components. The way in which these terms are approximated at the SCV faces is crucial to the overall accuracy and monotonicity of the solution (Truscott and Turner, 2005). The diffusive terms evident in Eq. 1.27 require the evaluation of the tensors $\overline{\overline{D}}_{\text{eff}}$, $\overline{\overline{D}}_b$ the secondary variable ρ_g and the gradients $\nabla \omega_v$, $\nabla \overline{p}_b$ at the SCV face. If $N_k^j(x,y)$ are the usual Lagrange polynomial C^0 basis functions then the gradient for the jth element associated with the SCV face can be written, for example for ω_v as $(\nabla \omega_v)_j = \sum_{k=1}^{3} \nabla N_k^j \omega_k^j$ (Dhatt and Touzot, 1984). One notes, however, that more accurate gradient approximations can be determined if one is prepared to overlook the additional complexities involved with the implementation (Jayantha and Turner, 2003; Moroney and Turner, 2006). The advective terms in Eq. 1.27 require that secondary variables such as ρ, μ and $\overline{\overline{\Lambda}}$ be approximated. For the latter tensor, this implies evaluation of the components of the liquid mobility tensor at the SCV face, which is the main focus of the next subsection.

1.4.2
Evaluation of the Tensor Terms at the CV Face

As can be seen from Eq. 1.27 it is necessary to define spatial averaging techniques for the evaluation of the tensor terms at the SCV faces. Typically, the components of the diffusion tensors $\overline{\overline{D}}_{\text{eff}}$ and $\overline{\overline{D}}_b$ are averaged. The correct treatment of the advection terms within the discrete conservation laws is essential to the accuracy and performance of the numerical scheme and ensures that smearing of the drying fronts is avoided (Turner and Perré, 2001). In *TransPore*, flux limiting (Sweby, 1984; Unger et al., 1996) is used as the spatial weighting scheme, which is more complicated than first-order upstream weighting, or upwinding. This technique allows coarse meshes to be employed for the simulations and maintains reasonably accurate results. This spatial weighting scheme is summarized for the components of the mobility tensors as:

$$(k)_{FL(p,nb_i)} = k_{ups} + \frac{\sigma(r)}{2}\left(k_{dwn} - k_{ups}\right) \tag{1.28}$$

where k_{ups} and k_{dwn} represent the values of the mobility-tensor components at the upstream and downstream points between nodes p and its neighbor nb_i, respectively, chosen according to the sign of the flow direction indicator (FDI) $(\mathbf{v} \cdot \mathbf{n})_{cvface}$ (see Truscott and Turner, 2005 for further details). Care must be taken with the construction of the FDI to avoid oscillatory behavior, and subsequent divergence of the outer nonlinear Newton iterations. Typically, the best results require that the FDI be constructed as the sum of fluxes through the two common SCV faces that share the same element edge, and that the most up-to-date iterate information must be used for this construction (Truscott and Turner, 2005).

The limiter $\sigma(r)$ is a function of the smoothness sensor $r_{(p,nb_i)}$ (Turner and Perré, 2001). The smoothness sensor $r_{(p,nb_i)}$ utilizes the ratio of the FDIs through the appropriately chosen SCV faces of the second upstream and upstream locations, respectively:

$$r_{(p,nb_i)} = \frac{(\mathbf{v} \cdot \mathbf{n})_{cvface_{2ups}}}{(\mathbf{v} \cdot \mathbf{n})_{cvface_{ups}}} \tag{1.29}$$

We favor the use of the van Leer flux limiter $\sigma(r) = \frac{r+|r|}{1+|r|}$ (van Leer, 1974) and the second upstream point (2ups) is the mesh node found by using the maximum flow method discussed elsewhere (Forsyth, 1994; Unger et al., 1996; Turner and Perré, 2001), which seeks to track the location of the streamline from the upstream point. The effect of the derivatives of the second upstream point variables has been ignored during the construction of the Jacobian matrix discussed below (Truscott and Turner, 2005).

1.4.3
Solution of the Nonlinear System

After integrating over each CV within the computational domain, a system of nonlinear equations results that has as its coordinates discrete analogues of the drying conservation equation. The resulting coupled nonlinear coordinate function set $f_w(\mathbf{u})$, $f_e(\mathbf{u})$ and $f_a(\mathbf{u})$ for a given control volume depends on the state variable triplet $(X, T, \bar{\rho}_a)$ at the node p and all of the neighboring nodes that surround the control volume under consideration. These coordinate functions combine the nonlinearities of the transport coefficients with the relevant geometric factors generated from the CV-FE discretization process. The solution vector $\mathbf{u}$ contains the primary variables (stored as triplets) for each CV within the mesh. Once the nonlinear functions have been assembled for each CV within the computational domain, the result is a large system of nonlinear equations, expressed here as $\mathbf{F}(\mathbf{u}) = 0$. This system then must be solved at each time step in order to advance all of the primary variables in time. The underlying iterative solution strategy is carried out in two distinct phases that are known as outer and inner iteration, each of which is elaborated upon in subsequent paragraphs.

1.4.3.1 Outer (Nonlinear) Iterations

Newton's method forms the basis for most modern methods of solving nonlinear systems of equations (Kelley, 1995). The method constructs a set of iterates of the following form:

$$\mathbf{u}^{(n+1)} = \mathbf{u}^{(n)} + \delta \mathbf{u}^{(n)}, \quad \mathbf{J}(\mathbf{u}^{(n)})\delta \mathbf{u}^{(n)} = -\mathbf{F}(\mathbf{u}^{(n)}) \quad (1.30)$$

The process commences with an initial approximation $\mathbf{u}^{(0)}$ and generates a sequence of iterates that converge to the desired solution provided the initial approximation is sufficiently close to the root. The process gives rapid (*quadratic*) convergence to the root in this case (see Kelley, 1995). The vector quantity $\delta \mathbf{u}^{(n)}$ is known as the *Newton step or Newton search direction*. Note that if $\mathbf{J}$ is *singular* at any $\mathbf{u}^{(n)}$ during the computations then $\delta \mathbf{u}^{(n)}$ is not be defined. Furthermore, the Newton algorithm when applied to the solution of the nonlinear system under consideration here, may not converge for a given time step due to the occurrence of nonphysical values of the primary variables, or due to the generation of a large Newton step throughout the iterations. In order to remedy these limitations, a globally convergent method is used (Jennings and McKeown, 1992; Press et al., 1992; Kelley, 1995), which seeks to minimize the square of the Euclidean norm of the residual of the error vector:

$$g(\mathbf{u}^{(n)}) = \| \mathbf{F}(\mathbf{u}^{(n)}) \|^2 = \sum_{i=1}^{N} f_i^2(\mathbf{u}^{(n)}). \quad (1.31)$$

The globally convergent Newton's method is summarized as follows:

$$\mathbf{u}^{(n+1)} = \mathbf{u}^{(n)} + \lambda^{(n)} \delta \mathbf{u}^{(n)}, \quad \mathbf{J}(\mathbf{u}^{(n)})\delta \mathbf{u}^{(n)} = -\mathbf{F}(\mathbf{u}^{(n)}). \quad (1.32)$$

A line-search strategy is used to ensure that a decrease in $g(\mathbf{u})$ occurs at each iteration. Since it can be shown that the Newton search direction is always a descent direction of $g(\mathbf{u})$, it follows that a small move from $\mathbf{u}^{(n)}$ along $\delta \mathbf{u}^{(n)}$ will always cause $g(\mathbf{u})$ to decrease (Kelley, 1995). The objective of the line search is to backtrack along the Newton search direction until a value of $\lambda^{(n)}$ is found such that $g(\mathbf{u}^{(n)} + \lambda^{(n)} \delta \mathbf{u}^{(n)})$ has decreased sufficiently. The full Newton step $\lambda^{(n)} = 1$ is trialled at first to determine if a sufficient decrease has occurred, which ensures quadratic convergence when $\mathbf{u}^{(n)}$ is sufficiently close to the zero of $\mathbf{F}(\mathbf{u})$. If this step is rejected then it is possible to backtrack along the Newton search direction using a variety of line-searching methods that range in levels of sophistication. In *TransPore*, a parabolic two-point line search followed by a cubic three point search (Jennings and McKeown, 1992; Press et al., 1992; Kelley, 1995), if necessary, is used to determine the correct value of $\lambda^{(n)}$ to use in Eq. 1.32. The *Armijo, Goldstein and Price* rules for selecting $\lambda^{(n)}$ are used to ensure that a sufficient decrease in g(**u**) has occurred during the line-search procedure (Jennings and McKeown, 1992).

1.4.3.2 Construction of the Jacobian

The Jacobian matrix must be generated and then the corresponding linearized system solved at each iteration of the Newton method. As a result of the CV-FE discretization process, the coordinate functions $f_w(\mathbf{u})$, $f_e(\mathbf{u})$ and $f_a(\mathbf{u})$ depend upon only a small subset of the entire solution vector. The construction of the Jacobian matrix is complicated and requires the derivatives of the accumulation term and all flux terms within the discrete conservation laws to be approximated. We calculate these derivatives using first-order finite-difference approximations (Turner and Perré, 1996; Truscott and Turner 2005; Moroney and Turner, 2006). The perturbation parameter used in this approximation is chosen large enough to avoid round off errors and small enough to elude a poor approximation for the derivative. This process requires that all terms in the nonlinear function be shifted in terms of the dependent (state) variables at each SCV face, which can be computationally demanding and efficient coding strategies must be devised to enable efficient simulations (Truscott and Turner 2005; Moroney and Turner, 2006).

In order to minimize possible sources of floating-point error in the Newton method, particularly for the construction of the approximate Jacobian, the system of nonlinear equations is scaled using diagonal matrices $\mathbf{D}_e^{-1}\mathbf{F}(\mathbf{D}_v\mathbf{v}) = \mathbf{0}, \mathbf{v} = \mathbf{D}_v^{-1}\mathbf{u}$. In particular, the conservation equations are scaled by the inverse of the diagonal matrix $\mathbf{D}_e = \mathrm{diag}(\rho_R, \Delta h_v, 1, \ldots, \rho_R, \Delta h_v, 1)$ to bring them all roughly to the same order of magnitude and the primary variables are scaled by the inverse of the diagonal matrix $\mathbf{D}_v = \mathrm{diag}(1, T_R, 1, \ldots, 1, T_R, 1)$ to ensure that they are all approximately O (1). ρ_R is a reference density representative of the solid-phase density, usually ρ_s for an homogeneous porous medium, and T_R a reference temperature representative of the drying conditions. This scaling enables the same shift parameter to be employed for the approximation of the Jacobian entries and also can aid in enhancing the orthogonality of the Krylov subspace basis vectors by controlling the impact of repeated matrix multiplication by the Jacobian throughout the Arnoldi process in GMRES (Saad, 2003), which is discussed next.

1.4.3.3 Inner (Linearized System) Iterations

As discussed above, each of the coordinate functions of $\mathbf{F}(\mathbf{u})$ depends only upon a small subset of all the discrete unknowns, the underlying Jacobian matrix has a block sparse structure, comprising 3×3 blocks. A reverse Cuthill–McKee (RCM) node reordering scheme (see George and Liu, 1981) is first applied to the matrix graph to reorder the unknowns to guarantee that the bandwidth of the Jacobian is tight. This preprocessing phase enables the Newton search direction to be determined at each iteration using an efficient block ILU(0) right preconditioned iterative GMRES solver (Saad, 2003), which assumes the form

$$\mathbf{J}^{(n)}\mathbf{M}^{-1}\mathbf{x} = -\mathbf{F}^{(n)}, \quad \mathbf{x} = \mathbf{M}\delta\mathbf{u}^{(n)} \tag{1.33}$$

Krylov methods are the favored means for solving the linearized system (1.30) because they naturally exploit the sparsity of the Jacobian. A recent error bound on the norm of the linear residual given by Ilić and Turner (2005) highlighted the role that eigenvalues, both small and large in magnitude, play on the convergence of Krylov subspace methods. To overcome the problems associated with residual stagnation, we use the block ILU(0) preconditioner in an attempt to cluster the (large) eigenvalues to around one and then utilize the implementation of GMRES-DR due to Morgan (2002) to deflate any remaining troublesome eigenvalues that are small in magnitude, from the spectrum $\sigma(\mathbf{JM}^{-1})$ during the iterations. To achieve this goal, GMRES-DR uses the augmented Krylov subspace (with superscripts omitted for clarity) given by

$$K_m = \text{span}\{\mathbf{v}_1, \mathbf{v}_2, \ldots, \mathbf{v}_k, \mathbf{JM}^{-1}\mathbf{r}_0, (\mathbf{JM}^{-1})^2\mathbf{r}_0, \ldots, (\mathbf{JM}^{-1})^{m-k}\mathbf{r}_0\}$$

where r_0 is the initial residual vector given by $r_0 = \mathbf{F} + \mathbf{JM}^{-1}\delta\mathbf{u}_0$, the matrix $\mathbf{M}^{-1}$ is the right block ILU(0) preconditioner, the m vectors are the usual Arnoldi vectors (Saad, 2003) and the $\mathbf{v}_i, i = 1, , k$ are approximate eigenvectors computed using a harmonic Ritz projection. The harmonic Ritz projection is preferred over a Ritz projection because it provides good approximations to the interior eigenvalues of the preconditioned Jacobian matrix $\mathbf{JM}^{-1}$ located near the origin (Morgan, 2002; Saad, 1997). The solution is then extracted from the affine space $\delta\mathbf{u}^{(n)} \in \delta\mathbf{u}_0 + \mathbf{M}^{-1}K_m$. The GMRES-DR algorithm first orthogonalizes the (current) harmonic Ritz vectors $\mathbf{V}_k = [\mathbf{v}_1, \mathbf{v}_2, \ldots, \mathbf{v}_k]$ to produce the matrix $\tilde{\mathbf{V}}_k$. Then $m - k$ iterations of the Arnoldi recurrence are performed while still maintaining orthogonality to $\tilde{\mathbf{V}}_k$ to produce the Arnoldi-like relation

$$\mathbf{JM}^{-1}[\tilde{\mathbf{V}}_k \; \mathbf{V}_{m-k}] = [\tilde{\mathbf{V}}_k \; \mathbf{V}_{m-k+1}]\overline{\mathbf{H}}_m, \tag{1.34}$$

in which $\overline{\mathbf{H}}_m$ is upper Hessenberg except for its leading dense $(k+1) \times (k+1)$ submatrix. The solution is then updated as

$$\delta\mathbf{u}^{(n)} = \delta\mathbf{u}_0 + \mathbf{M}^{-1}[\tilde{\mathbf{V}}_k\mathbf{V}_{m-k}]\mathbf{y}_m, \quad \mathbf{y}_m = \text{argmin}_y \left|[\tilde{\mathbf{V}}_k\mathbf{V}_{m-k+1}]^T\mathbf{r}_0 - \overline{\mathbf{H}}_m\mathbf{y}\right|_2$$

Next, GMRES-DR sees the harmonic Ritz vectors associated with the k smallest in magnitude harmonic Ritz values computed with the aid of Relation (1.34) and finally the process is restarted again using those vectors in the next cycle to hopefully improve the convergence rate offered by the classical restarted GMRES.

1.5
Possibilities Offered by this Modeling Approach: Convective Drying

1.5.1
High-temperature Convective Drying of Light Concrete

Light concrete, often designated by a brand name (Ytong), is one of the porous media that has been studied in detail, especially by Krischer and Kröll (1963). This is an isotropic medium that allows reproducible experiments, excellent for model validation.

Among the several tests performed, the results of the three most representative are discussed here [see Perré et al. 1993 for further details].

- Test 1 involved a superheated steam flow with velocity $V = 6$ m s^{-1} and dry-bulb temperature $T_{dry} = 145$ °C.
- Test 2 used moist air flow ($V = 4$ m s^{-1}), with a temperature difference $\Delta T_{dw} = T_{dry} - T_{wet} = 30$ K, where T_{dry} is the dry-bulb temperature and T_{wet} is the wet-bulb temperature. This test will be denoted as "soft".
- Test 3 also used moist air with $\Delta T_{dw} = 90$ K and a high velocity ($V = 10$ m s^{-1}). This test will be denoted as "severe".

Tests were carried out in a wind tunnel that can hold samples with a length of up to 1 m (Fig. 1.8). The fluid velocity is in the range 2–15 m s^{-1} and the dry air temperature is accurate to one or two degrees Celsius in the range 40–200 °C. An external boiler generating low-pressure steam allows regulation of the wet-bulb

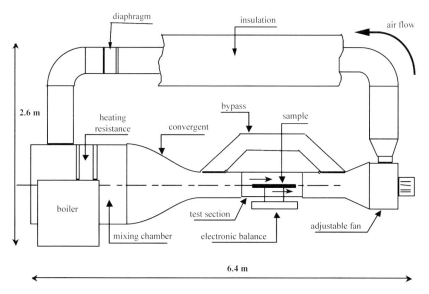

Fig. 1.8 Scheme of the wind-tunnel drier [adapted from Perré and Martin, 1994].

1.5.1.1 Test 1: Superheated Steam

By comparison with Test 2 (Fig. 1.11), it can be seen that the main effect of high-temperature convective drying is a significant reduction in the drying time (Figs. 1.9 and 1.10). Actually, when $T_{dry} = 145$ °C and $V = 6$ m s^{-1} seven hours of drying are sufficient to reduce the moisture content of a slab of light concrete (dimensions in Fig. 1.10) from 90% to around 2%.

The measured temperature versus time curves show a long plateau at the boiling temperature ($T_{boiling} = 100$ °C), followed by a more or less sudden increase towards the steam temperature. The deeper the location of the thermocouple within the slab, the longer is the duration of the plateau.

With superheated steam, the wet-bulb temperature equals the boiling point of water; therefore, the transition between the first drying period (evaporation from the exchange surface) and the second drying period (internal evaporation) has no discernible effect on the measured internal temperatures.

All the calculated results (kinetics, drying time, temperature at different locations) are very close to the experimental results. Note that due to the edge effect and because the medium is isotropic, the endpiece that is exposed to the flow dries faster than the half-length section of the sample.

The simulation shows that the long period at 100 °C observed during the experiment is, in fact, divided into two periods (Fig. 1.10):

- The first period is a real constant drying rate period, with liquid migration up to the surface, together with surface evaporation and a vapor flux in the boundary layer. In the numerical

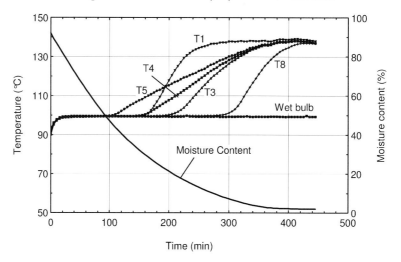

Fig. 1.9 Convective drying at high temperature of light concrete. Experimental results (superheated steam: Tdry ≈ 145 °C, V = 6 m s^{-1}).

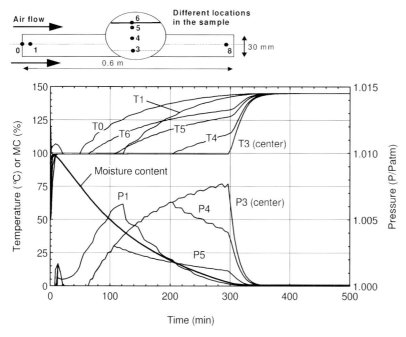

Fig. 1.10 Convective drying at high temperature of light concrete. Numerical simulation (superheated steam: $T_{dry} = 145\,°C$, $h = 30\,W\,m^{-2}\,K^{-1}$).

simulation, the surface temperature (T_6) is available, which allows the end of the constant drying rate period to be marked at the point when the surface temperature starts to increase above the boiling temperature, after around 60 min.
- In the second period, a drying front appears within the medium and internal evaporation becomes evident. Nevertheless, the gaseous permeability is high enough to evacuate the vapor flux by a slight overpressure. The temperature that produces this overpressure according to the saturated vapor curve is very close to 100 °C. The temperature increases only in the dry zone, where a thermal gradient is required in order to transport the energy of vaporization towards the front position.

The inward advance of the front can be followed by the successive breakaway of the different internal temperatures (T_5 and T_4). When the centre temperature (T_3) increases beyond 100 °C, the first drying period has long been completed. At this stage, the whole medium is in the hygroscopic range and the process is close to an end. The drying rate decreases sharply and temperatures increase rapidly.

Due to a rather high value of the intrinsic permeability, the internal pressure remains low during the entire process and is very difficult to detect. However, this information is available in the numerical simulation. The evolution of the internal pressure at different locations has been plotted in Fig. 1.10. The first peak of pressure, between 10 and 20 min, is due to the initial increase in temperature. Then, the level of pressure results in a pressure gradient that drives the vapor towards the exchange surface via a diffusive-convective mechanism. This explains why no overpressure is observed during the first drying period.

1.5.1.2 Tests 2 and 3: Moist Air, Soft and Severe Conditions

Due to the separation between the wet-bulb temperature and the boiling point, the case of moist air is more interesting to analyse. Because of their similarities, we treat the "soft" ($\Delta T_{dw} = 30$ K, Figs. 1.11 and 1.12) and the "severe" conditions ($\Delta T_{dw} = 90$ K, Figs. 1.13 and 1.14) together. It can be seen in Fig. 1.13 that, as a result of the large difference $T_{dry} - T_{wet}$, the temperature regulation is not as accurate in the wind-tunnel dryer.

For the "soft" operating conditions, the drying process is much longer (650 min for test 2 instead of 200 min for Test 3). The form of the temperature curves (Figs. 1.11 and 1.12) indicates three different phases:

- During the first two hours, the temperature curves show a plateau region at the wet-bulb temperature (85 °C), during which the slab temperature is uniform. This proves the existence of a constant drying rate period.
- After this period, the curves present a second plateau or a horizontal inflexion above T_{wet}, in the range 85–90 °C.
- Finally, the temperatures continue to increase and reach the flow temperature without any pronounced plateau at the boiling temperature $T_{boiling}$.

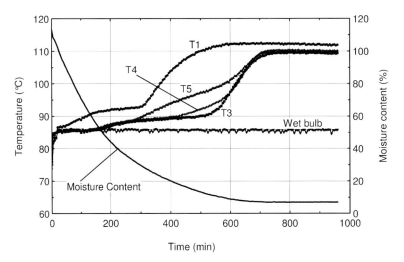

Fig. 1.11 Convective drying at high temperature of light concrete. Experimental results (moist air "soft conditions": $T_{dry} \approx 115$ °C, $T_{wet} \approx 85$ °C, $V = 4$ m s^{-1}).

1.5 Possibilities Offered by this Modeling Approach: Convective Drying

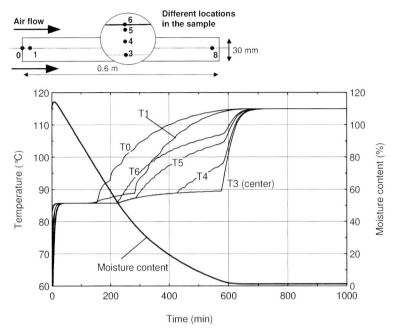

Fig. 1.12 Convective drying of light concrete. Numerical simulation (moist air "soft conditions": $T_{dry} = 115\ °C$, $T_{wet} = 85\ °C$, $h = 25\ W\ m^{-2}\ K^{-1}$).

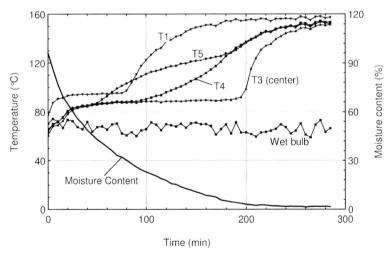

Fig. 1.13 Convective drying of light concrete. Experimental results (moist air "severe conditions": $T_{dry} \approx 160\ °C$, $T_{wet} \approx 70\ °C$, $V = 10\ m\ s^{-1}$).

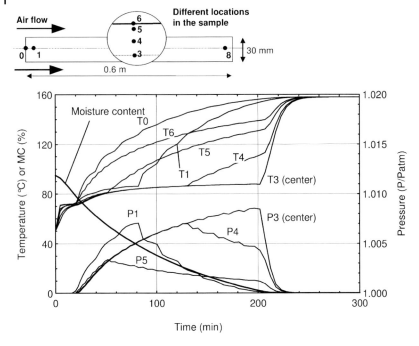

Fig. 1.14 Convective drying of light concrete. Numerical simulation (moist air "severe conditions": $T_{dry} = 158\ °C$, $T_{wet} = 70\ °C$, $h = 35\ W\ m^{-2}\ K^{-1}$).

In the case of "severe conditions" the essential point to note is the rapid drying process, the temperature variations being similar as in the previous test. The temperatures pass through a slight plateau at $T_{wet} = 70\ °C$, then increase up to a second plateau in the range 85–90 °C, beyond which they rise steeply through the boiling temperature until they attain the dry-bulb temperature of the moist air flow.

In both runs, the numerical curves show a first drying period for which all the temperatures remain close to the wet-bulb temperature. However, this period is much shorter ($\approx$20 min compared to 200 min) in the case of severe drying conditions (Figs. 1.13 and 1.14). A period of decreasing drying rate then appears. The successive departures of the temperatures from the value at the center clearly indicate that a drying front moves within the medium. Up to this point, the process remains similar to what was observed for superheated steam.

What is new here is the variation of the center temperature. At the end of the constant drying rate period, this value increases slowly from the wet-bulb temperature up to about 90 °C (i.e. 10 °C below the boiling point). It is to be noted that this evolution is the same for the two tests (low and severe drying conditions). This stage, at around 90 °C, comes from the dynamic balance that exists, in the dry layer, between the thermal flux driven by convection and the total vapor flux times the specific enthalpy of evaporation. In the case of light concrete, the convective flux induced by vapor diffusion allows the total vapor flux imposed by thermal conduction to be evacuated at a front temperature significantly below the boiling point of water. A simple analytical model allowed this phenomenon to be explained and predicted (Perré et al., 1993).

The same mechanism occurs for the temperature T_1 (at around one centimetre from the endpiece). However, this part of the slab, which receives the bulk of the incoming air flow directly, undergoes much more severe drying conditions. Consequently, the constant drying rate period does not exist in this region. The endpiece enters directly in the second stage of drying and tends towards the above-mentioned temperature, dictated by a dynamical balance between vapor flux and heat supplied.

In the simulated results, all of the trends are well predicted. The fact that the endpiece does exhibit a short constant drying rate period in the simulation is primarily because transfer coefficients are assumed to be uniform throughout the sample.

Once the front reaches the center of the slab all temperatures increase rapidly up to the dry-bulb temperature. This final period resembles that observed in the test with saturated vapor.

1.5.2
Typical Drying Behavior of Softwood: Difference Between Sapwood and Heartwood

In a freshly felled and sawn tree, it is easy to distinguish sapwood from heartwood by touch or by sight. This is not possible a few days later, because the surface loses moisture. Nevertheless, in the case of high-temperature drying, the increase of internal pressure gives rise to longitudinal migration of liquid towards the endpieces, provided that the permeability and the moisture content are high enough. This phenomenon can be observed in industrial kilns, and allows us to identify sapwood after a few hours of drying (Fig. 1.15).

To illustrate differences in the drying process, drying experiments were carried out with superheated steam at 150 °C on both sapwood and heartwood of Norway spruce

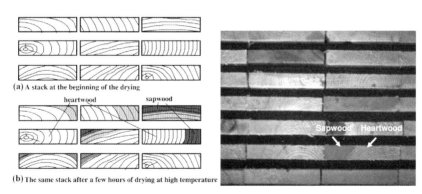

Fig. 1.15 A stack of boards during high-temperature drying (shaded areas indicate wet zones).

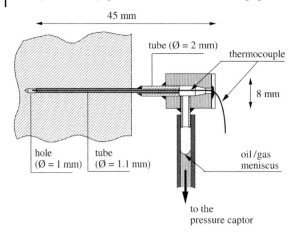

Fig. 1.16 Schematic diagram of airtight temperature-pressure sensor.

(*Picea abies*). For these experiments, small airtight sensors (Fig. 1.16) have been designed (Perré et al., 1993; Perré and Martin, 1994). This sensor of 1.1 mm diameter allows temperature and pressure to be measured simultaneously and at the same point. Note that the pressure gauge placed outside the test section is at room temperature. In order to prevent condensation into this captor, which would induce local evaporation within the board, the connecting tube was filled with oil, taking care to ensure an oil/gas interface inside the test chamber.

Representative experimental results are depicted in Fig. 1.17. The main advantage of high-temperature convective drying is an acceleration of internal moisture transfer due to the development of an overpressure in the gaseous phase inside the board. After the initial transient period, the constant drying-rate period takes place for a sapwood board. During this period, which lasts for several hours, all temperatures equilibrate at the wet-bulb temperature and the overpressure remains very small. At the beginning of the second drying period (after around 350 min), an important overpressure develops due to the temperature increase, which disappears only once the entire board enters the hygroscopic range. At this moment, all temperatures approach the dry-bulb temperature and moisture content tends towards the EMC (equilibrium moisture content). Indeed, at EMC, the water activity is such that the partial pressure of vapor equals the atmospheric pressure corresponding to when the board is at the dry-bulb temperature.

The results obtained for heartwood (lower plot of Fig. 1.17) are quite different. No constant drying-rate period can be observed. Just a short plateau at the boiling point is detectable at the rear end of the board (T_8). Consequently, the overpressure remains high (especially for the center pressure P_3) up to the end of drying. The maximum pressure is higher for heartwood than for sapwood. The differences in drying kinetics are of great interest. In spite of the high initial moisture content of sapwood (170% against 60%), the permeability of heartwood is so low that the curves cross each other

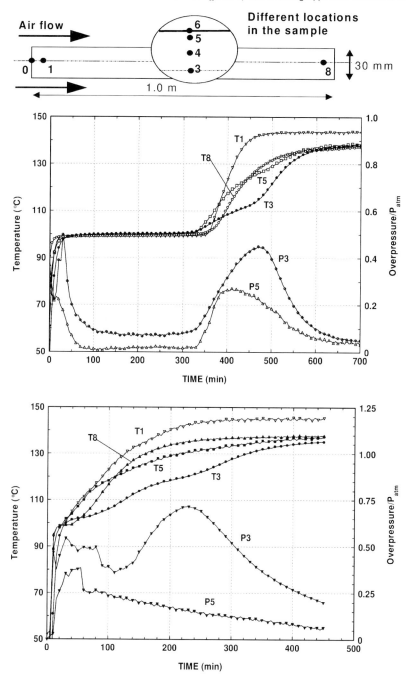

Fig. 1.17 Experiment on spruce (*Picea abies*) dried with superheated steam at 150 °C. For temperature and internal pressure at different locations note the difference between sapwood (top) and heartwood (bottom).

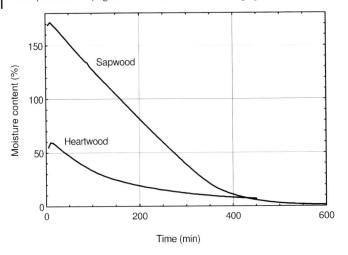

Fig. 1.18 Moisture-content loss obtained for sapwood and for heartwood (same experiments as shown in Fig. 1.17).

at 450 minutes of drying (Fig. 1.18). This is consistent with observations on entire stacks (Salin, 1989).

The strategy of simulating the differences between heartwood and sapwood lies in only two sets of parameters: permeability and initial moisture content (Tab. 1.1). The permeability values are based on considerations concerning pit aspiration.

Through the use of only these differences, all the trends observed for sapwood and heartwood were found in the simulated results (Perré and Martin, 1994; Perré and Turner, 1996). The most spectacular effect is the longitudinal flow that is driven by the overpressure (Fig. 1.19). After 5 h, in this case of high-temperature convective drying, the sapwood board delivers a large supply of water to the endpiece, while the heartwood endpiece is already within the hygroscopic range. These carpet plots should be compared with Fig. 1.15.

Tab. 1.1 Parameter values used in the modeling to distinguish sapwood and heartwood.

Parameters	Heartwood	Sapwood
Initial MC (dry basis, %)	70	180
Longitudinal permeability (m^2)		
gas	10^{-13}	2×10^{-13}
liquid	10^{-13}	10^{-12}
Transverse permeability (m^2)		
gas	10^{-16}	2×10^{-16}
liquid	10^{-16}	10^{-15}

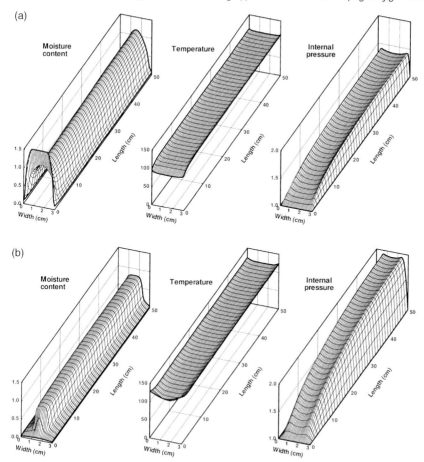

Fig. 1.19 High-temperature drying (140 °C/85 °C). Carpet plots after 5 h of drying. (a) sapwood: Evidence of internal overpressure, resaturation of the endpiece, thermal conduction along the thickness and endpiece close to the wet-bulb temperature. (b) heartwood: Absence of endpiece resaturation in spite of the high value of internal pressure.

1.6 Possibilities Offered by this Modeling Approach: Less-common Drying Configurations

1.6.1 Drying with Volumetric Heating

Through the source term Φ in the energy conservation Eq. 1.21, the proposed mathematical formulation is suitable for any drying process involving a volumetric heating (for example microwave drying or radio-frequency drying). In many cases it is also mandatory, because due to this thermal source term, it is very easy for the internal part of the product to attain and exceed the boiling point of water, so that the internal pressure must be computed and its effects on mass migration must be taken into account.

Moreover, one has to be aware that, in the domain of electromagnetic heating, intuition often produces misleading rules. Comprehensive models are therefore welcome and should be used widely to simulate such drying processes. For example, simple rules stipulate that the volumetric source term is proportional to the square of the averaged electromagnetic field E times the local material loss factor ε''. From this perspective one might deduce that wet boards in an entire stack should receive more energy during dielectric heating than other boards. In fact, what happens in practice is exactly the opposite because the actual electromagnetic field is very small in these boards (Bucki and Perré, 2003).

Two examples are depicted in this section: microwave and radio-frequency drying. In both cases, the source term field is obtained from a rigorous solution of Maxwell's equations (Perré and Turner, 1997; Bucki and Perré, 2003) and, the modeling approach is complex and quite tricky. Four variable fields are computed in these models, which enables the intricate coupling mechanisms that develop during drying to be captured during the simulations:

- The temperature and moisture-content fields allow the dielectric properties to be calculated within the sample. This part requires prior material characterization.
- Maxwell's equations are solved using this field together with the appropriate boundary conditions.
- The power field is used in the energy equation, which influences the temperature field.
- Temperature has a strong effect on mass migration, namely via the saturated pressure curve and as a result of the internal overpressure.
- New temperature and moisture content fields result from this coupling and the model then returns to the top to proceed in time.

Typical tests carried out with microwave heating highlight four drying periods (Fig. 1.20):

- A heating period: at first the supplied energy heats the medium with negligible mass loss.
- A "streaming" period: when the temperature reaches and overtakes the boiling point of water, the internal pressure becomes high enough to drain liquid water out of the medium via a pumping effect. The pressure level required to allow this drainage is a percolation pressure whose value strongly depends on the porous medium (Perré, 1995).
- An enthalpic period: when the moisture content decreases, vapor migration takes place instead of liquid migration. The

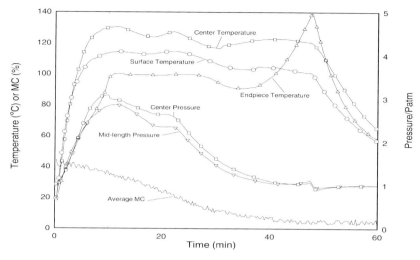

Fig. 1.20 Experimental results for combined microwave and convective drying of spruce heartwood (Experimental incident power approximately 100 W) (Perré and Turner, 1997).

mechanism becomes simpler: the vapor flux, sustained through vaporization by the amount of energy, leaves the medium by diffusion-convection (or pure convection).
– A possible "burning" period: finally, when there is insufficient water remaining inside the medium, the temperature increase can be very spectacular for some products. Indeed, the loss factor (imaginary permittivity) often increases with temperature. This means that the drier part of the medium becomes hotter and receives more energy, which can lead to a thermal runaway.

Figures 1.21 and 1.22 depict typical simulation examples of the evolution of this intricate coupling during the microwave drying of softwood. These carpet plots exhibit the four variable fields computed throughout the drying test: power, temperature, moisture content and gaseous pressure. The longitudinal and highly permeable board direction is along the largest dimension (7.5 cm). Half the board is simulated here: the symmetry plane is in the rear plane and the endpiece in the front plane. Figure 1.21 represents a sapwood board after 70 min of drying. The effect of the standing waves is evident on the power field, because the sample dimensions are comparable to the wavelength. This effect is probably one of the main disadvantages of microwave drying for large samples, which renders the task of obtaining rapid and uniform drying difficult. As a result of the thermal conduction and evaporation-condensation, the temperature field is more uniform than the power field. Because the temperature level is higher than the boiling point of water, an overpressure develops inside the sample. The gradient of the overpressure stretches along the permeable direction (the longitudinal wood direction) and gives rise to an important liquid migration that resaturates the endpiece. In the case of heartwood,

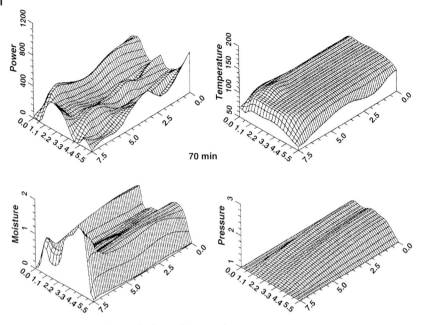

Fig. 1.21 Microwave drying of softwood. Carpet plots depicting the intricate coupling between the four variable fields. For this example of sapwood drying, the effect of the pressure field on longitudinal migration, which resaturates the endpiece, is evident (Perré and Turner, 1997).

the overpressure is even higher, but the low permeability and initial moisture content of heartwood do not produce this important longitudinal migration so that the endpiece remains dry. A location near the endpiece is even drier, which provokes a thermal runaway (Fig. 1.22), very similar to the one observed during the experiment.

To emphasize the differences between microwave and radio-frequency drying, Fig. 1.23 depicts the variable fields (volumetric power, temperature, moisture content and internal pressure) obtained after 2 and 20 h of convective drying of an oak section undergoing radio-frequency heating. Again, the power field is computed by solving Maxwell's equations (Bucki and Perré, 2003, 2004). The coupling between the four variables still exists in radio-frequency drying, however, standing waves cannot develop because the wavelength is much higher than the sample dimensions. In the case of radio-frequency heating, optimization of the process seems to be simpler. Nevertheless, as stated in the introduction, the problem lies here with the optimization of a whole stack of products (Bucki and Perré, 2003).

1.6.2
The Concept of Identity Drying Card (IDC)

Based on the results gained from measurement and numerical analysis, the dynamic distribution and development of local temperature and pressure inside a seasoned

1.6 Possibilities Offered by this Modeling Approach: Less-common Drying Configurations

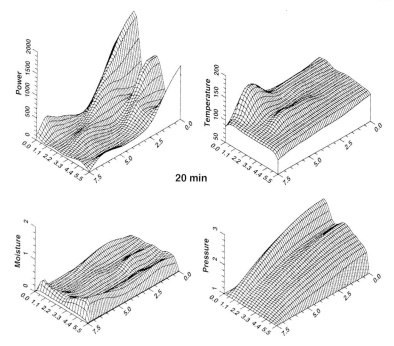

Fig. 1.22 Microwave drying of softwood. In this case of heartwood, a thermal run-away effect can be observed near the endpiece (Perré and Turner, 1997).

medium are coupled together by a temperature–pressure graph, herewith called an "identity drying card" (IDC), which is a concept introduced some time ago (Perré 1995). By using an IDC it is possible to visualize the internal profile of temperature and pressure, the dominant transport properties (permeability and diffusivity), the mechanism of transport (diffusion, convection or both) and the phase transitions during drying.

More specifically, the amount of dry air, the moisture content in the hygroscopic region or the danger due to internal mechanical loads of the handled materials can be identified with the aid of an IDC. Figure 1.24 depicts a typical IDC obtained for convective drying with superheated steam and microwave drying, which allows different features of these curves to be specified. Note that an IDC is a local value, specific to the sensor location (or to a numerical node in a computational mesh): all the comments concerning the global process (in particular the drying phases) are valid only if the chosen point is at the center of the product.

From these diagrams, the reader is able to explore phenomena such as increases in permeability; increases in diffusivity; decreases in the wet-bulb temperature (from superheated steam to moist air); and decreases in the external pressure. Hence, they are a very useful tool to understand the coupled mechanisms that occur during drying with internal vaporization.

34 | *1 Comprehensive Drying Models based on Volume Averaging*

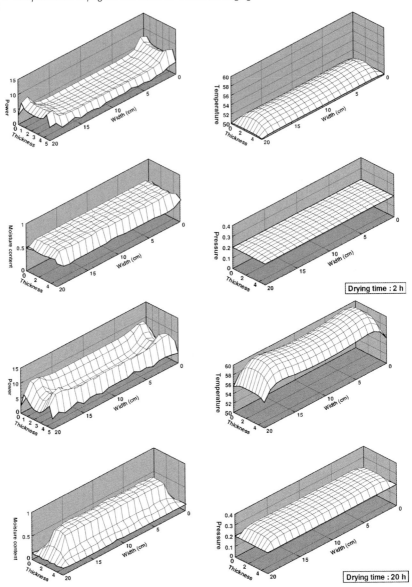

Fig. 1.23 Example of drying simulation: comprehensive modeling of convective drying with radio-frequency heating.

1.6.3
Drying of Highly Deformable Materials

Derived from a previous version of *TransPore*, the numerical model presented in this section has been developed in order to describe the drying of porous media that

1.6 Possibilities Offered by this Modeling Approach: Less-common Drying Configurations

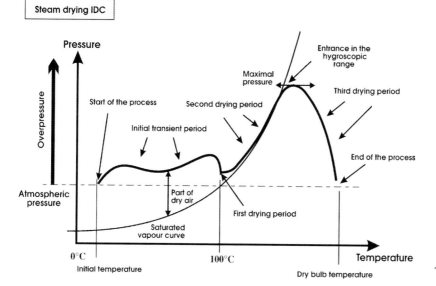

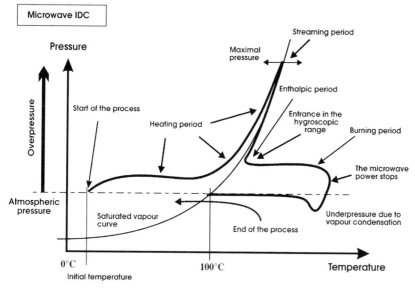

Fig. 1.24 Typical IDC diagrams observed for convective and microwave drying.

undergo large deformations due to shrinkage. For such products, the change of geometrical shape has to be taken into account when solving the balance equations.

The initial product shape is described using triangular elements. From this classical FE mesh, Control Volumes are constructed and used to solve the heat- and mass-transfer equations, while the initial FE mesh is used to solve the mechanical problem.

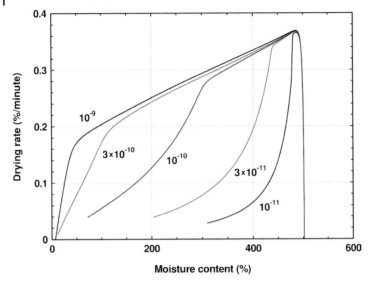

Fig. 1.25 Drying curves for different mass diffusivities (in $m^2\ s^{-1}$). Note the effect of this parameter on the duration of the first drying period (period at constant drying flux).

Due to the large shrinkage values, nonlinear problems arise from the large strain field generated during the process. In order to overcome this difficulty, the constitutive law is written on a local intermediate configuration in which rotation is chosen to oppose the rotation appearing in the polar decomposition of the deformation gradient tensor (Perré and May, 2001).

One typical example is presented here for potato. This product undergoes the majority of its shrinkage within the domain of free water. Consequently, the sample shape varies significantly although free water remains available at the exchange surface. Paradoxical results can be obtained when simulating low-temperature convective drying of this product: a decreasing drying rate is exhibited during the first drying period. This paradox, which has also been observed during experiments, is obviously due to the reduction of the exchange surface.

Global drying curves calculated for different values of the diffusion coefficient are depicted in Fig. 1.25. Each curve clearly shows two drying periods:

- Although the drying rate decreases due to the reduction in exchange surface, the first period is actually a first drying period (confirmed in the model by the temperature that remains at the wet-bulb temperature). During this period, the drying flux is constant.
- The second period is the classical falling rate period, also observed for rigid products. During this period, free water no longer exists at the surface.

Figure 1.26 depicts the shapes and moisture content fields calculated at different drying times for a mass diffusivity equal to $3 \times 10^{-10}\ m^2\ s^{-1}$. These plots allow us to

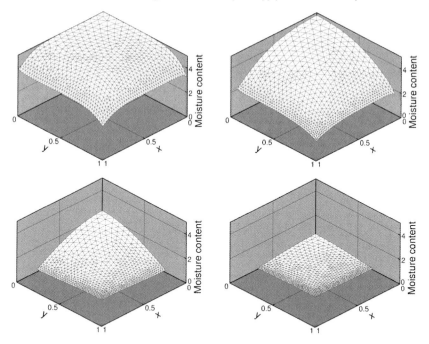

Fig. 1.26 Shape and moisture content field of a French fry during drying (2, 10, 25 and 40 h of drying). Mass diffusivity = 3×10^{-10} m^2 s^{-1}.

understand how the constant drying flux period lasts for a long time while the sample undergoes significant reduction in dimension, hence in exchange surface area.

1.7
Homogenization as a Way to Supply the Code with Physical Parameters

Although the set of macroscopic equations presented throughout the previous sections is a powerful foundation for the computational simulation of drying, it needs several physical parameters to be known, most of which are functions of both temperature and moisture content. Indeed, the Whitaker approach involves many "effective" parameters that have to be measured or predicted. Consequently, supplying the computer model with all physical characterizations is a tedious task, which restrains the use of modeling. The first goal for a scaling approach is to use modeling to predict one part of the parameters required at the macroscale. Homogenization is one of the mathematical tools that allows the macroscopic properties to be predicted from the microscopic description of a heterogeneous medium (Sanchez-Palencia, 1980; Suquet, 1985; Hornung, 1997). The principle of the method will be explained hereafter, using a simple parabolic equation as the reference problem:

$$\begin{aligned}
\frac{\partial u}{\partial t} &= \nabla .(a.\nabla u) + f \quad \text{in } [0, T] \times \Omega \\
u &= 0 \quad \text{on } [0, T] \times \partial\Omega \\
u(0, x) &= \varphi(x) \in L^2(\Omega)
\end{aligned} \quad (1.35)$$

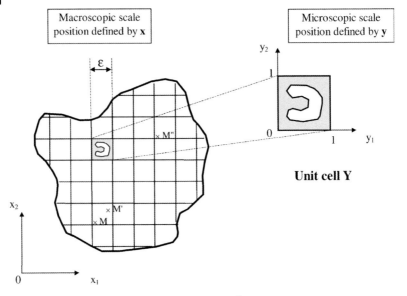

Fig. 1.27 Principle of double coordinates system used in periodic homogenization (after Sanchez-Palencia, 1980).

where Ω is the (bounded) domain of interest, $a(x)$ the diffusion coefficient (order 2 tensor), $u(t, x)$ the variable field (i.e. temperature for thermal diffusion or moisture content for mass diffusion) and φ a source term.

Now consider a heterogeneous and periodic medium, comprising a juxtaposition of unit cells (Fig. 1.27). A small parameter ε denotes the ratio between the macroscopic scale, denoted by vector x, and the microscopic scale, denoted by vector y. x is used to locate the point in the macroscopic domain Ω whereas y is used to locate the point within the unit cell Y. Using this new configuration, the reference problem becomes multiscale:

$$\frac{\partial u^\varepsilon}{\partial t} = \nabla \cdot (a^\varepsilon \cdot \nabla u^\varepsilon) + f \quad \text{in } [0, T] \times \Omega$$
$$u^\varepsilon = 0 \quad \text{on } [0, T] \times \partial\Omega \tag{1.36}$$
$$u^\varepsilon(0, x) = \varphi(x) \in L^2(\Omega)$$

Where $u^\varepsilon(t, x) = u(t, x, \frac{x}{\varepsilon}) = u(t, x, y)$ and $a^\varepsilon(x) = a(x, \frac{x}{\varepsilon}) = a(x, y)$, a uniformly elliptic, bounded and Y-periód set of functions in $\mathbb{R}^n$ (n = number of space dimensions). Although the homogenization procedure can be derived with the diffusion coefficient dependent on x, the unit cell is supposed to be dependent only upon y, namely

$$a^\varepsilon(x) = a(y) \tag{1.37}$$

1.7 Homogenization as a Way to Supply the Code with Physical Parameters

The homogenization theory indicates that

$$u^\varepsilon \xrightarrow[\varepsilon \to 0]{} u^0 \quad \text{weakly in } H_0^1(\Omega) \tag{1.38}$$

where u^0 is the solution of the homogenized problem

$$\begin{aligned}
\frac{\partial u^0}{\partial t} &= \nabla \cdot (A^0 \cdot \nabla u^0) + f \quad \text{in } [0, T] \times \Omega \\
u^0 &= 0 \quad \text{on } [0, T] \times \partial\Omega \\
u^0(0, x) &= \varphi(x) \in L^2(\Omega)
\end{aligned} \tag{1.39}$$

The macroscopic property, A^0, is given by

$$\underset{\substack{\text{Homogenised} \\ \text{coefficient}}}{A_{ij}^0} = \underset{\substack{\text{Average of the} \\ \text{microscopic coefficient}}}{\langle a_{ij}(y) \rangle} + \underset{\substack{\text{Corrective} \\ \text{term}}}{\sum_{k=1}^{n} \left\langle a_{ik}(y) \frac{\partial \xi^j}{\partial y_k}(y) \right\rangle} \tag{1.40}$$

In Eq. 1.40 the functions ξ^j are solutions of the following problems that must be solved over the unit cell Y

$$\sum_{i=1}^{n} \frac{\partial}{\partial y_i} \left(\sum_{k=1}^{n} a_{ik}(y) \frac{\partial \xi^j}{\partial y_k}(y) \right) = -\sum_{i=1}^{n} \frac{\partial a_{ij}(y)}{\partial y_i}, \quad j = 1 \cdots n \tag{1.41}$$

Equation 1.40 indicates that the macroscopic property consists of two contributions:
- the average of the microscopic properties, which accounts for the proportion and values of each phase in the unit cell,
- a corrective term, which accounts for the morphology of the constituents within the unit cell, thanks to the solution of the cell problems. This term might be very important. For example, the macroscopic stiffness of the earlywood part of softwood is only 5–10% of the averaged value of the microscopic properties, which means that the corrective term removed 90–95% of this averaged value (Farruggia, 1998).

Equations 1.40 and 1.41 can be derived either by the method of formal expansion or, more rigorously, by using adapted test functions in the variational form of Eq. 1.36 (Sanchez-Hubert and Sanchez-Palencia, 1992). To derive the limit problem in a formal way, the unknown function u^ε is developed as the following expansion

$$u^\varepsilon(t, x, y) = u_0(t, x) + \varepsilon u_1(t, x, y) + \varepsilon^2 u_2(t, x, y) + \cdots \tag{1.42}$$

Due to the rapid variation of properties inside Y, two independent space derivatives exist

$$\nabla(u) = \nabla_x(u) + \frac{1}{\varepsilon} \nabla_y(u) \tag{1.43}$$

Applying this derivative rule to problem (1.36) leads to a formal expansion of the parabolic equation in powers of ε. The term with ε^{-2} indicates that u_0 does not depend on y. Using this result, the term with ε^{-1} gives

$$\nabla_y \cdot (a(y)\nabla_y u_1(t, x, y)) = -\nabla_y \cdot (a(y)\nabla_x u_0(t, x)) \tag{1.44}$$

Equation 1.44 is linear, so u_1 can be expressed as a linear expression of the derivatives of u_0 with respect to x

$$\nabla_y u_1(t, x, y) = \sum_{j=1}^{n} \nabla_y \xi^j(y) \frac{\partial u_0(t, x)}{\partial x_j} \tag{1.45}$$

In Eq. 1.45 $\xi^j(y) \in H^1_{per}(\Omega)$ are Y-periodic functions, which are solutions of the following problems

$$\nabla_y \cdot (a(y)\nabla_y \xi^j(t, x, y)) = -\nabla_y \cdot (a(y)\vec{e}_j) \tag{1.46}$$

Note that Eq. 1.46 is just Eq. 1.41 written with the derivative notation defined in Eq. 1.43. $\vec{e}_j$ is the unit vector of axis j.

Finally, the term with ε^0 reads

$$\frac{\partial u_0}{\partial t} + \nabla_x \cdot (a(y)(\nabla_x u_0 + \nabla_y u_1)) + \nabla_y \cdot (a(y)(\nabla_x u_1 + \nabla_y u_2)) = f \tag{1.47}$$

Averaging Eq. 1.47 over the unit cell Y allows the Y-periodic terms to vanish

$$\frac{\partial u_0}{\partial t} + \nabla_x \cdot ((\langle a(y) \rangle + \langle b(y) \rangle)\nabla_x u_0) = f$$
$$\text{with} \quad b_{ij}(y) = \sum_{k=1}^{n} a_{ik}(y)\frac{\partial \xi^j(y)}{\partial y_k} \tag{1.48}$$

Equation 1.48 is just the homogenized problem and the rule to obtain the homogenized property A^0, as already formulated in Eq. 1.39 and Eq. 1.40, respectively.

The detail of the formal expansion in powers of ε for the mechanical problems encountered in drying can be found elsewhere, for elasticity (Sanchez-Palencia, 1980; Léné, 1984; Sanchez-Hubert and Sanchez-Palencia, 1992), for thermoelasticity (L'Hostis, 1996) and for shrinkage (Perré, 2002; Perré and Badel, 2003).

The homogenization formulation results in classical PDE problems. Moreover, owing to the assumption that the microscopic and macroscopic scales are independent, together with the simple physical formulation used in this work (elastic constitutive equation and shrinkage proportional to the change of moisture content), the problems are steady-state, linear and uncoupled.

However, the computer model must be able to handle any geometry and deal with properties that vary strongly in space. This is why the FE method is among the appropriate numerical strategies. Finally, the mesh must represent as closely as

possible the real morphology of the porous medium. The best strategy able to fulfil this requirement consists in building the mesh directly from a microscopic image of the porous medium. To address this demand, two numerical tools have been developed:
- *MeshPore*: a software developed to apply image-based meshing (Perré, 2005),
- *MorphoPore:* a code based on the well-known finite-element strategy, which is specifically devoted to solving homogenization problems and can deal with all kinds of boundary conditions encountered when solving the cell problems.

These codes are written in Fortran 95 and run on a PC as a classical Windows application, as a result of the graphical library used for the pre- and postprocessing (Winteracter 5.0).

Figure 1.28 depicts one typical set of solutions, which allows the macroscopic properties (stiffness and shrinkage) of one annual ring of oak to be calculated. In this figure, the solid lines represent the initial position of boundaries between different

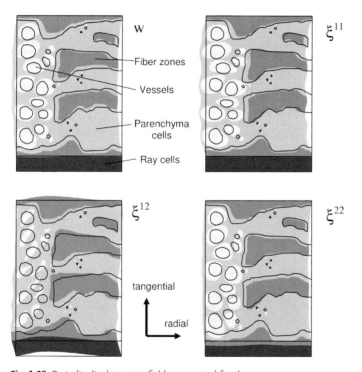

Fig. 1.28 Periodic displacement fields computed for the four problems to be solved over the representative cell. (W) shrinkage problem; (ξ_{11}) and (ξ_{22}) mechanical problems in the radial and tangential directions. (ξ_{12}) corresponds to the shear problem. Solid lines represent the initial contours of tissues (Perré and Badel, 2003). Color code, from dark to light: ray cells, fibre zones, parenchyma cells and vessels.

kinds of tissues (vessels, parenchyma cells, fibre and ray cells), while the colored zones represent the deformation of these zones as calculated for each elementary solution (an amplification factor is applied so that the deformation field can be observed easily). These solutions emphasize the complexity of the pore structure of oak, and its implication on mechanical behavior. For example, the fibre zones are strong enough to impose this shrinkage to the rest of the structure (problem w) and the ring porous zone is a weak part unable to transmit any tangential forces (in this part, the ray cell enlargement is negligible, problem 11) and prone to shear strain (problem 12).

This approach allowed us to quantify the effect of fibre proportion and fibre zone shape on the macroscopic values, or to predict the increase of rigidity and shrinkage coefficients due to the increase of the annual ring width (Perré and Badel, 2003).

Keeping in mind that this homogenization procedure was obtained by letting ε tend towards zero, the time variable undoubtedly disappears within the unit cell Y. Consequently, the macroscopic property has to be calculated only once and subsequently used in the homogenized problem. This is a typical sequential coupling.

1.8
The Multiscale Approach

1.8.1
Limitations of the Macroscopic Formulation

Up to this point, we have presented a set of macroscopic equations that allows numerous transfer and drying configurations to be computed and a mathematical method, homogenization, which allows the macroscopic properties (the "effective" parameters of the macroscopic set) to be computed from the constitution of the material at the microscopic level (concept of unit cell, or REV, representative elementary volume).

Figure 1.29 depicts two examples for which the previous approach fails: soaking samples of hardwood with water. In such a process, a typical dual-scale mechanism occurs: the water flows very rapidly in those vessels that are open and connected. In oak, this easy transfer occurs in the vessels without (or with a low amount of) thyloses, whereas the early part of each annual growth ring is very active in the case of beech. Then,

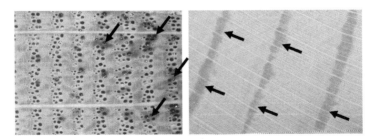

Fig. 1.29 Absence of local thermodynamic equilibrium when soaking wood samples, oak (left) and beech (right). View of the outlet face (S. Ghazil – LERMAB).

moisture needs more time to invade the remaining part of the structure, by liquid transfer in low permeable tissues or by bound water and water diffusion. The photographs in Fig. 1.29 prove that the local thermodynamic equilibrium cannot always be assumed. This occurs as soon as the macroscopic and the microscopic time scales have the same orders of magnitude. As a result, the microscopic field not only depends on the macroscopic field, but also on the history of this macroscopic field. Such media manifest microscopic storage with memory effects. Obviously in these instances, the previous formulations (macroscopic formulation and change of scale) must be discarded.

Different strategies (see Perré 2006, for a short survey) have been derived, ranging from simple global formulations to more comprehensive ones:
- Relatively simple global formulations are able to account for the microscopic delay but have to be supplied with experimental knowledge.
- A mesoscopic model consists in dealing with the microscopic detail at the macroscopic level. This is just a continuous model for which the physical parameters change rapidly in space. The complex geometrical structure of the heterogeneous medium has to be meshed and the grid to be refined enough for the small structure to be captured. Such models are obviously able to capture all of the subtle mechanisms and interaction between scales, but these are very demanding in terms of computer resources. They must be limited to cases for which the ratio between the microscale and the macroscale (ε) remains close to unity.
- Finally, homogenization can be extended to these configurations without local equilibrium. To do this, the microscopic transfer properties have to be scaled by ε^2 before calculating the limit as ε goes to zero (Hornung, 1997).

A strategy similar to mesoscopic models was adopted to obtain a realistic description of fluid flow in beech. In this case the x-z plane is the unit cell (microscopic cylindrical coordinates, with the vessel in the centre surrounded by the fibre zone) and the y-axis the macroscopic direction for fluid flow (Perré, 2004). Figure 1.30 depicts the moisture content fields computed at different soaking times. A fast wetting front invades the vessel at first. This phenomenon lasts about one minute. After this first step, one can see how moisture migrates in the fibre zone by two different mechanisms:
- a radial migration from the invaded vessel along the complete length of the sample (20 cm in this case),
- a significant macroscopic migration within the fibre zone from the injection plane (especially at 2 h).

1.8.2
The Stack Model: An Example of Multiscale Model

The multiscale approach considers simultaneously several scales in terms of time and dimension, and it demands much more computational time because at each

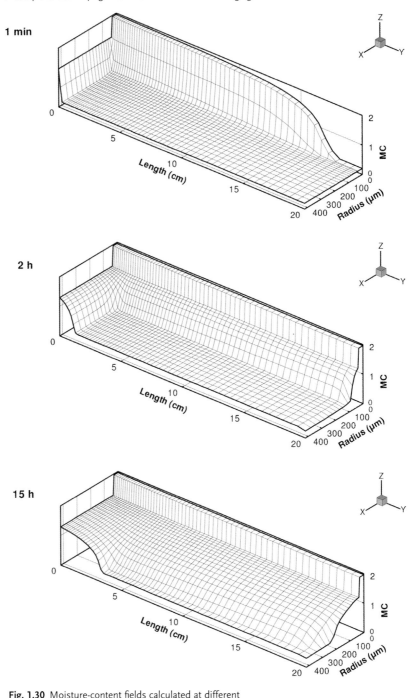

Fig. 1.30 Moisture-content fields calculated at different resaturation times using a mesoscopic model to simulate rapid fluid migration in the vessel and slow migration in the surrounding zone (Perré, 2004).

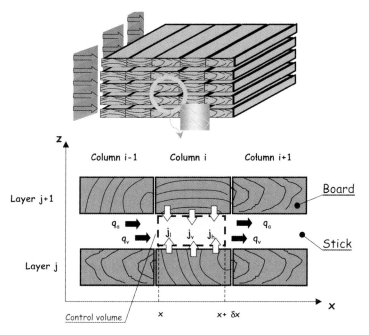

Fig. 1.31 Control volume in the airflow and fluxes at the control faces.

global time step the local variable field has to be updated for each point of the global grids. This approach is required as soon as the global time constant is driven by the local time constant, and the local variable field depends on both the global variable field and its history.

A batch lumber kiln is a perfect example of a drying process for which an intricate coupling exists between two scales. Boards are arranged in layers, and each board is placed edge to edge on the same tier (Fig. 1.31). Consequently, moisture migration and heat transfer take place only in the direction of the board thickness, which results primarily in one-dimensional moisture-content and temperature profiles. Board layers are spaced to enable airflow across their surfaces. The air parameters evolve along the airflow pathway between two adjacent board layers. Therefore, the drying conditions at the exchange surface of a board depend strongly on the fluxes of energy and moisture exchanged between the airflow and the boards.

Strong coupling between heat and mass transfer exists within each single board but also between boards, through the two-way interaction between the airflow and the boards. All boards of the stack can have different characteristics and are in close interaction as a result of their effect on the drying conditions within the stack. This complexity is dealt with via a dual-scale computational model that is summarized here and described in detail in Perré and Rémond (2006). The coupling between scales is realized by embedding the information gained from the fine-scale simulation into the coarser scale simulation.

1.8.2.1 Global Scale

At the global scale, the changes of air temperature and relative humidity along the air flow within the stack are determined from balance equations that involve the heat and mass fluxes computed for each board.

The vapor mass balance, applied to the global control volume depicted in (Fig. 1.31), leads to the relations:

$$\frac{dq_v}{dx} = J_v^{up,j} + J_w^{up,j} + J_v^{lo,j+1} + J_w^{lo,j+1}, \tag{1.49}$$

where q_v is the vapor mass flow (kg s^{-1} m^{-1}) in the airflow, and $J_v^{up,j}, J_w^{up,j}$ are the vapor and liquid water fluxes (kg s^{-1} m^{-2}) leaving the upper face of the board layer j (superscript lo, $j+1$ for example, represents the lower face of layer $j+1$).

When the exchanges that exist between the airflow and the porous medium are neglected, a similar approach leads to a simple relation for the balance of dry air:

$$\frac{dq_a}{dx} = 0 \tag{1.50}$$

where q_a is the mass flow of dried air (kg s^{-1} m^{-1}) passing through the control volume.

An enthalpy balance is applied to the control volume to produce:

$$\begin{aligned}q_v(x+dx)h_v(x+dx) - q_v(x)h_v(x) + q_a(x+dx)h_a(x+dx) - q_a(x)h_a(x) \\ = dx\{J_v^{up,j}h_v + J_w^{up,j}h_w + J_e^{up,j} + J_v^{lo,j+1}h_v + J_w^{lo,j+1}h_w + J_e^{lo,j+1}\}\end{aligned} \tag{1.51}$$

where h_a, h_v and h_w are the specific enthalpies of air, water vapor and liquid water, respectively, and J_e is the heat flux exchanged between the board and the airflow.

1.8.2.2 Local Scale

In all previous equations, local fluxes exchanged between the board and the airflow are calculated using *TransPore*, which solves the coupled heat and mass transfer equations within the board. The boundary conditions used for this convective exchange are depicted in Eq. (1.22).

1.8.2.3 Coupling Approach

The coupling between the stack and the board is a two-way process (see Fig. 1.32):

- From the fluxes of heat, vapor and liquid, leaving each face of the boards at time t, but also from the stack-inlet drying conditions, the air velocity and the stick thickness, the drying conditions (temperature and vapor pressure) throughout the stack are calculated.

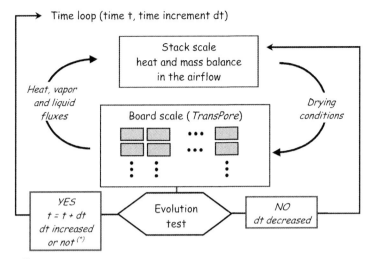

Fig. 1.32 The two-way coupling between the global and local scales in the board-stack model.

- The *TransPore* modules allow the drying of each single board to advance by the global time step d*t*, which enables the values of the heat, vapor and liquid fluxes to be known for time $t + dt$.

This coupling procedure is iterated until the end of the simulated drying time. Note that each of the *TransPore* modules has its own local time step, automatically adjusted during the calculation according to the convergence conditions of that board. In order to keep a good efficiency of the global simulation, all local time steps are controlled independently. Generally, the local time steps are different from one board to another and smaller than the global time step.

Such a dual-scale model is computationally time consuming and this is why the one-dimensional version of *TransPore* is preferred for the simulations.

1.8.2.4 Samples Simulations

The first computed test is focused on the effect of the stack configuration on the drying kinetics of the boards. All boards are identical here and the drying conditions are constant: $T_{dry} = 65\ °C$ and $RH = 76\%$ (hence $EMC = 12\%$, at the stack inlet). Flat-sawn $30 \times 200\ mm^2$ spruce heartwood boards were selected. The simulation results are presented in Fig. 1.33.

As expected, the boards close to the stack inlet dry much faster than those close to the outlet side. Indeed, the dry-bulb temperature decreases along the airflow because it has to supply the heat of vaporization, whereas the dew point increases due to the vapor flux coming from the boards. Consistently, the drying rate of the boards situated at the stack outlet increases only once the inlet boards are dry enough for their drying rate to decrease significantly. Increasing the air velocity and the stick

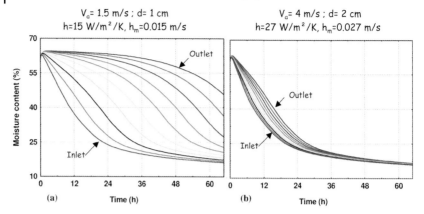

Fig. 1.33 MC evolution of 10 identical boards of the same layer. (a) Low air velocity (V_a) and small stick thickness (d). (b) High velocity and thick stick.

thickness dramatically increases the uniformity of drying conditions within the stack, see Fig. 1.33.

In order to reduce the effect of the board location on drying kinetics, industrial kilns typically periodically reverse the airflow during drying. Simulations were run on the previous stack of identical boards and the same external conditions but with an airflow direction reversed every two hours. As expected, the dispersion of moisture content curves is reduced considerably during all the drying process when the air flow is reversed periodically (Fig. 1.34). As a result of changing air parameters, this leads to the periodical application of a new drying gradient at the board surfaces. Boards located near both stack edges dry faster than those in the middle of the stack (Fig. 1.34a).

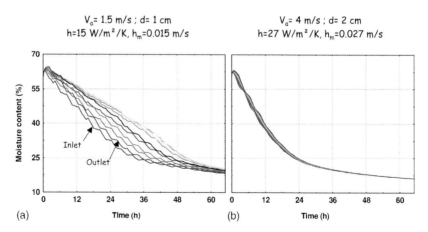

Fig. 1.34 Same problem and parameters as in Fig. 1.33, however, with reversal of airflow direction every two hours.

Note that the drying time at a given stack moisture content is at most equal to, or shorter than, when the airflow is reversed periodically. Indeed, oscillations in the air conditions delay the formation of the peripheral dry layer in the boards and slightly extend the first drying period for boards located at the stack inlet. During this period moisture migrates by capillarity, which is faster than diffusion. During the second drying period, oscillations of the drying gradient affect the moisture content only near the surface of the board, because the frequency of the oscillations is short in comparison to the times involved with mass transfer.

1.8.2.5 Accounting for Wood Variability

In drying models it is unrealistic to assume that all pieces of wood within the stack have the same average properties. Since boards come from several trees growing under different environmental conditions, they consist of different genetic composition. In addition, the boards may have been cut using differing sawing patterns within the log. Therefore, one can imagine that pieces of wood within the stack have dramatic variations in their properties and in their drying rate.

In order to account for this variability, a Monte-Carlo method has been implemented to generate the stack computed in the dual-scale model. The log diameter is generated as a random variable uniformly distributed between two values specified by the sawmill. A sample sawing pattern plane, which is used in the wood industry, has been applied to all logs according to their randomly generated diameters (Fig. 1.35). Sapwood and heartwood zones have been allocated a different initial moisture content and liquid permeability. It is clear that a one-dimensional model can not fully describe the drying behavior of a board that contains both heartwood and sapwood zones in its section. An approximate one-dimensional method is thus used. The moisture content and liquid permeability profiles across the board are calculated according to its proportion of heartwood and sapwood width using a simple linear relationship. Consequently, the boards are usually not symmetrical, see Fig. 1.35.

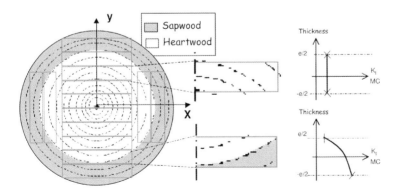

Fig. 1.35 Sawing-pattern plane, and profiles of moisture content (MC) and liquid permeability.

The oven dry density of each log is also generated as a random variable with a Gaussian distribution between two values depending on the forest station where the trees grew. This value modifies some other physical and mechanical parameters such as the porosity, the diffusion coefficient of bound water, thermal conductivity, Young's modulus and the Poisson coefficient.

1.8.2.6 Accounting for Drying Quality

During drying, stresses and deformations develop in the board through shrinkage as a result of the moisture and temperature field variations. The mechanical behavior is complex and combines elasticity, plasticity, viscoelasticity and mechanosorption. In order to take into account the drying quality in the dual-scale model, a rigorous one-dimensional mechanical formulation described in Rémond (2004, 2006) and based on previous works (Perré and Passard, 1995, 2004; Mauget and Perré, 1999), has been used for calculating the stress and deformation development during drying. This is fitted into a module and then added to each module of *TransPore* (Fig. 1.36).

When the deformation field induced by shrinkage does not fulfil the geometrical compatibility, a strain tensor

$$\overline{\overline{\varepsilon^{tot}}} = \overline{\overline{\varepsilon^{sh}}} + \overline{\overline{\varepsilon^{e}}} + \overline{\overline{\varepsilon^{ms}}} + \overline{\overline{\varepsilon^{ve}}} \tag{1.52}$$

related to the stress field is generated. The strain tensor of shrinkage $\overline{\overline{\varepsilon^{sh}}}$ is supposed to be proportional to variations in bound water content, the elastic strain is connected to actual stresses via Hooke's law, the viscoelastic strain tensor $\overline{\overline{\varepsilon^{ve}}}$ uses four thermally activated Kelvin elements, and the mechanosorptive strain tensor $\overline{\overline{\varepsilon^{ms}}}$ uses the concept of strain limitation (Salin 1992).

The mechanical problem is solved in terms of displacement. The total strain tensor is then simply deduced from the displacement field. Introducing Eq. 1.52 in the mechanical balance equations produces a set of equations that depend on the

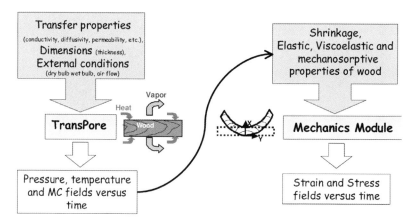

Fig. 1.36 Input/output data in the code of the local scale.

displacement field, as well as the temperature and moisture fields. At each time increment, temperature and moisture content fields are calculated by *TransPore*, and the set of mechanical equations is solved to estimate the displacement field (Fig. 1.36).

As a result of accounting for wood variability, wood quality and board location in the stack, the dual-scale model has great potential for predicting what is observed in an industrial kiln. Different sample simulations have been selected to emphasize the model potential, and to show that this comprehensive computational model is now able to link theory and practice. The wood stack contained 100 boards (10 × 10, width and height).

Figure 1.37 depicts the moisture content and stress evolutions of boards in the stack when the drying conditions are those of Fig. 1.33b. The drying simulation of this stack is completed in less than 40 s on a 2.8 GHz Xeon processor. The results show a broad distribution of the initial moisture content of boards so that the permeability value also varies significantly. As a consequence, the moisture-content evolution of the boards in the stack also has a great heterogeneity; some boards reach the equilibrium moisture content after 24 h of drying, whereas others attain it after 60 h. At the end of the first drying period, zones of the board section close to the exchange surface enter the hygroscopic range, which induces shrinkage and gives rise to tensile stress. This instant varies significantly from 6 h up to 55 h for boards according to their initial moisture content, their location in the stack and their transfer properties. Surface cracks might appear during this stage. Stress reversal due to the memory effect is not exhibited for any board in 65 h of drying. After 65 h, some boards still have an important moisture content gradient, and the "postkiln" drying could be responsible of internal cracks for these boards (see Rémond et al., 2006).

The foregoing dual-scale computational model is able to simulate simultaneously the drying of hundreds of boards, accounting for wood variability and stack interaction. As a result of its speed, this code is a real operational tool for finding a subtle compromise between drying time, product quality and process cost.

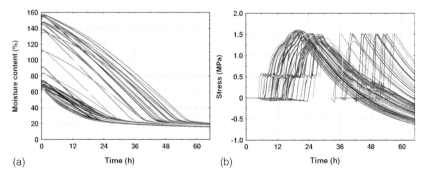

Fig. 1.37 (a) Moisture-content evolution of every individual out of a 10 × 10 stack of wood boards (b) Stress evolution of these boards at the exchange surface.

1.9
Conclusion

This chapter outlines the derivation of a comprehensive macroscopic drying model by volume averaging. The macroscopic formulation has been used to simulate a wide cross section of important drying processes. We have presented succinct explanations of the physics embedded within the mathematical formulation and given a brief overview of the computational strategies used to ensure the efficient solution of the coupled, nonlinear drying conservation equations. A number of drying case studies that are particularly relevant to industry have been presented to elucidate the physics of drying, thus providing the drying practitioner with insight into the underlying heat, mass and momentum transfer phenomena evolving within the porous medium. The often neglected role of overpressure in the product has been analysed. We have discussed the limitations of the macroscopic formulation and provided solutions to overcome these limitations, which include homogenization and a novel multiscale approach.

The next chapter builds upon our exposition in a natural manner, by introducing the concept of the discrete modeling of drying phenomena. This approach tries to remove some of the discussed limitations of continuous macroscopic models.

References

Bear J., Corapcioglu, M. Y., 1987. *Advances in Transport Phenomena in Porous Media*. Martinus Nijhoff Publishers, Lancaster.

Bimbenet, J. J., Daudin, J. D., Wolf, E., 1985. Air drying kinetics of biological particles. In: *Proceedings of the 4th International Drying Symposium*. Kyoto, Japan.

Bird R. B., Stewart W. E., Lightfoot, E. N., 1960. *Transport Phenomena*. John Wiley, New York.

Bojan N., Easymesh – mesh generation software, niceno@univ.trieste.it

Bucki M., Perré P., 2003. Physical formulation and numerical modeling of high frequency heating of wood. *Drying Technology Journal* **21**: 1151–1172.

Carbonell, R. G., Whitaker, S., 1983. Dispersion in pulsed systems. II: Theoretical developments for passive dispersion in porous media. *Chem. Eng. Sci.* **38**: 1795–1802.

Carbonell, R. G., Whitaker, S., 1984. Heat and Mass Transfer in Porous Media. In: *Fundamentals of Transport Phenomena in Porous Media*, edited by J. Bear and M. Y. Corapcioglu, Martinus Nijoff Publishers, Dordrecht, The Netherlands, pp. 123–198.

Dhatt G., Touzot G., 1984. The Finite Element Method Displayed. Translation of: Une présentation de la méthode des éléments finis, Wiley-Interscience, New York, pp. 120–121.

Dullien F. A. L., Dong W., 1996. Experimental determination of the flow transport coefficients in the coupled equations of two-phase flow in porous media. *Transport in Porous Media* **25**: 97–120.

Farruggia F., 1998. Détermination du comportement élastique d'un ensemble de fibres de bois à partir de son organisation cellulaire et d'essais mécaniques sous microscope. Thèse de Doctorat-ENGREF, Nancy, France.

Ferguson W. J., Turner I. W., 1996. A Control Volume Finite Element Numerical Simulation of the Drying of Spruce. *J. Comp. Phy.* **125**: 59–70.

Forsyth P. A., 1994. Three-Dimensional Modeling of Steam Flush for DNAPL Site

Remediation. *Int. J. Num. Methods in Fluids* **19**: 1055–1081.

Glatzmaier G. C., Ramirez W. F., 1988. Use of volume averaging for the modeling of thermal properties of porous materials. *Chem Eng. Sci.* **43**(12): 3157–3169.

George A., Liu J. W. H., 1981. *Computer solution of large sparse positive definite systems*, Prentice-Hall, Englewood Cliffs, NJ, USA.

Gray W. G., 1975. A derivation of the equations for multiphase transport, *Chem. Eng. Sci.* **30**: 229–233.

Hornung U., 1997. *Homogenization and porous media*, Springer-Verlag, New York.

Ilic M., Turner I. W., 2005. Krylov subspaces and the analytic grade. *Numerical Linear Algebra with Applications* **12**: 55–76.

Jason, A. C., 1958. A study of evaporation and diffusion processes in the drying of fish muscle. In: *Fundamental Aspects of the Dehydration of Food Products*. Society of Chemical Industry, London, and The Macmillan Co., New York, pp. 103–135.

Jayantha, P. A., Turner, I. W., 2003. A Second Order Finite Volume Technique for Simulating Transport in Anisotropic Media. *The International Journal of Numerical Methods for Heat and Fluid Flow* **13**(1): 31–56.

Jennings A., McKeown J. J., 1992. *Matrix Computation*, 2nd Edition, John Wiley, New York.

Jung, S. K., Choi, Y. H., Shon, T. H., Choi, J. U., 1986. The drying characteristics of apples at various drying conditions. *Korean Journal of Food Science and Technology* **18**(1): 61–65.

Kannan, D., Bandyopadhyay, S., 1995. Drying characteristics of a tropical marine fish slab. *Journal of Food Science and Technology* **32**(1): 13–16.

Kamke, F. A., Casey, L. J., 1988. Fundamentals of flakeboard manufacture: internal-mat conditions. *Forest Products Journal* **38**(6): 38–44.

Kelley C. T., 1995. Iterative methods for linear and nonlinear equations, number 16. In: *Frontiers in Applied Mathematics*, SIAM, Philadelphia.

Krischer, O., Kröll, K., 1963. *Die wissenschaflichen Grundlagen der Trocknungstechnik*. Springer-Verlag, Berlin.

Léné F., 1984. Contribution à l'étude des matériaux composite et de leur endomagement. Thèse d' état, Université Pierre et Marie Curie, Paris.

L'Hostis G., 1996. Contribution à la conception et à l'étude de structures composites thermoélastique. Thèse de Doctorat, Université Paris 6, Paris.

Lowery D. P., 1979. Vapor pressure generated in wood during drying, *Wood Science* **5**: 73–80.

Magee, T. R. A., Wilkinson, C. P. D., 1992. Influence of process variables on the drying of potato slices. *International Journal of Food Science and Technology* **27**: 541–549.

Marle C., 1982. On macroscopic equations governing multiphase flow with diffusion and chemical reactions in porous media. *Int. J. Eng. Sci.* **20**: 643–662.

Mauget, B., Perré, P., 1999. A large displacement formulation for anisotropic constitutive laws, *European Journal of Mechanics – A/Solids* **18**: 859–877.

May, B. K., Perré P., 2002. The importance of considering exchange surface area reduction to exhibit a constant drying flux period in foodstuffs. *Journal of Food Engineering* **54**(33): 87–98.

Morgan R. B., 2002. GMRES with deflated restarting. *SIAM J. Sci. Comput.* **24**(1): 20–37.

Moroney T. J., Turner I. W., 2006. A finite volume method based on radial basis functions for two-dimensional nonlinear diffusion equations. *J. Applied Mathematical Modeling* **30**: 1118–1133.

Patankar S. V., 1980. *Numerical Heat Transfer and Fluid Flow*, Hemisphere publishing Corporation, New York.

Perré P., 1995. Drying with Internal Vaporization: Introducing the Concept of Identity Drying Card. *Drying Technology Journal* **13**: 1077–1097.

Perré P., 1996. The Numerical Modeling of Physical and Mechanical Phenomena Involved in Wood Drying: an Excellent Tool for Assisting with the Study of New Processes. Tutorial, 5th Int. IUFRO Wood Drying Conference, Québec, Canada, pp. 9–38.

Perré P., 2002. Wood as a multi-scale porous medium: Observation, Experiment, and Modeling. Proceedings of the First International Conference of the European Society for Wood Mechanics, EPFL, Lausanne, Switzerland, pp. 365–384.

Perré P., 2004. Evidence of dual scale porous mechanisms during fluid migration in hardwood species: Part II: A dual scale computational model able to describe the experimental results. *Chinese J. Chem. Eng.* **12**: 783–791.

Perré P., 2006. Multiscale aspects of heat and mass transfer during drying, Transport in Porous Media, **66**(1–2): 59–76.

Perré P., Badel E., 2003. Properties of oak wood predicted from X-ray inspection: representation, homogenisation and localisation. Part II: Computation of macroscopic properties and microscopic stress fields. *Annals of Forest Science* **60**: 247–257.

Perré P., Bucki M., 2004. High-frequency/vacuum drying of oak: modeling and experiment. 5th Workshop of COST Action E15, Athens, Greece, 10 pages.

Perré P., Degiovanni A., 1990. Simulations par volumes finis des transferts couplés en milieu poreux anisotropes: séchage du bois à basse et à haute température, *Int. J. Heat and Mass Transfer* **33**: 2463–2478.

Perré, P., Martin, M., 1994. Drying at high temperature of heartwood and sapwood: theory, experiment and practical consequence on kiln control. *Drying Technology* **12**(8): 1915–1941.

Perré, P., May B., 2001. A numerical drying model that accounts for the coupling between transfers and solid mechanics: Case of highly deformable products. *Drying Technology Journal* **19**: 1629–1643.

Perré, P., May B., 2007. The existence of a first drying stage for potato proved by two independent methods, *Journal of Food Engineering* **78**: 1134–1140.

Perré, P., Moser, M., Martin, M., 1993. Advances in transport phenomena during convective drying with superheated steam or moist air. *Int. J. Heat and Mass Transfer* **36**: 2725–2746.

Perré, P. and Passard, J., 1995. A control-volume procedure compared with the finite-element method for calculating stress and strain during wood drying. *Drying Technology* **13**(3): 635–660.

Perré, P., Passard, J., 2004. A physical and mechanical model able to predict the stress field in wood over a wide range of drying conditions. *Drying Technology Journal* **22**: 27–44.

Perré P., Rémond R., 2006. A dual scale computational model of kiln wood drying including single board and stack level simulation. *Drying Technology Journal* **24**: 1069–1074.

Perré P., Turner I. W., 1996. The use of macroscopic equations to simulate heat and mass transfer in porous media. In: *Mathematical Modeling and Numerical Techniques in Drying Technology*, Chap. 2, edited by Turner, I. and Mujumdar, A. S., Marcel Dekker, New York, pp. 83–156.

Perré P., Turner I. W., 1997. Microwave drying of softwood in an oversized waveguide. *AIChE Journal* **43**: 2579–2595.

Perré P., Turner I. W., 2000. An efficient two-dimensional cv-fe drying model developed for heterogeneous and anisotropic materials. 12th International Drying Symposium, Drying' 2000, proceedings on CD-Rom, paper 262, 10 pages, The Netherlands.

Perré P., Turner I. W., 2001. Determination of the Material Property Variations Across the Growth Ring of Softwood for Use in a Heterogeneous Drying Model. *Holzforschung* **55**, 318–323, 417–425.

Perré P., I. W. Turner, 2002. A Heterogeneous Wood Drying Computational Model that accounts for Material Property Variation Across Growth Rings. *Chemical Engineering Journal* **86**: 117–131.

Press W. H., Teukolsky S. A., Vetterling W. T., Flannery B. P., 1992. *Numerical Recipes in Fortran – The Art of Scientific Computing*. Cambridge University Press, Cambridge.

Quintard M., Whitaker S., 1993. One and two-equation models for transient diffusion processes in two-phase systems. In: *Advances in Heat Transfer* **23**, Academic Press, New York, pp. 369–465.

Quintard M., Whitaker S., 1994. Transport in ordered and disordered porous media. *Transport in Porous Media* **14**: 163–206.

Quintard M., Whitaker S., 1995. Local thermal equilibrium for transient heat conduction: theory and comparison with numerical experiments. *Int. J. Heat and Mass Trans.* **38**: 2779–2796.

Rémond R., 2004. Approche déterministe du séchage des avivés de résineux de fortes épaisseurs pour proposer des conduites industrielles adaptées, PhD thesis, 209 p., ENGREF, Nancy, France.

Rémond, R., Passard, J., Perré, P., 2006. The effect of temperature and moisture content on the mechanical behavior of wood: comprehensive model applied to drying and bending. *European Journal of Mechanics – A/Solids*, **26**(3), p. 558–572.

Saad Y., 1997. Analysis of augmented Krylov subspace techniques. *SIAM Journal of Matrix Analysis and Applications* **18**: 435–449.

Saad Y., 2003. *Iterative methods for sparse linear systems.* Second Edition, SIAM, Philadelphia.

Salin J. G., 1989. *Remarks on the influence of heartwood content in pine boards on final moisture content and degrade.* IUFRO Wood Drying Symposium, Seattle.

Salin, J. G., 1992. Numerical prediction of checking during timber drying and a new mechanosorptive creep model. In: *Holz als Roh- und Werkstoff*, Vol. **50**(5), pp. 195–200.

Sanchez-Palencia, E., 1980. Non-homogeneous media and vibration theory. *Lecture Notes in Physics* **127**, ed. by E. Sanchez-Palencia and A. Zaoui, Springer-Verlag, Berlin.

Sanchez-Hubert, J.Sanchez-Palencia, E., 1992. *Introduction aux méthodes asymptotiques et à l' homogénéisation*, Masson, Paris.

Saravacos, G. D., Charm, S. E., 1962. A study of the mechanism of fruit and vegetable dehydration. *Food Technology* **15**: 78–80.

Sereno, A. M., Medeiros, G. L., 1990. A simplified model for the prediction of drying rates for foods. *Journal of Food Engineering* **12**: 1–11.

Slattery J. C., 1967. Flow of viscoelastic fluids through porous media. *Am. Inst. Chem. Eng. J.* **13**: 1066–1071.

Slattery, J. C., 1972. *Momentum, Energy and Mass Transfer in Continua.* McGraw-Hill, New York.

Suquet, P. M., 1985. Element of homogenization for inelastic solid mechanics. In: *Homogenization Techniques for Composite Media. Lecture Notes in Physics* **272**, ed. by E.Sanchez-Palencia and A. Zaoui, Springer-Verlag, Berlin.

Sweby P., 1984. High resolution schemes using flux limiters for hyperbolic conservation laws. *SIAM J. Numer. Anal.* **21**: 995–1011.

Truscott S., Turner I. W., 2005. A Heterogeneous Wood Drying Computational Model. *J. Applied Mathematical Modeling* **29**: 381–410.

Turner I. W., Perré P., 1996. A synopsis of the strategies and efficient resolution techniques used for modeling and numerically simulating the drying process. In: *Numerical Methods and Mathematical Modeling of the Drying Process.* Chap. 1, ed. by I. W.Turner and A.Mujumdar, Marcel Dekker, New York, pp. 1–82.

Turner I. W., Perré P., 2001. The use of implicit flux limiting schemes in the simulation of the drying process: A new maximum flow sensor applied to phase mobilities, *Journal of Applied Mathematical Modeling* **25**: 513–540.

Unger A. J. A., Forsyth P. A., Sudicky E. A., 1996. Variable spatial and temporal weighting schemes for use in multi-phase compositional problems, *Adv. Water Res.* **19**(1): 1–27.

Van Leer B., 1974. Towards the ultimate Conservative Difference Scheme, II. Monotonicity and Conservation Combined in a Second Order Scheme. *J. Comp. Phys.* **14**: 361–370.

Whitaker S., 1977. Simultaneous Heat, Mass, and Momentum Transfer in Porous Media: A Theory of Drying. *Advances in Heat Transfer* **13**: 119–203.

Whitaker S., 1998. Coupled Transport in Multiphase Systems: A Theory of Drying. *Advances in Heat Transfer* **31**: 1–104.

Zhao, Y., Poulsen K. P., 1988. Diffusion in potato drying. *Journal of Food Engineering* **7**: 249–262.

2
Pore-network Models: A Powerful Tool to Study Drying at the Pore Level and Understand the Influence of Structure on Drying Kinetics

Thomas Metzger, Evangelos Tsotsas, Marc Prat

2.1
Introduction

Pore-network models are network representations of the structure of porous bodies. Typically, one distinguishes between pores, which are voids of relatively large section, and throats, the narrow segments between them. The network is built from pore nodes connected by throats.

In the field of multiphase flow through porous media, pore-network models are a classical tool (Blunt et al. 2002); they are employed in two approaches, which can recently also be found in the context of drying (Prat 2002): on the one hand, they can help to study macroscopic transport properties, like relative permeabilities or capillary pressure as a function of saturation, which are used in continuum models; this approach has been taken by Nowicki et al. (1992) who seemingly developed the first pore-network drying model. On the other hand, they enable the fundamental investigation of phenomena directly at the sample scale. Such modeling permits situations to be addressed where averaged models are questionable, like near a percolation threshold. A drying model of this type was proposed by Prat (1993) one year later.

This model takes into account capillary and gravity effects (thanks to the invasion percolation algorithm) and the transport of vapor by diffusion in the gas phase (see Section 2.2). Since this first attempt, increasingly sophisticated models have been developed to account for viscous effects (Yiotis et al. 2001; Irawan et al. 2006), thermal effects (Huinink et al. 2002; Plourde and Prat 2003; Surasani et al. 2006), film effects (Yiotis et al., 2003, 2004, 2005; Prat 2006; Camassel et al. 2005) and binary liquids (de Freitas and Prat 2000). Three-dimensional versions, though not including all effects, have also been presented (Le Bray and Prat 1999; Segura and Toledo 2005; Metzger et al. 2005; Yiotis et al. 2006). Experiments with etched networks have led to satisfactory comparisons in terms of drying patterns (Laurindo and Prat 1996) and could show the strong influence of film flows on drying rates (Laurindo and Prat 1998).

Modern Drying Technology. Edited by Evangelos Tsotsas and Arun S. Mujumdar
Copyright © 2007 WILEY-VCH Verlag GmbH & Co. KGaA. All rights reserved.
ISBN: 978-3-527-31556-7

The analysis of drying patterns can be performed using invasion percolation concepts as proposed by Tsimpanogiannis et al. (1999), Prat and Bouleux (1999), notably in relation with the experimental results reported by Shaw (1987) indicating that viscous effects stabilize the invasion. Interestingly, invasion percolation concepts were also proven appropriate for analyzing the drying of sapwood (Salin 2006).

More recently, pore-network models have proven useful for analyzing the impact of changes at the pore or microstructure scale on drying behavior, namely wettability, pore shapes, pore-size distribution, pore structure (Prat 2006; Segura and Toledo 2005; Metzger et al. 2005; Irawan et al. 2005).

In this chapter, we first present the models in detail, starting with a basic isothermal version and presenting thereafter upgrades to account for further transport phenomena, and then apply them to study the influence of (micro) structure on drying.

2.2
Isothermal Drying Model

Isothermal modeling is a reasonable approximation only for very moderate drying conditions. However, it did greatly contribute to the development of pore-network drying models and their use in a systematic analysis of the involved transport phenomena. Nonisothermal pore-network investigations of drying are still rare (Huinink et al. 2002; Plourde and Prat 2003); one attempt to model convective drying with heat transfer (Surasani et al. 2006) will be presented in Section 2.3.1.

A second important assumption (throughout this chapter) is constant gas pressure. For vapor transfer in the gas-filled part of the porous medium, we use the logarithmic law describing combined diffusion and convection next to an evaporating liquid/gas interface. However, viscous effects to the convective flow of gas are not modeled. For high drying rates in large networks of small pores, this would no longer be a good approximation.

In our model, the Kelvin effect, i.e. the reduction of equilibrium vapor pressure over a curved meniscus (with the possibility of capillary condensation), and the Knudsen effect, i.e. modified diffusion at pore sizes as small as the mean free path of gas molecules, are neglected. This is justified for most of our investigations with pores larger than 100 μm.

Like in all network models of drying, we will not describe adsorption but restrict ourselves to capillary-porous media containing only free liquid, either in saturated pores or in corner films of empty pores. The liquid is, with the exception of Section 2.3.2, pure water.

2.2.1
Model Description

In this section, the network geometry as well as the basic drying algorithm will be presented in detail. This basic version requires further assumptions that will be abandoned later on. First, we assume that capillary effects dominate over viscous effects in liquid flow. As we will see, this is a good approximation for large pores and not too narrow pore-size distributions. Additionally, we neglect gravity, which is reasonable for small network size or two-dimensional networks with horizontal orientation. Furthermore, films in empty pores are not modeled at this stage. Finally, we assume a perfectly wetting liquid phase. The generalization to a different contact angle is straightforward as long as it is below and not too close to 90°; the drying behavior of hydrophobic networks (contact angles greater than 90°) is, however, dramatically different (see Section 2.2.5).

2.2.1.1 Network Geometry and Corresponding Data Structures

At first, we introduce the general geometry of the network and how it can be described by appropriate data structures. Though other geometries have been used in the literature, we restrict ourselves – unless otherwise mentioned – to the following: the pores are represented by nodes without volume, but they give the spatial grid of the pore space; and the throats connecting these nodes shall contain all pore volume and are cylindrical tubes, for simplicity. The length of such a cylinder equals the distance between the connected pores, hence leading to an overlap that is not yet accounted for (Fig. 2.1a). Cylinders are easier in modeling (no corner flow) than throats of rectangular cross section, but harder to produce for experiments and may not be very representative of real porous media (see Section 2.2.4).

The throats may have different radii and lengths, and the number of throats connected to one pore node may also vary, in order to reasonably represent a real pore structure with a pore-size distribution and variable connectivity of pores. However, we will only discuss regular networks with uniform throat length, starting with two-dimensional square ones.

The function of the throats is manifold, they serve as conductors for vapor and liquid transport (length and radius playing a role), define by their radius distribution the liquid pressure field and the order of invasion during drying, and determine by their volume the corresponding time scale. The nodes only serve to compute and store at discrete space points values of vapor pressure, liquid pressure and temperature.

We will use data arrays to describe the whole network concerning positions of nodes, throat connections and neighboring relationships between throats and pores. This is very convenient for an efficient drying algorithm and general enough to be used for irregular networks. These data structures are illustrated by the example of Fig. 2.1b. As a first step, all pores are numbered and their position in the two- or three-dimensional space is stored in an array. Next, throat connections between the pores are defined and, by this, implicitly numbered. For each throat, the neighboring pores are stored in a line of the array

$$\mathrm{TNP} = \begin{bmatrix} 1 & 2 \\ 1 & 3 \\ 1 & 4 \\ 2 & 5 \\ \vdots & \vdots \end{bmatrix} \qquad (2.1)$$

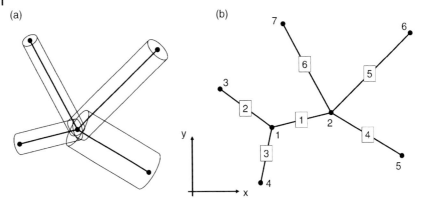

Fig. 2.1 Part of an irregular pore network showing (a) how pore nodes are connected by cylindrical throats of random radius and indicating (b) the numbering of pores and throats.

In principle, this defines the network completely and all additional arrays describing the neighboring relationships are only defined for convenience in the drying algorithm. For each pore, its neighboring pores are stored in PNP and the neighboring throats connecting it to these pores in PNT (respecting the same order):

$$\text{PNP} = \begin{bmatrix} 2 & 3 & 4 & 0 \\ 1 & 5 & 6 & 7 \\ \vdots & \vdots & \vdots & \vdots \end{bmatrix}, \quad \text{PNT} = \begin{bmatrix} 1 & 2 & 3 & 0 \\ 1 & 4 & 5 & 6 \\ \vdots & \vdots & \vdots & \vdots \end{bmatrix} \quad (2.2)$$

For each throat, the neighboring throats are set into TNT, and the pores by which they are linked are set (respecting the same order) into TNP2, which is an expansion of TNP:

$$\text{TNT} = \begin{bmatrix} 2 & 3 & 4 & 5 & 6 \\ \vdots & \vdots & \vdots & \vdots & \vdots \end{bmatrix}, \quad \text{TNP2} = \begin{bmatrix} 1 & 1 & 2 & 2 & 2 \\ \vdots & \vdots & \vdots & \vdots & \vdots \end{bmatrix} \quad (2.3)$$

Now, a radius r_{ij} is attributed to each throat according to a given random distribution; throat length can be computed from the distance between connected pores, but in this work we will take a uniform value L. As secondary properties, throat cross-sectional area A_{ij} as well as throat volume V_{ij} can be obtained. In the following, one index will denote pores, and two indices throats (connecting the indicated pores).

2.2.1.2 Boundary-layer Modeling

Only convective drying with air (kinematic viscosity ν) flowing at velocity u will be considered here. At low Reynolds numbers $\text{Re} = uL_\text{network}/\nu$, a laminar boundary layer will develop at the planar surface of the network, so that heat and mass transfer between this surface and the bulk of drying air can be described by heat conduction or vapor diffusion through the boundary layer. Correlations for transfer coefficients are available in form of dimensionless numbers; for mass transfer, the Sherwood number is

$$\text{Sh} = \frac{\beta L_{\text{network}}}{\delta} = 0.664\,\text{Re}^{1/2}\text{Sc}^{1/3} \tag{2.4}$$

with vapor diffusivity δ and Schmidt number $\text{Sc} = \delta/\nu$. Similar correlations are available for the more realistic case of a turbulent boundary layer. Note, however, that Eq. 2.4 was developed for a plate at uniform partial pressure, which is not necessarily the case in drying (Masmoudi and Prat 1991). Traditionally, vapor transfer from a surface node to the bulk air was described by a local mass-transfer coefficient β:

$$\dot{M}_{v,i} = \beta A_i \frac{P\tilde{M}_v}{\tilde{R}T} \ln\left(\frac{P - p_{v,\infty}}{P - p_{v,i}}\right) \approx \beta A_i \frac{\tilde{M}_v}{\tilde{R}T} \left(p_{v,i} - p_{v,\infty}\right) \tag{2.5}$$

As illustrated in Fig. 2.2a, this will induce a decrease in overall drying rate as soon as surface throats empty, because an additional resistance to vapor transfer is immediately introduced for these throats. As a consequence, the first drying period, in which a porous medium dries at the same rate as a completely wet surface, can not be obtained by the model.

Since one reason for the constant rate period lies in lateral diffusion within the boundary layer (Schlünder, 1988a), this has to be included into the modeling. Figure 2.2b shows how the grid of nodes is extended into the gas phase to this purpose. The connections between these nodes are treated as throats that are completely characterized by their cross-sectional area A_{ij} and length L. For square

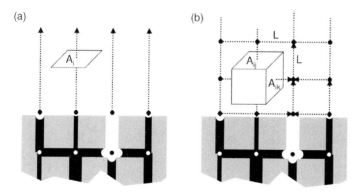

Fig. 2.2 Modelling of vapor transfer through boundary layer by (a) local mass-transfer coefficient or (b) discretized description of diffusion.

(or cubic) networks, the cross-sectional area is $A_{ij} = L^2$, except for horizontal throats along the network surface where half this value is taken. The number of nodes in the boundary layer is given by its thickness $s = \delta/\beta$ and the distance between the nodes L. In order to compute s, we may also take a value greater than L_{network} – as the length of the porous body in air flow direction – if the network only represents part of the object.

In Section 2.2.6, we will see that this description of the boundary layer will enable a first drying period to be reproduced for certain pore structures.

2.2.1.3 Saturation of Pores and Throats

All simulations start with a pore space completely saturated with liquid. During drying, the liquid/air interface (in the form of menisci) will move into the pore network and network saturation will decrease; the change of local saturation must be modeled and stored in respective data structures. To this purpose, we introduce throat saturations S_{ij}, which can take values from zero to one. Figure 2.3 shows a partially filled network; for clarity, throats and pores are drawn separately. Throats can either be empty ($S_{ij} = 0$), partially filled, when they contain one or two menisci ($0 < S_{ij} < 1$), or completely full ($S_{ij} = 1$). Pore saturation S_i can only take discrete values: for $S_i = 0$, we have a gas pore, and for $S_i = 1$, a liquid pore. The saturation of the neighboring throats decides: only if all neighboring throats are (at least partially) filled, i.e. $S_{ij} > 0$, the pore is a liquid pore. This is clear, since a pore is invaded by the gas phase when the first of its neighbor throats is completely emptied. Then, the phase boundary, represented by menisci, will move into the remaining liquid-filled throats.

Depending on throat saturation, transport mechanisms are different. In empty throats, vapor diffusion can occur due to differences in vapor pressure; in full throats, liquid can be pumped due to differences in liquid pressure. In partially filled throats, liquid may evaporate at the menisci and/or flow due to liquid pressure differences.

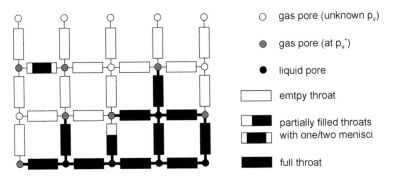

Fig. 2.3 Small partially saturated network with filling states of pores and throats.

2.2.1.4 Vapor Transfer

Since gas is the invading phase, it is always continuous. During drying, vapor is transferred from the liquid/gas phase boundary through the already empty pores and the boundary layer to the bulk of drying air. This vapor diffusion is modeled by a quasisteady approach: the motion of menisci is not continuously tracked, but time is discretized by the complete emptying of a single throat. During these natural time steps, we assume constant vapor flow rates without accounting for local accumulation of vapor in the gas phase or continuously increasing mass-transfer resistances. Additionally, we assume that in gas pores next to (partially) filled throats the gas is saturated with vapor (local equilibrium), hence neglecting any resistance in the empty section of partially saturated throats (a reason for this is given in Section 2.2.3 when viscous forces are introduced). Then, we can set up the following mass balances for all gas pores of unknown vapor pressure $p_{v,i}$ (in network and boundary layer).

$$\sum_j \dot{M}_{v,ij} = \sum_j A_{ij} \frac{\delta}{L_{ij}} \frac{P \tilde{M}_v}{\tilde{R}T} \ln\left(\frac{P - p_{v,i}}{P - p_{v,j}}\right) = 0 \qquad (2.6)$$

The sum is over all neighboring gas pores that are connected by an empty throat. The boundary conditions to this set of equations are given by equilibrium vapor pressure in pores adjacent to (partially) filled throats ($p_{v,j} = p_v^*$) and by the vapor pressure of drying air at the far end of the boundary layer ($p_{v,j} = p_{v,\infty}$) as shown in Fig. 2.4.

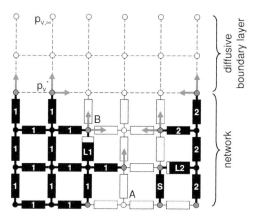

Fig. 2.4 Clusters and local evaporation rates during drying of a small network.

The set of equations can be transformed into a linear system by replacing $p_{v,i}$ with the variable $x_i = \ln(1 - p_{v,i}/P)$ and introducing vapor conductances g_{ij}.

$$\mathbf{A} \cdot \mathbf{x} = \begin{pmatrix} & \vdots & & \vdots & \\ \cdots & \sum g_{ij} & \cdots & -g_{ik} & \cdots \\ & \vdots & & \vdots & \\ \cdots & -g_{ik} & \cdots & \sum g_{ki} & \cdots \\ & \vdots & & \vdots & \end{pmatrix} \cdot \begin{pmatrix} \vdots \\ x_i \\ \vdots \\ x_k \\ \vdots \end{pmatrix} = \mathbf{b} = \begin{pmatrix} \vdots \\ \sum g_{il} x^* \\ \vdots \\ g_{kl} x_\infty \\ \vdots \end{pmatrix} \quad (2.7)$$

In this context, the convenience of the data structures PNP and PNT in combination with pore saturations S_i and throat saturations S_{ij} becomes obvious. First, all gas pores with unknown vapor pressure (building up the vector of unknowns **x**) can be easily identified by scanning the saturation of their neighboring throats. Likewise, each row of matrix **A** is determined by the saturations of pore and throat neighbors. And finally, the vector of boundary conditions **b** gets contributions from neighbor pores with known vapor pressure.

From the solution of Eq. 2.7 we directly know the vapor flow rates in any empty throat. During the drying of a network, it will happen that whole regions of gas pores are at the saturation vapor pressure even if they are not direct neighbors of a meniscus. Accordingly, a net vapor flow will not occur in every empty throat (see throats to pore A in Fig. 2.4).

2.2.1.5 Capillary Pumping of Liquid

Within the liquid phase, which is initially continuous, mass transfer is by convection due to differences in capillary pressure (with surface tension σ)

$$P_c(r_{ij}) = \frac{2\sigma \cos \theta}{r_{ij}} \quad (2.8)$$

which describes by how much liquid pressure can be less than gas pressure if the meniscus is fully developed (with radius of curvature r_{ij}); in the following, we will assume zero contact angle θ. Since viscous effects are neglected, liquid can always be pumped from the meniscus in the largest throat (where liquid pressure is highest) to all other menisci. As a consequence, meniscus throats will empty in the order of decreasing radius. Since the radius is randomly distributed, the liquid phase will rapidly split up into disconnected clusters. Figure 2.4 gives an example of a drying network with two such clusters and one single liquid throat (S). Within a cluster, always the largest meniscus throat will empty (L1 and L2). The rate at which this happens is given by the overall evaporation rate from the cluster surface, i.e. the sum of local vapor flow rates from every meniscus belonging to the cluster. If several menisci are connected to a gas pore (node B), the total vapor flow away from this gas pore is distributed to the menisci according to the respective throat cross sections (again, no difference is made between partially and completely filled throats).

Clusters can also be temporarily trapped, i.e. no liquid will evaporate from any of their menisci (such as for single throat S in our example).

The time step of the drying algorithm is determined by the cluster that can first evaporate all remaining liquid in its largest throat. After this time, throat and pore saturations are updated, changing the linear system for vapor diffusion, the choice of menisci per cluster, or even the number of liquid clusters and cluster affiliation of liquid throats.

2.2.1.6 Cluster Labeling

It has become obvious that at any instant we must know which liquid throats – specifically those containing a meniscus – are connected, in order to find for each cluster which candidate will empty next and in which time. To this purpose all liquid throats are given labels assigning them to one or the other cluster. (Alternatively, liquid pores could be labeled.) We apply a variant of the Hoshen–Kopelman algorithm (Al-Futaisi and Patzek 2003; Metzger et al. 2006) in order to update these labels after a throat has emptied, using the old cluster information.

The Hoshen–Kopelman algorithm scans all liquid throats one by one, checking for already labeled connected liquid throats (connection is by a liquid pore). If, during this process, labels are found to be identical, they are not immediately corrected, but the information is stored in a separate array. Only after all liquid throats have been labeled will these labels be corrected. Single liquid throats do not participate in capillary flow and are labeled zero.

During each time step, normally only one throat empties, so that changes are possible only within the corresponding cluster, in the sense that it splits up into several clusters or single throats. Therefore, all updating can be restricted to one cluster. It is even better to start the relabeling process at the emptied throat and define suitable stopping criteria.

Figure 2.5 shows the situation when the first throat in the network has emptied, leading to the emptying of a pore and local separation of liquid by creation of new

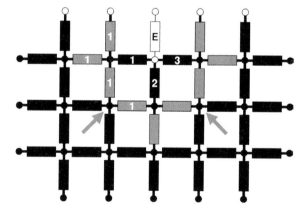

Fig. 2.5 Updating of cluster labels after emptying of one throat (E).

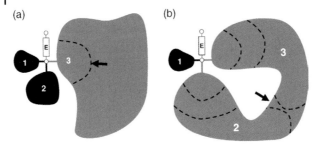

Fig. 2.6 Stopping criteria for cluster labeling after emptying of one throat (E).

menisci. Hence, we must check if the respective liquid throats still belong to the same cluster. First, the former neighbors of the emptied throat are given different labels. Then, liquid throats connected to them are identified; we may call these their *children*. These children are labeled with the method of Hoshen–Kopelman, always attributing the minimum label already existing in the neighborhood. If two labels are found to be identical, this information is stored. In our example, all new labels are in fact identical; the arrows indicate that this is discovered when labeling the first generation of children.

In general, several generations of children must be relabeled before the algorithm stops because one of the following criteria is fulfilled:

1. all throats of the old cluster have been relabeled.
2. all members of the new generation have been given the *same* label because:
    - all except one of the newly created clusters are completely labeled;
    - all remaining different clusters are found to be connected.

Figure 2.6 illustrates the two variants of case 2; the second example can only happen in three dimensions where liquid loops may provide a far connection for locally separated throats.

At the end, like in the usual Hoshen–Kopelman algorithm, these intermediate labels are updated. If new clusters are created, old labels need to be shifted to accommodate them.

The algorithm employs previous information so that usually only the direct neighborhood of the emptied throat has to be scanned, reducing computation time drastically.

2.2.1.7 Drying Algorithm

All discussed elements are combined into an algorithm to compute the drying of a pore network according to the flow chart of Fig. 2.7. First, network geometry is defined and the boundary layer discretized (step 1); the boundary-layer thickness is computed from the characteristic length of the porous body and velocity of drying air.

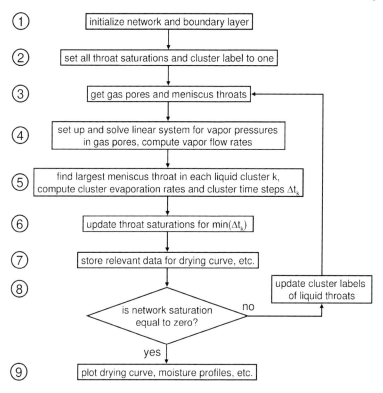

Fig. 2.7 Flow chart for isothermal convective drying of pore network when gravity and viscous effects can be neglected.

Then, all throats are filled by liquid and the initial (unique) cluster label is defined (step 2). From throat saturations, pore saturations are derived and meniscus throats identified (step 3). For gas pores (initially only in the boundary layer), the vapor pressure field is obtained, using the boundary conditions in the bulk drying air and in pores neighbor to a meniscus throat; from this, a vapor flow rate is computed for every throat (step 4). For each cluster (initially only one), the meniscus throat with the highest liquid pressure is found; in the absence of gravity and temperature gradients, this is simply the largest one. Vapor flow rates away from the cluster are summed up to the cluster evaporation rate, and the time to empty the largest throat is computed (step 5). From these cluster time steps, the minimum is chosen and used to update throat saturations (step 6). Time step, total evaporation rate, network saturation, throat saturations and moisture profiles are stored (step 7) for later use to plot the drying curve or evolution of phase distributions (step 9). As long as the network is not completely dry, cluster labels are updated (step 8), and quasisteady evaporation and invasion of the next meniscus throat is computed again.

This drying algorithm will be used in the following section and extended in later sections in order to additionally account for gravity, liquid viscosity, etc.

2.2.2
Simulation Results and Experimental Validation

As a first application of the above-described drying model, we look at the drying of a square 50 × 50 network with monomodal throat-size distribution. The radius of each throat is chosen independently according to the normal probability density function

$$f(r) = \frac{1}{\sqrt{2\pi}\sigma_0} \exp\left[-\frac{(r-r_0)^2}{2\sigma_0^2}\right] \tag{2.9}$$

with mean $r_0 = 50$ μm and standard deviation $\sigma_0 = 10$ μm; throat length is uniformly 500 μm so that $L_{\text{network}} = 25$ mm. The drying air contains no moisture ($p_{v,\infty} = 0$) and is at room temperature ($T = 20$ °C) and atmospheric pressure ($P = 1$ bar); the mass-transfer coefficient is $\beta = 0.5$ mm s^{-1}. Figure 2.8 shows how the phase distribution evolves during drying. Clearly, the results depend strongly on random network generation, and, for better illustration, we chose an example with well-pronounced capillary fingering. Very similar drying patterns have been observed in experiments with etched micromodels of comparable dimensions but with throats of rectangular cross section obeying a log-normal size-distribution law (Laurindo and Prat 1996).

During drying, liquid is removed by evaporation and throats are invaded by gas from the surface; due to capillary pumping, large throats are invaded first. This process can initially be described by invasion percolation rules; with time, a fractal liquid/gas phase boundary will develop that may extend far into the network. As a consequence, besides the main liquid cluster, numerous small disconnected liquid

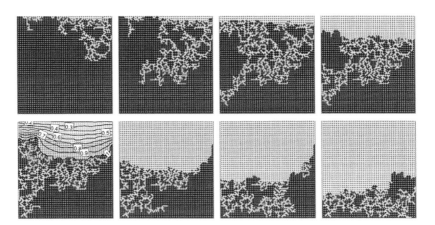

Fig. 2.8 Phase distributions during drying of a 50 × 50 network for overall saturations S=0.9,0.8,...,0.2 (liquid-filled throats are in black, gas-filled throats in white). Levels of constant vapor pressure are shown for the half-saturated network (values are fractions of saturation).

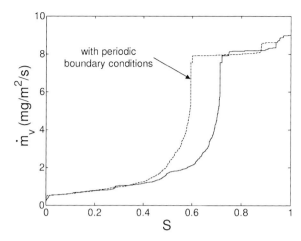

Fig. 2.9 Drying curve for 50 × 50 network in Fig. 2.8 and influence of periodic boundary conditions in the lateral direction.

clusters may form. Contrary to usual invasion percolation, these clusters are only temporarily trapped. If they have meniscus throats near the surface, they can be invaded; but they may also be screened from evaporation by other clusters, if they are deeper inside the network. Once the main liquid cluster is disconnected, a receding front of evaporating small clusters can be observed. In Fig. 2.8, the vapor pressure distribution in the gas-filled region is shown for a network saturation of $S = 0.5$. The line of saturated air marks the evaporation front that is the envelope of the liquid cluster region.

Figure 2.9 depicts the drying rates for the same network; of course, they are closely related to the phase distributions that determine the vapor-transfer resistance in the network. Additionally, boundary-layer thickness plays an important role. Here, we chose a very thick boundary layer ($s = 50$ mm, discretized by 10 vertical nodes) allowing better illustration of transport phenomena inside the network because mass-transfer resistances in the boundary layer and network are comparable. As a result, the evaporation rate will not drop significantly until the whole surface region of the network is dry. As long as part of the surface is still saturated (for $S > 0.75$), lateral diffusion in the boundary layer keeps the drying rate almost at the initial level.

For a thinner boundary layer, overall drying rates are increased and the shape of the drying curve changes because, then, the evaporation rate is more sensitive to surface saturation of the network. Every time surface throats empty, the drying rate will drop significantly, and a sequence of quasiconstant drying periods is obtained (for an example, see Fig. 2.23). In contrast to the drying curve, phase-distribution patterns depend little on boundary-layer thickness, since cluster formation is dominated by the spatial throat radius distribution and only cluster evaporation rates are influenced significantly.

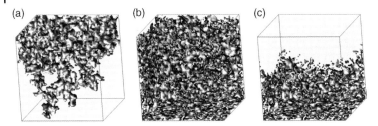

Fig. 2.10 Evolution of phase boundary for three-dimensional network (adapted from Le Bray and Prat 1999).

If we change to periodic boundary conditions by connecting the right and the left-hand side of the network, we can eliminate border effects and increase the effective network size. As a consequence, capillary pumping may be more efficient (first row in Fig. 2.16) so that drying rates stay high for a wider range of saturations (dashed curve in Fig. 2.9).

Three-dimensional results for the drying of a $51 \times 51 \times 51$ network (Le Bray and Prat 1999) are shown in Fig. 2.10 as the gas/liquid phase boundary (a) at breakthrough, i.e. when the gas reaches the bottom of the network, (b) when the main liquid cluster is disconnected from the network surface and (c) during the receding front period. As compared to two-dimensional networks, long-distance capillary flow paths are more easily maintained due to the additional dimension. Consequently, the main liquid cluster stays connected to the network surface for longer.

In the work from Le Bray and Prat (1999), vapor transfer through the boundary layer is modeled by a constant local transfer coefficient (not accounting for lateral diffusion), so that the drying rate is almost proportional to surface saturation. Therefore, the drying rate drops drastically at the beginning, and then stabilizes when capillary pumping becomes efficient (Fig. 2.11). Recent simulations show that the drying rate can be maintained at its high initial value over this whole period if

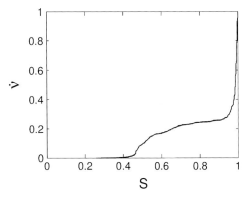

Fig. 2.11 Dimensionless drying curve of three-dimensional network (adapted from Le Bray and Prat 1999).

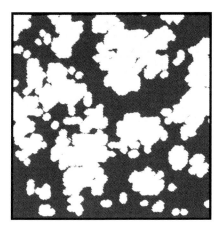

Fig. 2.12 Distribution of dry patches (in white) on network surface (adapted from Le Bray and Prat 1999).

lateral vapor transfer in the boundary layer is modeled (Yiotis et al. 2006). In this context, it is also worth mentioning that network modeling can reproduce the characteristic wet and dry patches at the surface of a porous medium that are larger than individual pores (Fig. 2.12).

2.2.3
Gravity and Liquid Viscosity – Stabilized Drying Front

In this section, the isothermal drying model is extended by gravitational and viscous effects acting on the liquid phase. In principle, gravity can stabilize or destabilize the fractal drying front – depending on the orientation of the gravitational field with respect to the open side of the network; in contrast, liquid viscosity will always be stabilizing. In our drying simulations, we will restrict ourselves to stabilizing gravity (for the destabilizing case, refer to Yiotis et al. 2006; Laurindo and Prat 1996); this is given for a vertically oriented network that is open to evaporation at its top. In the following, we will separately present the necessary model extensions; then, we use a series of drying simulations to point out similarities and differences of the two effects.

2.2.3.1 Modeling Gravity
Gravity affects liquid pressure in the network so that, even in the absence of menisci, it would be a function of position

$$P_\mathrm{w}(z) = -(\rho_\mathrm{w} - \rho_\mathrm{a})gz \qquad (2.10)$$

where z is the height coordinate, ρ_w and ρ_a are densities of liquid water and air (the small vapor fraction is neglected), respectively, and g the gravitational acceleration.

Now, if menisci are present, then capillary pressure, Eq. 2.8, counteracts these gravitational pressure differences. Assuming that the liquid is in vertical equilibrium, gravity and capillary forces balance each other. Concerning the order of emptying, in the case of uniform throat radius, throats are strictly invaded from top to bottom. If throat radius is randomly distributed, it is convenient to introduce the potential (Wilkinson 1984):

$$\Phi(r_{ij}, z) = -\frac{2\sigma}{r_{ij}} + (\rho_w - \rho_a)gz \tag{2.11}$$

Then, throats with a higher potential Φ, i.e. having a larger radius or being located higher in the field of gravity, will be invaded first. Oscillations in phase distributions during drying (refilling of already empty throats) might occur, if Φ is not negative everywhere; this case is, however, not considered here.

From this, we see that the only change necessary to include gravity into the model is in step 5 of the drying algorithm where we must find the meniscus throats with the highest throat potential instead of those with the largest radius (Fig. 2.7).

2.2.3.2 Modeling Liquid Viscosity

Viscous effects in the liquid phase were first modeled by Nowicki et al. (1992), who, unfortunately, do not report on drying behavior at the sample scale.

Yortsos and coworkers (Yiotis et al. 2001) proposed a model including gas and liquid viscosity but with different network geometry: spherical pores contain the liquid or gas, throats serve as conductors and capillary barriers. They studied the drying of a pore network connected to a small fracture. For high gas flow rates through the fracture, significant pressure gradients develop in the gas phase (also within the network); the invasion of pores is strongly influenced leading to a piston-like displacement of the liquid with a stabilized phase boundary. For low gas flow rates, the process is controlled by capillary fingering and phase distributions are similar to those in Fig. 2.8. Whereas in the investigation by Yortsos gas viscosity played the major role, we will study the influence of friction in the liquid phase.

In the following, we will first illustrate the new situation due to liquid viscous effects and, then, introduce an appropriate algorithm to describe it for our network geometry (Irawan et al. 2006). When friction is non-negligible, differences in capillary pressure may not be enough to pump liquid from the meniscus in the largest throat to all other menisci in the same cluster for keeping them stationary. As a result, we may have many partially filled throats with moving menisci. Taking a small liquid cluster, namely cluster 2 in Fig. 2.4, we can explain what may happen.

Figure 2.13 shows the relative size of throat radii ($r_2 < r_1 = r_4 < r_3 < r_5$) as well as the places of evaporation (gray arrows). Initially, throat 3 is the largest to contain a meniscus and therefore has the highest liquid pressure $P_{w,3} = P - 2\sigma/r_3$; consequently, it must empty. Additionally, throat 1 will empty, because its meniscus cannot produce a low enough liquid pressure to supply liquid at the local evaporation rate. In

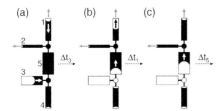

Fig. 2.13 Simultaneous emptying and refilling of throats in the presence of liquid viscosity.

contrast, the meniscus in throat 2, which can produce a lower pressure and has a lower evaporation rate, as well as the meniscus in throat 4, which need not supply any liquid because there is no evaporation, can remain stationary.

For simplicity, we do not aim to resolve the motion of single menisci in time; instead, we assume quasisteady flow rates for vapor and liquid, and hence constant meniscus velocities (white arrows in Fig. 2.13), until another throat has been completely emptied. In our example, this is throat 3 after Δt_3. Then, cluster size is reduced by separation of a single throat 4 (now trapped), leaving throat 5 with the largest meniscus radius. Since its liquid pressure is higher than the former $P_{w,3}$ and since the distance to the meniscus in throat 1 has been reduced, the pressure difference can now pump more liquid to throat 1 than is evaporated. As a consequence, this throat refills in Δt_1, again assuming quasisteady transport. Then, its meniscus is stationary, until throat 5 has emptied.

Strictly speaking, the partial emptying and refilling of throat 1 would be accompanied by a drop and increase in the local evaporation rate, because of the additional vapor diffusion resistance in the empty throat section. In combination with the quasisteady approach, this would lead to oscillations in the overall drying rate. This unwanted behavior is avoided by neglecting the resistance to vapor transfer in partially filled throats altogether (see above).

In order to correctly identify all moving menisci in a cluster and obtain their respective velocities, one must solve the balance for liquid mass flow rates (assuming Poiseuille flow) at each liquid pore i

$$\sum_j \dot{M}_{w,ij} = \sum_j \frac{\rho_w \pi r_{ij}^4}{8\eta L_{ij}}(P_{w,i} - P_{w,j}) = 0 \qquad (2.12)$$

where $0 < L_{ij} \leq 1$ is the liquid-filled length of the throat, η dynamic viscosity of water, and $P_{w,j}$ liquid pressure at the neighboring meniscus or node. (The sum is over all neighbor throats because a pore node is only liquid if all connected throats contain liquid.) What makes this task tricky is the fact that the boundary conditions to Eq. 2.12 are given at the menisci and depend on whether they are moving or stationary.

In a full throat, the meniscus is stationary, if it can produce a low enough liquid pressure to supply water at the evaporation rate; in this case, a second kind of boundary condition is applied: the liquid mass flow in the throat equals the

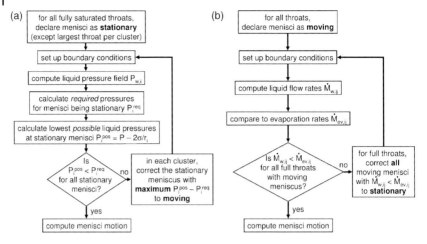

Fig. 2.14 Flow charts of possible algorithms to identify moving and stationary menisci.

evaporation rate at the meniscus, $\dot{M}_{w,ij} = \dot{M}_{ev,ij}$. On the other hand, menisci are moving in the largest throat of a cluster, in all partially filled throats and in full throats where liquid cannot be supplied at the evaporation rate; here we assume the meniscus to be fully developed, no matter if the throat is filling or emptying, so that the first kind boundary condition $P_{w,j} = P - 2\sigma/r_{ij}$ is applied.

The unique set of boundary conditions for the liquid phase must be found iteratively; two possible algorithms are given in Fig. 2.14. The first (a) starts with the situation as in the nonviscous case where only the largest throat of each cluster has a moving meniscus. Then, the menisci that were declared stationary but cannot provide a low enough liquid pressure are, one by one, corrected to moving, starting with the one having the largest discrepancy. The procedure must be stepwise because the correction of one boundary condition will raise the pressure in the cluster – possibly making further corrections unnecessary.

The second algorithm (b) starts from the opposite assumption that all menisci are moving (high-viscosity limit). Then, all menisci that can pump enough liquid are corrected to become stationary. This rise in cluster pressure will allow more menisci to be stationary, and iteration is necessary until the boundary conditions are without contradiction. A first comparison of the algorithms suggests the second one to be faster, but a systematic investigation is still due.

Once the right boundary conditions are found and the linear system, Eq. 2.12, is solved, meniscus velocities are obtained from the difference between evaporation and liquid flow rates. From this, the times required for complete filling or emptying of throats are computed to choose the global time step, just as in the basic drying model.

The phenomenon of liquid redistribution after a meniscus has passed from a narrow to a wide throat (as in Fig. 2.13) is known as a Haines jump.

2.2.3.3 Dimensionless Numbers and Length Scales

It is useful to define dimensionless numbers for assessing the relative importance of gravitational or viscous effects in comparison to capillarity.

Assuming a narrow radius distribution ($\sigma_0 \ll r_0$), a typical difference in capillary pressure is

$$\Delta P_c = 2\sigma \left(\frac{1}{r_0 - \sigma_0/2} - \frac{1}{r_0 + \sigma_0/2} \right) \approx \frac{2\sigma\sigma_0}{r_0^2} \tag{2.13}$$

Similarly, gravitational effects may be characterized by the pressure difference

$$\Delta P_{\text{gravity}} = (\rho_w - \rho_a) g \Delta L \tag{2.14}$$

where ΔL is a *vertical* distance; choosing it as the throat length L, the ratio between gravitational and capillary forces defines the Bond number

$$B = \frac{(\rho_w - \rho_a) g L r_0^2}{2\sigma\sigma_0} \tag{2.15}$$

Alternatively, one may define, by setting $\Delta P_{\text{gravity}} = \Delta P_c$, a characteristic *vertical* distance

$$L_{\text{gravity}} = \frac{2\sigma}{(\rho_w - \rho_a)g} \frac{\sigma_0}{r_0} \frac{1}{r_0} \tag{2.16}$$

over which capillary forces may overcome gravity. One may expect gravitational effects if the network size $L_{\text{network}} = nL$ (n number of throats) is greater than L_{gravity}. For constant relative distribution width σ_0/r_0, we find that gravitational effects will become more important with increasing pore size (corresponding to an increase of Bond number). For constant aspect ratio of the cylindrical pores r_0/L and network size n, we find that $L_{\text{network}} \sim r_0$, whereas $L_{\text{gravity}} \sim r_0^{-1}$; this behavior is shown by curves in Fig. 2.15. In order to investigate the influence of gravity, we choose large-sized pores ($r_0 = 50$ μm) and vary the relative distribution width from 2 to 20% (open circles).

Concerning viscous effects, a typical pressure difference due to friction may be given by

$$\Delta P_{\text{friction}} = \frac{8\eta v \Delta L}{r_0^2} \tag{2.17}$$

where ΔL is a distance for liquid transport (in *any* direction) and v the throat average of liquid velocity; its maximum can be computed from initial evaporation rate and surface porosity of the network

$$v \leq \frac{\dot{m}_{v,I}}{\rho_w} \frac{L^2}{\pi r_0^2} \quad \text{with} \quad \dot{m}_{v,I} \approx \beta \frac{\tilde{M}_v p_v^*}{\tilde{R}T} \tag{2.18}$$

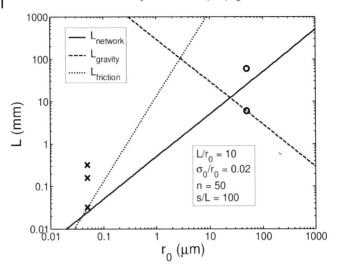

Fig. 2.15 Characteristic lengths for gravity and friction as compared to network size. (For the studied cases, circles and crosses give characteristic lengths of gravity and friction, respectively, whereas the solid line indicates network size).

Taking $\Delta L = L$, the ratio between viscous and capillary forces defines the Capillary number

$$\mathrm{Ca} = \frac{4\eta v L}{\sigma \sigma_0} \qquad (2.19)$$

A characteristic distance

$$L_\mathrm{friction} = \frac{\sigma}{4\eta v}\frac{\sigma_0}{r_0} r_0 \qquad (2.20)$$

over which capillary forces may pump liquid at velocity v, is obtained by setting $\Delta P_\mathrm{friction} = \Delta P_c$. Assuming, for simplicity, a linear dependency between boundary-layer thickness and network size (constant s/L), we have $\beta \sim 1/L$ and $L_\mathrm{friction} \sim r_0^2$ as shown in Fig. 2.15. Therefore, we will choose small pores ($r_0 = 50$ nm) and vary the relative distribution width from 2 to 20% for our investigation on liquid viscosity (three crosses). Note that we still neglect the Kelvin effect (the reduction of equilibrium vapor pressure at the meniscus is 2%) and the Knudsen effect (although the mean free path of air and vapor molecules is roughly equal to r_0).

Because of small network size ($n = 50$), pores will have to be chosen unrealistically small for a capillary porous medium (and pore-size distribution unrealistically narrow) to observe significant viscous effects. For realistic pore size (distributions), similar effects will only occur in large networks that cannot be simulated yet.

Gravitational and viscous forces will be important at the same time only for large objects with small pores.

2.2.3.4 Phase Distributions and Drying Curves

First, we take the 50 × 50 network (Fig. 2.8), with periodic boundary conditions, to investigate gravitational effects. To this purpose, the network is oriented in vertical direction and the distribution width of radius is rescaled (from $\sigma_0 = 10$ μm to $\sigma_0 = 1$ μm) – leaving the spatial distribution in the network unchanged.

Secondly, both network structure and boundary-layer thickness are shrunk by a factor of 1000 (so that the boundary-layer thickness is 50 μm) to study the influence of liquid viscosity; here again the radius distribution width is varied ($\sigma_0 = 10$, 5 and 1 nm). The results are given in terms of phase distributions (Fig. 2.16) and corresponding drying curves (Fig. 2.17).

Additionally, the reference case of zero gravity and liquid viscosity ($g = 0, \eta = 0$) is given; due to geometric similarity, we obtain identical phase-distribution patterns for small- and large-sized pores, only drying rates differ by one order of magnitude. The other extreme of infinite gravity or viscosity leads to a sharp receding front; then, the drying rate is given by

$$\dot{m}_v = -\left(\frac{s}{\delta} + \frac{z_f}{\delta_{eff}}\right)^{-1} \frac{P\tilde{M}_v}{\tilde{R}T} \ln\left(1 - \frac{p_v^*}{P}\right) \qquad (2.21)$$

where front position depends on saturation $z_f = L_{network}(1 - S)$ and the effective diffusivity in the network is obtained from the reduced cross section to $\delta_{eff} = \delta \pi r_0^2 / L^2$. The corresponding drying curves are given in Fig. 2.17.

As a major result we may see that an increase in the Bond or Capillary number – achieved by a narrower radius distribution – leads to the stabilization of the fractal phase boundary to a receding front that is accompanied by lower drying rates. In the gravity-controlled case, the overall extent of this front remains more or less constant during the drying process and it scales as $B^{-0.52}$ for two-dimensional networks (Prat and Bouleux 1999).

The behavior is slightly different for viscous effects: friction has no preferential direction so that drying behavior is significantly changed even for $L_{friction} > L_{network}$, whereas for gravity $L_{gravity} < L_{network}$ was required (Fig. 2.15). This can be illustrated at the limit of a very narrow pore-size distribution ($\sigma_0 = 1$ nm). Figure 2.18 shows the main cluster of the drying network; for clarity, disconnected small clusters are not plotted. On the right-hand side, the liquid pressure distribution is given. Partially filled throats are plotted as thick lines. All of them are of very similar radius, except those where menisci were created in the previous time step (51.98 nm and 51.47 nm). This leads to a high local pressure that will allow refilling of a nearby throat (50.27 nm). However, pressure variations are localized so that partially filled throats are competing for liquid over long distances with little pressure difference available (50.07 nm and 49.94 nm).

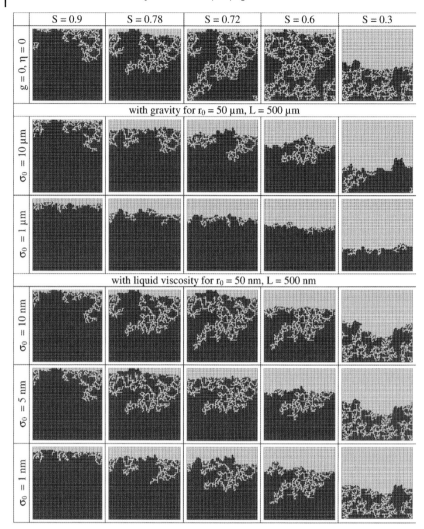

Fig. 2.16 Influence of gravity and liquid viscosity on phase distributions for a periodic 50 × 50 network with $s/L = 100$. For all cases, the relative size distribution of throat radii is identical, only the mean r_0 and the standard deviation σ_0 of this distribution are varied.

Furthermore, the Capillary number depends on the drying rate with the consequence that the drying front widens up as drying proceeds. From Fig. 2.16, we see that, at low network saturations, phase distributions differ only little from the nonviscous case, even if viscosity was important at the start of drying.

Concerning the drying rates for both gravitational and viscous effects (Fig. 2.17), they may locally be higher even if, at a global scale, drying is slower due to less

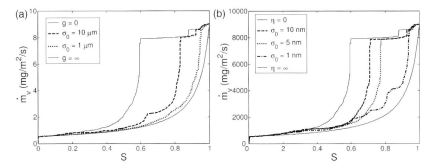

Fig. 2.17 (a) Influence of gravity, (b) Influence of liquid viscosity on drying curve (as in **Fig. 2.16**).

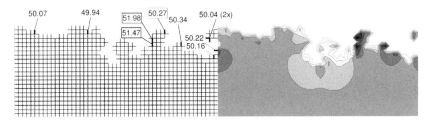

Fig. 2.18 Main cluster at $S = 0.957$ with pressure field (case of last row in **Fig. 2.16**): throat radii (in nm) are given for partially filled throats that are plotted as thick lines; ten dimensionless pressure levels are given with higher pressures indicated as darker.

efficient capillary pumping. This can be explained by a change in the order of throat emptying and resulting different phase distributions for a given saturation.

2.2.4
Film Flow

Unfortunately, the mentioned good agreement of experimental phase distributions with the simulated ones breaks down where drying rates are concerned: here the theoretical values are less than the experimental ones by a factor of about six (Laurindo and Prat 1998). To deal with this deviation, several authors have proposed empirical calibration parameters; for example, the external mass-transfer coefficient has been expressed as a function of porous surface saturation in many macroscopic models (Chen and Pei 1989).

A more fundamental approach explains the enhancement of drying rate by liquid film flows. In reality, liquid films are also expected to have some influence on drying patterns since the associated higher drying rates lead to higher viscous effects.

Yortsos and coworkers (Yiotis et al., 2003, 2004, 2005) were the first to propose a pore-network model of isothermal drying including the effect of liquid films. Then, a similar model was used by Prat (2006) to study the influence of contact angle and pore shape on drying rates (see Sections 2.2.6 and 2.4.1). These models are successful in the sense that a major impact on drying rates is predicted, which is consistent with the available experimental data. However, they need to be improved as discussed at the end of this section.

The films in these models consist of liquid trapped by capillarity along wedges and corners of the pore space; thus, they cannot exist in cylindrical throats. In contrast to such "thick" films, thin films of a few nanometres thickness can develop on a smooth surface (Churaev 2000). Transport in thin films only plays a role in very small pores because of their very low hydraulic conductivity. Flow in thick films is driven by capillarity and can, therefore, be much more efficient then vapor diffusion.

A simple but instructive introduction to thick films is to consider evaporation from a single capillary tube of square cross section. As sketched in Fig. 2.19, air invades the tube bulk as a result of evaporation, but liquid remains trapped within the corners owing to capillary effects. Liquid is transported to the tube entrance by the corner films, maintaining an evaporation flux much higher than in a circular tube. The films tend to thin out as the main meniscus recedes into the tube because of gravity and/or viscous effects. Eventually, the film tip also recedes into the tube, leading to the development of a dry zone between the film tip and tube entrance in which transfer is by vapor diffusion. The interested reader can refer to Prat (2006), Camassel et al. (2005), Coquard et al. (2005), for more details on the different evaporation regimes that are expected in tubes of polygonal cross section and to Eijkel et al. (2005) for some experimental evidence of the effect of corner films on evaporation in straight channels.

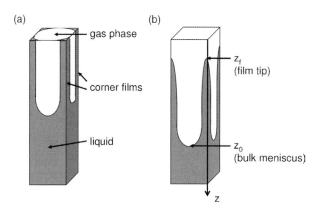

Fig. 2.19 Schematic of drying of a square capillary tube with corner films (a) up to the tube entrance and (b) receding into the tube.

A simple model of evaporation in polygonal tubes is based on the following assumptions (Yiotis et al., 2003, 2004, 2005; Prat 2006):

1. vapor diffusion in the gas phase is neglected in the region of the film (water-vapor pressure being essentially at equilibrium);
2. gravity effects in the films are negligible compared to viscous and capillary effects;
3. the external transfer length scale ($\sim$boundary-layer thickness s) is much larger than the tube hydraulic diameter.

The film is characterized by the radius of curvature R of the liquid/gas interface in the tube cross-section, the curvature of the film interface along the tube being negligible. Under these circumstances, the equation governing R can be expressed as, (Yiotis et al., 2003, 2004, 2005; Prat 2006),

$$\frac{d^2 R^3}{dz^2} = 0 \qquad (2.22)$$

with boundary conditions

$$\frac{N\rho_w \kappa \sigma R^2}{\eta} \frac{dR}{dz} = A \frac{\delta}{s+z_f} \cdot \frac{\tilde{M}_v}{\tilde{R}T} \cdot (p_v^* - p_{v,\infty}) \quad \text{at} \quad z = z_f \qquad (2.23)$$

$$R = \frac{R_0}{\chi} \quad \text{at} \quad z = z_0 \qquad (2.24)$$

where N is number of tube sides ($N = 4$ for square tube), κ a dimensionless viscous resistance factor, A is tube cross-sectional area, z_f is film–tip position, R_0 is radius of the largest inscribed sphere in the capillary and χ is a numerical factor depending on contact angle, critical contact angle θ_c (see below) and pore shape. Equation 2.24 expresses the continuity of the capillary pressure between bulk meniscus and corner menisci at z_0, whereas Eq. 2.23 assures that the mass flow rate within the film is equal to the evaporation rate. To extend the above model to a pore network the following main assumptions are made, (Yiotis et al., 2003, 2004, 2005; Prat 2006):

A1. Pore- and throat-size variability is neglected as far as transport in the film and vapor diffusion in the gas phase are concerned (though taken into account in invasion rules).
A2. An identical and constant film thickness is assumed at all clusters. So far, this is obtained from the invaded throat (other possibilities are given in Yiotis et al. 2004).

In the presence of films, one distinguishes three main regions as sketched in Fig. 2.20: the dry zone, in which transport is by gas-phase diffusion, the film region in which transport is in the liquid phase through films, as well as the clusters, which are fully saturated regions. In the film region, because of assumption A1, the governing

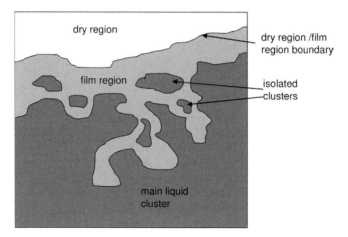

Fig. 2.20 Schematic of phase distribution during drying in the presence of liquid films.

transport equation takes the form

$$\nabla^2 R^3 = 0 \tag{2.25}$$

(with the Laplace operator ∇^2) which is a straightforward extension of Eq. 2.22. In the dry region, the governing equation takes the simple form (again thanks to assumption A1)

$$\nabla^2 p_v = 0 \tag{2.26}$$

These equations are completed by the continuity of mass flow at the boundary between film region and dry zone (analogous to Eq. 2.23), the boundary condition at the cluster boundary (analogous to Eq. 2.24 using assumption A2) and the boundary condition at the network surface; for details see, e.g., Prat (2006). It is clear that the boundary between the dry region and the film region evolves during drying. A nice way to avoid tracking this interface explicitly is to introduce a composite variable of the form $\phi = R^3 + \xi p_v$, where ξ is deduced from the equation expressing mass flow continuity at the film region/dry region boundary. As a result, the transport equation in both regions is simply

$$\nabla^2 \phi = 0 \tag{2.27}$$

which is analogous to the diffusion equation in the basic model version (see Section 2.2.1). With this formulation, it is therefore fairly easy to include the film effect in the pore-network model.

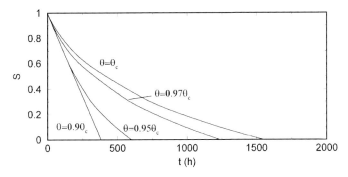

Fig. 2.21 Temporal evolution of network saturation for different contact angle θ when liquid films are present, in a 40 × 40 square network with throats of square cross section.

As an illustration, Fig. 2.21 shows the temporal evolution of network saturation for different values of contact angle θ in a 40×40 network ($L = 1$ mm, $s = 5$ μm). For $\theta \geq \theta_c$ ($= 45°$ in square throats, see Section 2.4.1), films cannot develop for geometrical reasons (Prat 2006). This corresponds to the model of Section 2.2.1. For $\theta < \theta_c$, films develop and their extent increases with $\cos\theta$. The results plotted in Fig. 2.21 are consistent with the experiments reported in Laurindo and Prat (1998). Much faster drying is predicted when thick films are present, and a small change in contact angle leads to markedly different drying times.

To conclude this section, it is clear that pores in real materials are generally more complex than the straight tubes considered here. Therefore, the presented model should be understood as describing the effect of film flows in an average sense only. Among the assumptions made, the most questionable is assumption A2. In fact, the invasion threshold varies from one cluster to the other and changes also as drying proceeds for a given cluster. This opens the possibility of intercluster film flow and also poses numerical problems since the boundary condition at the clusters for the film region fluctuates in time. This aspect deserves to be investigated in detail. Also quantitative comparisons with experimental data are yet to be made.

2.2.5
Wettability Effects

Not surprisingly, wettability conditions can greatly affect drying since capillarity is a dominant mechanism under normal circumstances. Here, we discuss the impact of wettability on drying patterns and drying rates by assuming that the whole sample can be characterized by a certain equilibrium contact angle θ taken in the liquid. The porous medium is referred to as hydrophilic when $\theta < 90°$ and hydrophobic when $\theta > 90°$. We further assume that only capillary forces control the invasion (except in films, if any, where viscous effects are present). It is not difficult to anticipate the additional effects of viscosity or gravity in this context, based on the results of Section 2.2.3. It should be pointed out, however, that the Capillary number increases with contact angle. Hence, for otherwise identical drying conditions, viscous effects will necessarily affect the drying pattern as the contact angle approaches 90°.

In the case of hydrophilic porous media there is a major effect on drying rates because of the liquid films that develop below the critical contact angle θ_c and have an extent strongly dependent on contact angle ((Prat 2006), Sections 2.2.4 and 2.4.1). Phase distributions are, however, not affected significantly, as experiments (Laurindo and Prat 1996) and numerical simulations (Prat 2006) indicate. This is consistent with the invasion percolation rule stating that the largest meniscus throat is selected in each cluster regardless of the value of contact angle.

Major changes due to wettability occur when the porous medium is hydrophobic. This case has been explored recently for teflonized porous media in relation with the study of evaporation in some components of fuel cells (Chapuis et al. 2006; Chapuis 2006). Here, the contact angle is typically 110° to 120°. Since gas is now the wetting fluid, the invasion is no longer analogous to a drainage process but to an imbibition one, e.g. Dullien (1992). Invasion rules in quasistatic imbibition are significantly more complex than in drainage (Lenormand and Zarcone 1984; Blunt and Scher 1995). First, one must consider the filling of throats and the filling of pores as two distinct steps. Secondly, the pore-filling invasion threshold depends on the number of surrounding throats filled with wetting fluid and also on pore geometry. Once the throat and pore invasion thresholds have been derived for a given geometry, it is, however, not difficult to develop a drying algorithm in the same spirit as for hydrophilic networks (Chapuis et al. 2006; Chapuis 2006). For the square networks studied in Chapuis (2006), the change in the invasion rules leads to invasion patterns dramatically different from the ones in a hydrophilic network, as illustrated in Fig. 2.22. It is obvious from the hydrophobic pattern that drying is faster in a hydrophilic network (Chapuis et al. 2006).

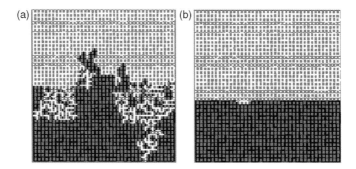

Fig. 2.22 Phase distributions in drying networks with identical microstructure but different wettability:
(a) hydrophilic ($\theta \approx 70°$) and (b) hydrophobic ($\theta \approx 120°$).

2.2.6
First Drying Period

One major challenge for network drying models is to obtain a first drying period, i.e. a stage at the beginning of the drying process during which the evaporation rate remains constant at the value corresponding to a completely wet surface. For real porous media, this high constant rate may be maintained down to quite low saturations.

Up to now, we saw that quasiconstant drying periods can be observed if, for some time, throat saturation near the network surface does not change. This is possible if the main cluster extends over most of the network, and throats in the depth of the network are invaded while water is pumped to (near) surface throats (Fig. 2.16 and Fig. 2.17, case of $g = 0, \eta = 0$). As already mentioned, the effect is more pronounced in three dimensions.

However, the evaporation rates during these quasiconstant periods are not equal to the high initial drying rate. This is because part of the surface dries out rapidly, and overall drying rate depends on the distribution of wet surface throats and on the transfer properties in the gas-side boundary layer. If lateral transfer in the boundary layer is not accounted for, the drying rate is (more or less) proportional to surface saturation. For lateral diffusive transfer, the boundary-layer thickness and distribution of wet surface throats play the decisive role.

Figure 2.23a shows for the two-dimensional case, how lateral transfer and boundary-layer thickness s modify the drying curve; for a better comparison, a normalized form has been chosen (initial drying rate is inversely proportional to s). Normalized rates during quasiconstant periods are increased by lateral transfer, which is more efficient for a thicker boundary layer. The fine structure of the curves is similar, reflecting the fact that the main influence of boundary-layer thickness is on dynamics and not on the order of throat emptying.

However, a true first drying period can be obtained, if – by a correlated pore structure – the wet throats are well distributed on the network surface. This can be simulated by a network of bimodal throat radius distribution where the large throats

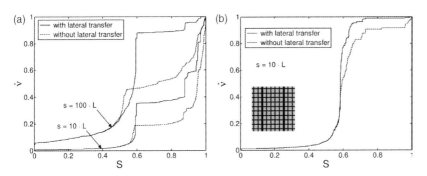

Fig. 2.23 Influence of boundary layer on drying curves for 50×50 network ($L = 500$ μm): (a) monomodal radius distribution (50 ± 10 μm), (b) bimodal radius distribution with macrochannels (100 ± 10 μm, 30% of pore volume).

form long channels towards the surface. Then, these macrochannels will dry out first, whereas most of the surface stays wet; in combination with lateral transfer in the boundary layer this leads to a first drying period, the duration of which is determined by the volume fraction of large throats.

Figure 2.23b shows the drying curve of a bimodal network that was obtained by rescaling every fifth vertical throat of the monomodal network to double its radius. Even for this favorable network structure, lateral transfer is important to obtain a first drying period (Irawan et al. 2005).

An alternative to explain and model the existence of a first drying period can be seen in (corner) film flows as discussed in the previous section. Then, we assume that surface throats do not really dry out when the bulk meniscus recedes, but that corner flow can still provide liquid for a while, in analogy to the above role of macrochannels and microthroats. Again, good lateral vapor transfer is assumed, this time over the short distance of a throat diameter. As a consequence, the high initial evaporation rate can be maintained over a significant period (Fig. 2.21).

Hence, several phenomena can contribute to the existence of a constant rate period: invasion percolation in 3D, film flow, external boundary-layer redistribution, biporous structures. Furthermore (Prat 2006), invasion percolation in a destabilizing gradient (IPDG, Laurindo and Prat (1996, 1998), Prat and Bouleux (1999)) can be significant. For example, IPDG patterns can be induced by gravity (Laurindo and Prat 1996), selective evaporation (de Freitas and Prat 2000), or thermal gradient (Plourde and Prat 2003). As illustrated in Fig. 2.24, IPDG is characterized by the preferential invasion of the region away from the surface of the porous medium and, therefore, by constant surface saturation over a long period of drying, which leads to a constant evaporation rate.

It may be concluded that:
- In the absence of major structural effects (e.g. bimodal pore-size distribution), a constant drying period cannot be observed in 2D under normal convective drying conditions (thin external mass boundary layer), unless by film flow.

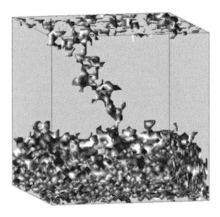

Fig. 2.24 Example of liquid/gas interface within a cubic network during drying under IPDG conditions (evaporation at top surface).

- A constant rate period can be observed in 3D even without films and lateral transfer in the external boundary layer only after a sharp decrease of the initial evaporation flux.
- Film effects are of major importance for the occurrence of a constant rate period both in 2D and 3D. With the films, the constant-rate period predicted by the pore networks is consistent with experiments, at least qualitatively (Prat 2006).
- The homogenization effect of the external boundary layer (lateral transfer effect) contributes to the occurrence and duration of constant rate period.

2.3 Model Extensions

2.3.1 Heat Transfer

We will now abandon the isothermal condition and model heat transfer in convective drying. So far, little work has been done in this field, only considering the influence of (imposed) thermal gradients on phase distributions (Huinink et al. 2002; Plourde and Prat 2006). Since this is a first modeling attempt (Surasani et al. 2006), we go back to the basic isothermal model of Section 2.2.1 (no gravitational or viscous effects) and assume that heat is transferred only by conduction in liquid and solid phase under conditions of local thermal equilibrium between the phases.

For discrete modeling of heat transfer, we attribute a control volume V_i to each pore (Fig. 2.25). The enthalpy contained in this volume is proportional to its temperature T_i

$$H_i = V_i (\rho c_p)_i T_i \tag{2.28}$$

where the heat capacity $(\rho c_p)_i$ depends on the radius and saturation of the neighboring throats

$$V_i (\rho c_p)_i = \left(V_i - \frac{L}{2} \sum_j \pi r_{ij}^2 \right) (\rho c_p)_s + \frac{L}{2} \sum_j \pi r_{ij}^2 S_{ij} (\rho c_p)_w \tag{2.29}$$

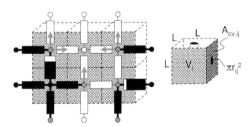

Fig. 2.25 Control volumes and exchange areas for modeling of heat transfer.

Each throat participates equally in two control volumes; for simplicity, the position of the liquid in partially filled throats is not resolved. The heat-flow rate between neighboring pores is

$$\dot{Q}_{ij} = \frac{A_{cv,ij}\lambda_{ij}}{L}(T_i - T_j) \tag{2.30}$$

with the exchange area $A_{cv,ij} = L^2$ (because both voids and solid can contribute to heat transfer). The effective thermal conductivity λ_{ij} is a function of throat saturation

$$A_{cv,ij}\lambda_{ij} = \left(A_{cv,ij} - \pi r_{ij}^2\right)\lambda_s + \pi r_{ij}^2 S_{ij}\lambda_w \tag{2.31}$$

where λ_s and λ_w are the thermal conductivities of solid and liquid, respectively. As a boundary condition, we take a local heat-transfer coefficient

$$\alpha = \beta \frac{\lambda_g}{\delta}\left(\frac{Pr}{Sc}\right)^{1/3} \tag{2.32}$$

from drying air to network surface, where β is the respective mass-transfer coefficient, λ_g the thermal conductivity and $Pr = \lambda_g/(\delta \rho c_p)$ the Prandtl number of air. The thermal boundary layer is not discretized. The other sides of the network are impervious to heat.

Additionally, we have heat sinks where water is evaporated and heat sources where it condenses. Altogether, the dynamic enthalpy balance for a control volume writes

$$V_i(\rho c_p)_i \frac{dT_i}{dt} = -\sum_j \dot{Q}_{ij} - \Delta h_v(T_i) \sum_j \dot{M}_{ev,ij} \tag{2.33}$$

Δh_v is temperature dependent evaporation enthalpy and $\dot{M}_{ev,ij}$ are evaporation rates at menisci in the control volume (negative for condensation).

Though evaporation is the major coupling factor between heat and mass transfer, two other effects must also be pointed out: since equilibrium vapor pressure is highly temperature dependent, the temperature distribution in the network strongly influences vapor diffusion (Huinink et al. 2002); and, because of the temperature dependency of surface tension, capillary pressure becomes temperature dependent as well

$$P_c(r_{ij}, T_{ij}) = \frac{2\sigma(T_{ij})}{r_{ij}} \tag{2.34}$$

so that the order of throat emptying will differ from the isothermal case. As indicated in Plourde and Prat (2003), this last effect should lead to a stabilization of the drying front in convective drying, if pore-size distribution is relatively narrow and temperature gradients are high.

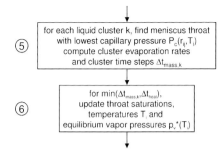

⑤ for each liquid cluster k, find meniscus throat with lowest capillary pressure $P_c(r_{ij}, T_i)$ compute cluster evaporation rates and cluster time steps $\Delta t_{mass,k}$

⑥ for min($\Delta t_{mass,k}, \Delta t_{heat}$), update throat saturations, temperatures T_i and equilibrium vapor pressures $p_v^*(T_i)$

Fig. 2.26 Modification of flow sheet for nonisothermal model (cf. **Fig. 2.7**).

For solving the coupled set of equations for heat and mass transfer, we chose to update the temperature field by the explicit scheme

$$T_i' = T_i + \frac{\Delta t_{heat}}{V_i(\rho c_p)_i} \sum_j \left(-\frac{A_{cv,ij}\lambda_{ij}}{L}(T_i - T_j) - \Delta h_v(T_i)\dot{M}_{ev,ij} \right) \quad (2.35)$$

where the thermal time step must fulfil the condition

$$\Delta t_{heat} < \frac{(\rho c_p)_i L^2}{\sum_j \lambda_{ij}} \quad (2.36)$$

in order to obtain a stable solution. Time discretization must also respect the restrictions for the mass transfer time step, i.e. the complete emptying of a throat. Figure 2.26 illustrates how the drying algorithm of the isothermal model (Fig. 2.7) must be adapted to include the described thermal effects.

In this first model version, the condensation of vapor is not fully accounted for in terms of mass conservation: only partially filled throats may be refilled, but not already empty ones. Though the respective error was less than 0.2 % for the given example, the model will be improved in this sense in the future.

The nonisothermal model was applied to convection drying of 41 × 41 networks (initially at 20 °C) with absolutely dry air at 80 °C; the throat length was 500 μm and boundary-layer thickness 3.5 mm. The thermal properties of the solid are chosen as for glass: $(\rho c_p)_s = 1.7 \times 10^6$ Jm^{-3}K^{-1} and $\lambda_s = 1$ Wm^{-1}K^{-1}. Drying of two networks

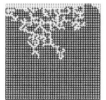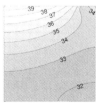

Fig. 2.27 Phase and temperature distributions for monomodal case at $S = 0.86$ and 0.55.

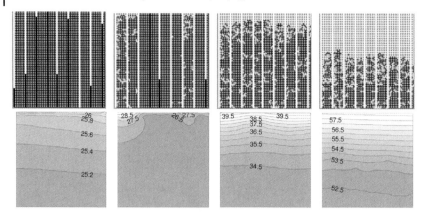

Fig. 2.28 Phase and temperature distributions for bimodal case at $S = 0.8$, 0.6, 0.4 and 0.2.

is presented: one has a monomodal throat radius distribution of mean 40 μm, the other a bimodal one, in which the large throats have a mean radius of 100 μm and account for 44% of pore volume. The relative standard deviation of each mode is 2%. Figures 2.27 and 2.28 show phase distributions and temperature fields. Drying curves are given in Fig. 2.29, together with corresponding isothermal results (at $T = 20\ °C$).

For the monomodal network structure, a first drying period does not exist, but the drying rate drops from the beginning, as parts of the surface dry out. For isothermal drying, quasiconstant periods can be observed between these events. However, in nonisothermal modeling, drying out of a surface region and the resulting decrease in local evaporation rate lead to warming up; as a consequence saturation vapor pressure

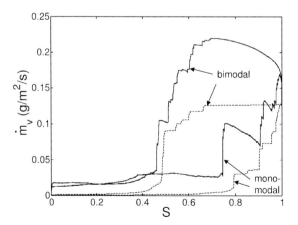

Fig. 2.29 Drying curves for mono- and bimodal radius distribution: comparison between nonisothermal (solid lines) and isothermal (dashed lines) modeling.

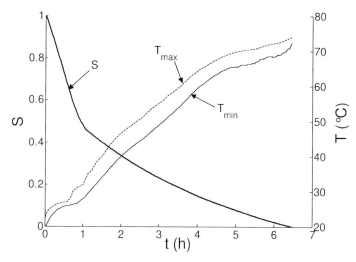

Fig. 2.30 Evolution of saturation and temperature during drying of bimodal network.

and drying rate will increase (Fig. 2.29), towards a new equilibrium between convective heating and evaporative cooling. The local cooling effect of surface evaporation can be seen in Fig. 2.27 (for $S = 0.86$). When the surface is completely dry, the differences in surface temperature are negligible and a unidirectional temperature gradient is obtained ($S = 0.55$). From Fig. 2.29 we can see that the gradual warming up of the network leads to much higher drying rates at low saturations than in isothermal modeling. The fine structure of the drying curve is also slightly different, reflecting the fact that, in a temperature gradient, throats empty in a slightly different order because capillary pressure, Eq. 2.34, does not depend on throat radius alone.

The bimodal network behaves very differently. During the emptying of macrothroats, surface saturation changes only little, and warming up towards wet-bulb temperature can be observed (Fig. 2.30). For the present choice of network properties, constant values for network temperature and drying rate are not quite attained, before microthroats start to empty and the falling rate period commences (see also Fig. 2.28, $S = 0.6$). As long as some macrothroats still contain liquid, they are preferably invaded and temperature has little influence on the order of throat emptying. In the second drying period, microthroats dry out with little capillary pumping, and the evaporation front recedes into the network ($S = 0.4$). Substantial heating maintains the drying rate at an elevated level very similar to that of the monomodal case. In the dry part of the network, steeper gradients develop than in the wet part because of the heat sink at the evaporation front and the saturation-dependent effective thermal conductivity ($S = 0.2$).

Even though the model still needs extension (full account of condensation, convective heat transfer, etc.), the major characteristics of convective drying can already be described such as warming up to the wet-bulb temperature, the first drying period or hot spots on the surface. This is promising for future development.

2.3.2
Multicomponent Liquid

Drying of porous media containing a multicomponent liquid occurs in pharmaceutical production (Schlünder 1988b), soil remediation (Ho and Udell 1995) and oil recovery (Morel et al. 1990). The interesting new feature is that the different components generally do not evaporate at the same rate. This induces changes in local liquid composition, surface tension and, possibly, contact angle. When a sufficient surface-tension gradient builds up over the porous sample, the invasion pattern is modified and a stabilization of the drying front is expected.

Freitas and Prat (2000) developed a pore-network model and applied it to evaporation of a 2-propanol/water mixture in air. The simulations indicate an accumulation of liquid near the open edge of the network, in accordance with experimental results (Le Romancer et al. 1994). The model takes into account capillary effects, ternary diffusion in the gas phase and binary convection-diffusion in the liquid. A special numerical procedure was developed in order to satisfy the assumption of local thermodynamic equilibrium at the liquid/gas interface. Simulations were restricted to two-dimensional networks. Film effects as well as changes of contact angle with composition were ignored. Among other things, these network simulations revealed that the gas phase is not in thermodynamic equilibrium with the liquid at the scale of a representative elementary volume (REV). This suggests that rather nonequilibrium than equilibrium assumptions should be used in continuum models to simulate this type of process.

2.4
Influence of Pore Structure

In the remainder of this chapter we will use basic versions of the network model to study the influence of pore structure on drying, taking advantage of the fact that the presented drying algorithms are versatile and can be applied to various network structures and pore shapes.

2.4.1
Pore Shapes

Figure 2.31 shows some typical pore shapes considered in network models, mainly in relation to oil recovery (Blunt et al. 2001). In our context, such "pore" shapes must be understood as the shape of the throat cross section. In drying, Prat (2006) considered triangular, square and hexagonal cross sections and found that drying rates can be greatly affect when film effects are significant. In contrast, pore shape has little influence (for hydrophilic systems and negligible viscosity) when liquid films cannot develop since vapor diffusion in the gas phase is not sensitive to it.

Fig. 2.31 Some shapes of pore cross section considered in pore-network models.

The important effect on drying rates can be easily understood by recalling that liquid films can only develop for $\theta < \theta_c$ (see Sections 2.2.4 and 2.2.5) and stating that this critical contact angle θ_c depends on tube shape. For a regular polygonal cross sections of N sides, it is simply given by $\theta_c = \pi/N$. Thus, corner flows cannot exist for $\theta \geq 60°$ in a triangular pore, $\theta \geq 45°$ in a square pore and $\theta \geq 30°$ in a hexagonal one. Assuming the same pore size (given by the radius of the largest inscribed circle), drying will be faster for shapes having greater critical contact angle because of a higher hydraulic conductivity for the films. Furthermore, films can be present for certain shapes and not for others, depending on contact angle.

More precisely (Prat 2006), three main regimes can be identified. Suppose we consider a limited range of pore shapes with corners, such as for example those depicted in Fig. 2.31. Under certain circumstances depending on contact angle, external flow conditions and sample size, films can extend up to the network surface during the whole drying process for all these pore shapes. Then, the system is not sensitive to pore shape, since the drying rate is essentially controlled by external transfer. As mentioned before, the system is also not sensitive to pore shape if the contact angle is greater than the largest critical contact angle of the range of pore shapes. In this case, no films develop and transport within the invaded region is by gas diffusion only. As to the influence of pore shape, the interesting regime is, therefore, the intermediate regime, in which films exist but recede into the network during drying.

Figure 2.32 exemplifies the evolution of saturation with time for three different pore shapes and the two-dimensional pore network with films described in

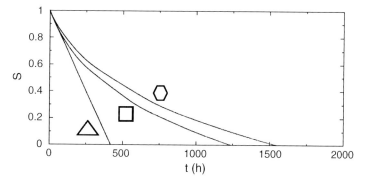

Fig. 2.32 Saturation as a function of time for three different pore shapes. Simulations for water at 20 °C and $\theta = 43°$ in a 40 × 40 square network.

(Prat 2006). As can be seen, the effect of pore shape can be very significant. For the specific example, drying occurs in the intermediate regime with the square shape and in the regime controlled by external conditions with the triangular shape (films extent up to the open edge of network); no films are present for the hexagonal shape.

2.4.2
Coordination Number

Concerning network structure, connectivity of pores is a major parameter. In a first step, regular monomodal networks of different coordination number Z have been investigated (Metzger et al. 2007). In two dimensions, throats were arranged in hexagonal, square and triangular structures corresponding to $Z = 3$, 4 and 6, respectively. Three-dimensional networks were built from tetrahedrons, cubes or octahedrons ($Z = 4$, 6 and 8). This choice allows deviating from the previous square and cubic networks to lower and higher coordination numbers and covering a relatively wide range.

Starting with a square 48×51 network and a cubic $15 \times 15 \times 16$ network (periodic boundary conditions, cylindrical throats of length $L = 500$ µm), the other networks were constructed to have (approximately) the same size, pore volume and porosity. As a consequence, length and total number of throats had to be varied. Parts of the two-dimensional networks are shown in Fig. 2.33. For all cases, the throat radius is normally distributed with a mean $r_0 = 40$ µm and standard deviation $\sigma_0 = 2$ µm. Drying parameters are the same (absolutely dry air at 20 °C, 5 mm boundary layer) so that we can study the influence of pore connectivity for otherwise identical conditions. The basic isothermal model of Section 2.2.1 is applied; viscous effects can be neglected for the selected conditions.

In order to eliminate the influence of random network generation, a number of Monte-Carlo (MC) simulations have been performed. The results are shown in Fig. 2.33 and 2.34 as cumulative distributions of normalized drying rate over network saturation and average saturation profiles over normalized network depth ζ (the average being taken for a given network saturation, and not for a given drying time). Note that an individual drying curve (dashed line) can be quite different from the cumulative results.

For all simulations, we find that gas first penetrates with a relatively flat saturation gradient, before a more or less sharp evaporation front recedes into the network. By comparing saturation profiles for the smallest and highest coordination number ($Z = 3$ and 6 in 2D, $Z = 4$ and 8 in 3D), it can be seen that for lower Z the initial gas-penetration gradients are steeper so that breakthrough occurs at lower network saturation, and that the drying front is smoother and starts to recede at lower overall saturation. As a consequence, the initial drop in drying rate occurs earlier for lower Z, but then drying rates stay elevated down to lower saturations.

The saturation profiles obtained for square and cubic networks with intermediate coordination number confirm the above-described tendency, though drying rates are

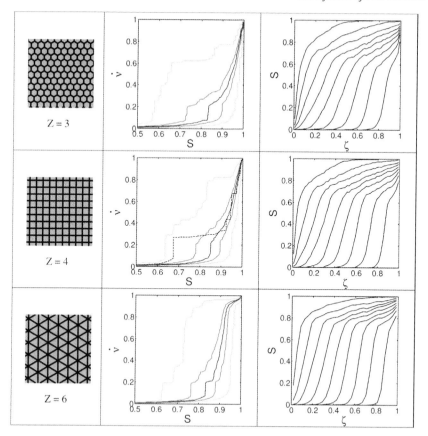

Fig. 2.33 Influence of coordination number on normalized drying curves (shown as levels of 25% of cumulative distribution) and saturation profiles (averages for multiples of 10% of network saturation) for two-dimensional networks (100 MC runs).

lower than expected (see especially the initial drop for the cubic network in Fig. 2.34). Further investigation is expected to cast more light on the influence of coordination number, with specific reference to orientation of throats (with respect to network surface), network size, the role of pores as links between throats, and throat length.

2.4.3
Bimodal Pore-size Distributions

In the foregoing, square networks with macrochannels perpendicular to the network surface have already been introduced. Here, we investigate in more detail the drying of pore networks with spatially highly correlated bimodal throat-size distributions

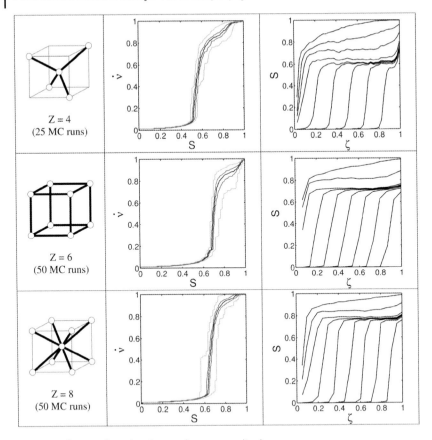

Fig. 2.34 Influence of coordination number on normalized drying curves and saturation profiles for three-dimensional networks (representation as in **Fig. 2.33**).

(Metzger et al. 2007). Three two-dimensional network structures will be discussed differing in the arrangement of large throats as:

S1. long channels perpendicular to the network surface (Fig. 2.35);
S2. long channels in both space directions (Fig. 2.36);
S3. regularly distributed small clusters (Fig. 2.37).

Structure 1 represents the case of two continuous "phases", one of micropores and one of macropores (though continuity cannot be strictly achieved in 2D). Structure 2 can be viewed as a simple representation of an agglomerate of microporous particles. In structure 3, a porous medium is modeled with distributed large pores; here only the microphase is continuous.

For all three structures, drying of 100 Monte Carlo realizations of a 48 × 51 network with periodic boundary conditions is simulated by the isothermal model of Section 2.2.1. Both small and large throats (again cylindrical) have a normal distribution of mean 40 μm or 100 μm and standard deviation 2 μm or 5 μm. The volume fraction

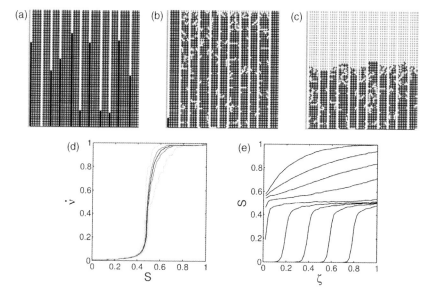

Fig. 2.35 Bimodal network structure 1: typical phase distributions for network saturations (a) 0.75, (b) 0.5 and (c) 0.25; (d) distribution of normalized drying curves and (e) average saturation profiles (with normalized network depth ζ) for 100 network realizations.

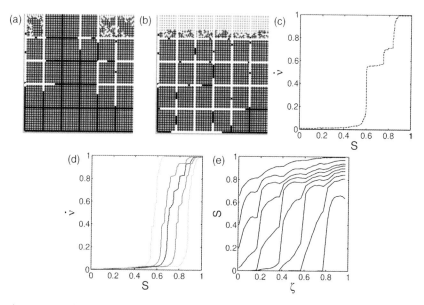

Fig. 2.36 Bimodal network structure 2: phase distributions at saturations (a) 0.8 and (b) 0.55 as well as (c) normalized drying curve for one MC simulation; (d) distribution of drying curves and (e) average saturation profiles for 100 network realizations.

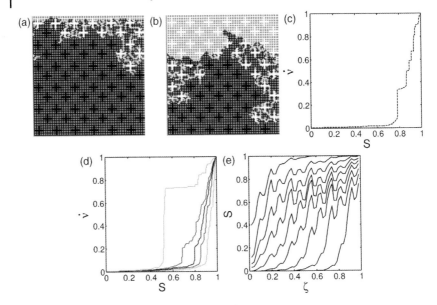

Fig. 2.37 Bimodal network structure 3: phase distributions at saturations (a) 0.85 and (b) 0.54 as well as (c) normalized drying curve for one MC simulation; (d) distribution of normalized drying curves and (e) average saturation profiles for 100 network realizations.

of large pores is almost constant between 44.7% for S1 and 43.6% for S3, so that all networks have the same porosity and pore-size distribution. Throat length and drying conditions are as in the last section, allowing for direct comparison with the square monomodal network (Fig. 2.33). The results are shown as typical phase distributions, distributions of normalized drying curves and average saturation profiles in Figs. 2.35–2.37.

As already discussed above, for network structure 1 the macrochannels empty first (almost completely) while liquid is pumped by capillary forces to the small throats at the network surface; the good distribution of wet surface throats in combination with lateral vapor transfer in the boundary layer can (almost) maintain the initial drying rate. This first drying period lasts down to network saturations roughly corresponding to the volume fraction of small throats. Then, the small throats dry out with less-efficient capillary pumping; an evaporation front recedes into the network. The drying curves in Fig. 2.35 have a very narrow distribution. Saturation profiles clearly show the two separate periods: emptying of macrothroats to a uniform saturation and receding front. Note that viscous forces will have little effect – even for smaller pore dimensions – because the capillary pressure difference between micro- and macro-pores is usually sufficiently large (Irawan et al. 2006).

It is stressed that preferential invasion of macrothroats and capillary pumping through microthroats to the surface can only be assured if both the macro- and microphase are continuous. This can easily be seen by looking at the other two network structures.

In structure 2, the random emptying of lateral macrochannels cuts off capillary flow paths to the network surface so that whole surface regions dry out and the first drying period will in general be short. Drying of the network will occur layer by layer because of the screening effect of disconnected liquid regions. In fact, the random emptying of microporous regions is quite similar to the drying of single throats in the monomodal case. Only their relative size is very different, leading to quasiconstant periods of longer duration, a wider range of drying curves and average saturation profiles displaying the spatial separation of the microporous regions (Fig. 2.36). In an average sense, the saturation profiles may be considered equivalent to those of the monomodal network (Fig. 2.33).

In the drying behavior of structure 3 the random process of invasion percolation is only temporarily overruled if a large throat region is reached: then, all of its members are first emptied completely before the random process continues (Fig. 2.37). Consequently, the drying curves and saturation profiles are very similar to those of the monomodal network, except for the fine structure due to regular location of macrothroats that have a higher probability of being emptied than the surrounding microthroats.

Finally, three-dimensional extensions of network structures 1 and 2 are considered (Fig. 2.38), by introducing macrochannels (20.7 % of the total pore volume) into the cubic network of Section 2.4.2. Due to the third dimension, the micropores form a continuous phase for both cases; therefore, the orientation of the macrochannels does not matter and the drying curves are very similar. In contrast to the monomodal network (Fig. 2.34) a first drying period can be observed (due to more and better distributed wet surface throats). Saturation profiles are very flat down to $S = 0.6$,

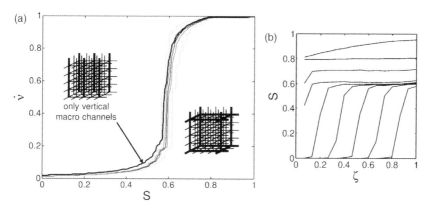

Fig. 2.38 Three-dimensional bimodal networks: (a) distribution of normalized drying curves (for the case of vertical macrochannels, only the median of the distribution is shown) and (b) average saturation profiles (for vertical macrochannels) for 50 network realizations.

because when all large throats are empty there is a large and uniformly distributed number of small-meniscus throats that are candidates for invasion. The evaporation front starts to recede at lower saturations.

2.4.4
Outlook

The above investigations of the influence of pore structure on drying kinetics are just a start, but they can demonstrate how powerful pore-network modeling can be in this context. In the future, networks with random pore locations, randomly chosen throat connections and a spatially correlated radius distribution may be investigated.

For a limited range of porous materials, pores are large enough that the necessary information might be obtained from imaging experiments. Otherwise, only averaged quantities are available, usually in the form of an experimental pore-volume distribution.

An interesting question in this framework will be how much structural information – such as porosity, pore-size distribution, coordination number (distribution) – is necessary to characterize a network in its drying behavior. It is also important to investigate how, for a predefined set of macroscopic structural properties, a pore network can be generated for use in drying simulations.

2.5
Towards an Assessment of Continuous Models

In all the above drying simulations, the pore network was considered to represent the porous medium at the sample scale. This approach obviously faces a problem since real porous bodies usually have too high a total number of pores to be dealt with numerically. As mentioned in the introduction, traditional drying models are continuous with effective parameters, and pore-network modeling has already been used to compute such macroscopic properties by taking the network as a representative elementary volume (REV) (Nowicki et al. 1992; Segura and Toledo 2005).

If we consider the discrete pore network approach as being more fundamental, it may help to investigate under what conditions continuous modeling – with the generalized Darcy law for convective flow, effective vapor diffusion coefficient and effective thermal conductivity – can correctly describe the transport phenomena. For this investigation, the macroscopic transport coefficients must be computed for a given network structure at the REV scale, as functions of average saturation. It will be interesting to study the influence of the spatial distribution of a given amount of liquid on the transport parameters, since we have already seen that phase distributions depend on drying conditions. The comparison of the two approaches is to be carried out at the sample scale using the effective parameters in the continuum model and simulating drying of a large pore network with the same structural characteristics. For a full comparison, the network drying model still needs to be extended to include gas convection and adsorption.

If both approaches are indeed equivalent, pore networks can help to provide the parameters of continuous modeling that can otherwise only be obtained by independent experiments.

References

Al-Futaisi, A., Patzek, T. W., 2003. Extension of Hoshen–Kopelman algorithm to non-lattice environments. *Physica* **A321**: 665–678.

Blunt, M. J., Scher, H., 1995. Pore-level modeling of wetting. *Phys. Rev.* **E52**: 6387–6403.

Blunt, M. J., Jackson, M. D., Piri, M., Valvatne, P. H., 2002. Detailed physics, predictive capabilities and macroscopic consequences for pore-network models of multiphase flow. *Adv. Water Resour.* **25**: 1069–1089.

Camassel, B., Sghaier, N., Prat, M., Ben Nasrallah, S., 2005. Ions transport during evaporation in capillary tubes of polygonal cross section. *Chem. Eng. Sci.* **60**: 815–826.

Chapuis, O., 2006. *Déplacements quasi-statiques eau-air et évaporation en milieux poreux modèles hydrophiles ou hydrophobes*, Ph.D Thesis, INP Toulouse, France.

Chapuis, O., Prat, M., 2007.Influence of wettability conditions on slow evaporation.*Phys. Rev.* **E75**: 1–11.

Chen, P., Pei, D.C.T., 1989. A mathematical model of drying process. *Int. J. Heat Mass Transfer* **18**: 297–310.

Churaev, N. V., 2000. Liquid and vapor flow in porous bodies: surface phenomena. In: *Topics in Chemical Engineering* **13**, Gordon and Breach, New York.

Coquard, T., Camassel, B., Prat, M., 2005. Evaporation in capillary tubes of square cross section. *Proceedings HT2005, ASME Heat Transfer Conference July 17–22*, San Francisco, CA, USA

deFreitas, D. S., Prat, M., 2000. Pore network simulation of evaporation of a binary liquid from a capillary porous medium. *Transport in Porous Media* **40**: 1–25.

Dullien, F. A. L., 1992. *Porous Media, Fluid Transport and Pore Structure* 2nd ed., Academic Press, San Diago, California.

Eijkel, J.C.T., Dan, B., Reemeijer, H. W., Hermes, D. C., Bomer, J. G., van denBerg, A., 2005. Strongly accelerated and humidity independent drying of nanochannels induced by sharp corners. *Phys. Rev. Lett.* **95**: 256107–1/4

Ho, C. K., Udell, K. S., 1995. Mass transfer limited drying of porous media containing an immobile binary liquid mixture. *Int. J. Heat Mass Transfer* **38**: 339–350.

Huinink, H. P., Pel, L., Michels, M.A.J., Prat, M., 2002. Drying processes in the presence of temperature gradients. Pore-scale modelling. *Eur. Physical J.* **E9**: 487–498.

Irawan, A., Metzger, T., Tsotsas, E., 2005. Pore-network modeling of drying: combination with a boundary layer model to capture the first drying period. *7th World Congress of Chemical Engineering*, Glasgow, Scotland.

Irawan, A., Metzger, T., Tsotsas, E., 2007. Isothermal drying of pore networks: Influence of friction for different pore structures. *Drying Technol.* **25**: 47–57.

Laurindo, J. B., Prat, M., 1996. Numerical and experimental network study of evaporation in capillary porous media – phase distributions. *Chem. Eng. Sci.* **51**: 5171–5185.

Laurindo, J. B., Prat, M., 1998. Numerical and experimental network study of evaporation in capillary porous media – drying rates. *Chem. Eng. Sci.* **53**: 2257–2269.

LeBray, Y., Prat, M., 1999. Three dimensional pore network simulation of drying in capillary porous media. *Int. J. Heat Mass Transfer* **42**: 4207–4224.

La Romancer, J.F., Defives, D.F., Fernandes, G., 1994. Mechanism of oil recovery by gas diffusion in fractured reservoir in presence of water. SPE/DOE 27746, pp. 99–111.

Lenormand, R., Zarcone, C., 1984. Role of roughness and edges during imbibition in square capillaries. SPE 13264.

Masmoudi, W., Prat, M., 1991. Heat and Mass Transfer between a Porous Medium and a Parallel External Flow – Application to Drying of Capillary Porous Materials. *Int. J. Heat Mass Transfer* **34**: 1975–1989.

Metzger, T., Irawan, A., Tsotsas, E., 2005. Discrete modeling of drying kinetics of porous media. *3rd Nordic Drying Conference*, Karlstad, Sweden.

Metzger, T., Irawan, A., Tsotsas, E., 2006. Remarks on the paper "Extension of Hoshen–Kopelman algorithm to non-lattice environments". by A. Al-Futaisi and T. W. Patzek, 2003, *Physica* **A321**: 665–678. *Physica* **A363**: 558–560.

Metzger, T., Irawan, A., Tsotsas, E., 2007. Influence of pore structure on drying kinetics: a pore network study. *Submitted to AIChE J.*

Morel, D.D., Bourbiaux, B., Latil, M., Thiebot, B., 1990. Diffusion effects in gas flooded light oil fractured reservoirs, SPE paper 20516, pp. 433–446.

Nowicki, S. C., Davis, H. T., Scriven, L. E., 1992. Microscopic determination of transport parameters in drying porous media. *Drying Technol.* **10**: 925–946.

Plourde, F., Prat, M., 2003. Pore network simulations of drying of capillary media. Influence of thermal gradients. *Int. J. Heat Mass Transfer* **46**: 1293–1307.

Prat, M., 1993. Percolation model of drying under isothermal conditions in porous media. *Int. J. Multiphase Flow* **19**: 691–704.

Prat, M., 2002. Recent advances in pore-scale models for drying of porous media. *Chem. Eng. J.* **86**: 153–164.

Prat, M., 2007. On the influence of pore shape, contact angle and film flows on drying of capillary porous media. *Int. J. Heat Mass Transfer* **50**: 1455–1468.

Prat, M., Bouleux, F., 1999. Drying of capillary porous media with stabilized front in two-dimensions. *Phys. Rev.* **E60**: 5647–5656.

Salin, J. G., 2006. Drying of sapwood analyzed as an invasion percolation process. *Maderas: Cienca y Technologia* **8**: 149–158.

Schlünder, E.-U., 1988a. On the mechanism of the constant drying rate period and its relevance to diffusion controlled catalytic gas phase reactions. *Chem. Eng. Sci.* **43**: 2685–2688.

Schlünder, E.-U., 1988b. Selective drying of mixture-containing products. *6th Intern. Drying Symposium*, Versailles, France. Sept. 5–8, 1, KL 9–23.

Segura, L. A., Toledo, P. G., 2005. Pore-level modeling of isothermal drying of pore networks. Effects of gravity and pore shape and size distributions on saturation and transport parameters. *Chem. Eng. J.* **111**: 237–252.

Shaw, T. M., 1987. Drying as an immiscible displacement process with fluid counterflow. *Phys. Rev. Lett.* **59**: 1671–1674.

Surasani, V. K., Metzger, T., Tsotsas, E., Towards a complete pore-network drying model: first steps to include heat transfer. *Proceedings 15th International Drying Symposium (IDS2006)*, Budapest, Hungary, 125–132.

Tsimpanogiannis, I. N., Yortsos, Y. C., Poulou, S., Kanellopoulos, N., Stubos, A. K., 1999. Scaling theory of drying in porous media. *Phys. Rev.* **E59**: 4353–4365.

Wilkinson, D., 1984. Percolation model of immiscible displacement in the presence of buoyancy forces. *Phys. Rev.* **A30**: 520–531.

Yiotis, A. G., Boudouvis, A. G., Stubos, A. K., Tsimpanogiannis, I. N., Yortsos, Y. C., 2003. Effect of liquid films on the isothermal drying of porous media. *Phys. Rev.* **E68**: 037303-1/4

Yiotis, A. G., Boudouvis, A. G., Stubos, A. K., Tsimpanogiannis, I. N., Yortsos, Y. C., 2004. The effect of liquid films on the drying of porous media. *AIChE J.* **50**: 2721–2737.

Yiotis, A. G., Stubos, A. K., Boudouvis, A. G., Yortsos, Y. C., 2001. A 2-D pore-network model of the drying of single-component liquids in porous media. *Adv. Water Resour.* **24**: 439–460.

Yiotis, A. G., Stubos, A. K., Boudouvis, A. G., Tsimpanogiannis, I. N., Yortsos, Y. C., 2005. Pore-network modeling of isothermal drying in porous media. *Transport in Porous Media* **58**: 63–86.

Yiotis, A. G., Tsimpanogiannis, I. N., Stubos, A. K., Yortsos, Y. C., 2006. Pore-network study of the characteristic periods in the drying of porous materials. *J. Colloid Interf. Sci.* **297**: 738–748.

3
Continuous Thermomechanical Models using Volume-averaging Theory

Frédéric Couture, Philippe Bernada, Michel A. Roques

3.1
Introduction

Of the roughly four million known substances, about 60 000 are processed and sold, and many of these must be dried. Drying can be defined as an operation in which a liquid/solid separation is accomplished by the supply of heat, with separation resulting from the evaporation of water or solvent. Nearly every industry has wet products that must be dried. Most of these are powders, grains, crystals, waste or colloids. The cost of drying is estimated to be about 15% of the total manufacturing cost of dry products.

To predict the possibility of mechanical damage due to cracks provides the incentive to model the transport phenomena that occur in highly deformable multiphase media during drying. Whatever the modeling approach (averaging theory, irreversible thermodynamics, intuitive equivalence to a continuous medium, etc.), a mathematical description that consists of the heat, mass and momentum conservation laws for each phase is needed. Although these equations are intimately coupled, numerical resolution requires subdivision of the model into two distinct parts. This separation is performed rigorously by replacing the momentum conservation equation of the solid phase by the sum of the momentum conservations equations for each phase. Let us call the resulting equation *total momentum conservation* as it depends on the rheological behavior of the total multiphase system. Indeed, under the commonly used assumption that the flow of each phase is quasisteady, this relation represents the mechanical equilibrium of the global multiphase medium subjected to the load induced by shrinkage of the wet solid due to solvent removal. Finally, the two distinct parts of the modeling can be summarized as follows:
  1. heat, mass and momentum conservation equation for each phase except the momentum conservation equation for solid,
  2. total momentum conservation.

At this stage, one can distinguish two approaches in the recent literature that correspond to two different rheological behaviors of the wet solid: elastic and viscoelastic.

Modern Drying Technology. Edited by Evangelos Tsotsas and Arun S. Mujumdar
Copyright © 2007 WILEY-VCH Verlag GmbH & Co. KGaA. All rights reserved.
ISBN: 978-3-527-31556-7

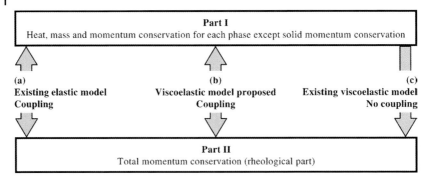

Fig. 3.1 Principle of modeling of the drying of deformable media: existing and proposed modeling. (a) Elastic case: rheological influence (part II) on transports (part I) is taken into account in the literature. (b) Viscoelastic case: rheological influence (part II) on transports (part I) is taken into account in the present work. (c) Viscoelastic case: rheological influence (part II) on transport is neglected in the literature.

As illustrated in Fig. 3.1a, the scheme of the numerical resolution implemented in order to simulate the drying of an elastic product ensures that the existing coupling effects between the total momentum conservation (part II) and the other equations describing heat and mass transport (part I) are taken into account (Jomaa and Puiggali 1991; Mrani et al. 1995; Mercier et al. 1998). The rheological part II is solved for unknown displacements using the principle of virtual work combined with a finite-element approach. The velocity of the solid phase is deduced from the displacements and is introduced in part I. The solution depends on solvent and temperature fields as the stress-strain relations depend on these fields that are themselves provided by the numerical solution of the governing heat, mass and momentum transport equations of part I. Thus, we see that the solid velocity is the coupling factor between the two above-mentioned parts of the model.

The theory of elasticity accounts for materials when neglecting the time effects is permitted (Sih et al. 1986). Deformation in many real products is found to be described as a result of their viscoelastic behavior. The theory of viscoelasticity is adequately described in Flügge (1967), Christensen (1971) and Ferry (1970). To modeling drying, experimental determination of the stress relaxation modulus from the tensile stress-strain curves at various temperatures and moisture contents has been the subject of several research projects. For instance, Hammerle et al. (1971) and Hammerle (1972) establish a stress based failure criterion for corn endosperm subjected to temperature and moisture gradients assuming the material to be linear viscoelastic. A solution for stresses in a viscoelastic sphere under the influence of radial temperature and moisture distribution is proposed by Rao et al. (1975). The stress-strain relations used are those for a Maxwell viscoelastic body with temperature- and moisture-dependent properties. Using the Boltzmann superposition principle, an analytical solution was obtained, whereas later finite-element formulations became more common for the solution of such problems (Rao 1989).

Nevertheless, viscoelasticity during drying is generally being discarded in favor of elasticity. Indeed, time effects on the rheological behavior discouraged authors from numerically solving the complete problem. The first steps have been undertaken but without coupling the total momentum conservation (part II) to the other laws (part I), as illustrated in Fig. 3.1c.

Classical approaches solve the heat- and mass-transport problem based on a rigid skeleton (assumed to be the part I previously defined), and with the predicted moisture and temperature fields, the rheological part II is tackled to predict stress and strain fields. Within this framework, some authors formulate a finite-element solution that can be used to analyse stress-crack formation in viscoelastic materials (Haghighi and Segerlind 1988). Others propose a fundamental theoretical analysis to predict the deformation characteristics of clay during drying in ceramic production (Hasatani et al. 1992). Linear viscoelasticity is assumed for the strain-stress analysis to account for the effect of creep. A large tensional stress, which may generate a crack, is observed initially around the surface. It is found that the time history of the volume change of the formed clay is significantly influenced by the drying conditions and / or by the drying rate. A theoretical basis for the nonlinear thermo-hydro-viscoelasticity is applied to simulate the stress states during the drying of grain kernels (Irudayaraj and Haghighi 1993). More recently, Itaya (1997) developed a transient three-dimensional analysis of the drying of a ceramic slab. The reader can find a review of the works on strain and stress in materials during drying in Hasatani and Itaya (1996).

We will consider drying of colloidal media either particulate or macromolecular with the aim of deriving generic rules in order to incorporate the real rheological behavior of products into the modeling of heat and mass transport. Most of them are viscoelastic and can be considered reasonably as a two-phase system over a very long time during processing. The emphasis here, is on describing mathematically and simulating the physics involved during the drying of a viscoelastic two-phase medium. As shown schematically in Fig. 3.1b, taking into account the influence of strains and stresses (part II) with respect to the internal transport (part I) constitutes the main novelty of this approach.

The macroscopic governing equations are derived using a volume-averaging method. Numerical solution enables two-dimensional drying of a viscoelastic macromolecular gel to be simulated. It is concluded that overcoming the numerical difficulties introduced by viscoelasticity is highly justified by the underestimated stresses resulting from an elastic assumption. However, we note that the mathematical description of the liquid momentum has to be improved to provide a closer representation of the real phenomena.

3.2
Modeling

A necessary preliminary to the mathematical formulation is to define precisely the range of the products that are covered by the proposed model.

3.2.1
Nature of Product Class

As previously mentioned, the media under investigation here are colloidal materials that exhibit a viscoelastic behavior. Not only is the academic problem challenging but also of great economic interest. Indeed, this class of materials contains numerous industrial products known as gels. Two types of gels can be distinguished as follows.
- *Particulate gels* can be described as stacking and assembling of microscopic domains (1–100 µm). A schematic representation is proposed in Fig. 3.2a. This subgroup contains materials such as paints, silica, alumina, clays and many other oxides.
- *Macromolecular gels* are such materials as agarose, gelatine, starch, polyacrylates, glues and reticulating varnishes. As illustrated in Fig. 3.2b, they consist of long interwoven chains of macromolecules. The interlocking can be fixed covalent bonding as in polyacrylamide or mobile sliding as in gelatine.

The complexity and our limited knowledge of the above-described microstructures contrast with our aim of proposing a modeling valid over a wide range of materials. This implies the use of a suitable homogenization method that permits elimination of the inherent microscopic heterogeneity of the medium. Among all the available methods of homogenization (see Chapter 1), we chose the volume-averaging technique (Marle 1967; Slattery 1981; Whitaker, 1977, 1986). However, applying this theory implies the following assumptions: the sophisticated microstructure of the gel is "simplified" to two continuous immiscible phases, a liquid and a solid one (see Fig. 3.2c). Indeed, the macroscopic description (or local-scale description) is provided by integration over a representative volume of the microscopic conservation laws. These laws and the associated boundary conditions linking the two phases are written for each phase on the basis of the continuum physics. Given the previous definition of particulate and macromolecular gels, consideration as a two-phase

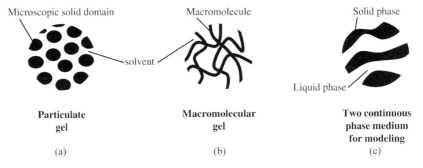

Fig. 3.2 Schematic representation showing (a), (b) the microstructure of gels and (c) identification into two continuous phases for modelling.

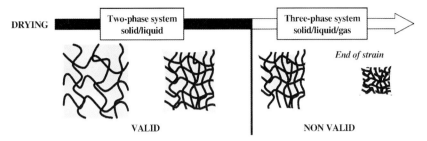

Fig. 3.3 Validity domain of proposed two-phase model.

system may be the most important limitation of our entire analysis. The solid microscopic domains or chains of macromolecules possibly form a discontinuous solid phase. Here, we assume that approximation of the gel as a two-phase system is adequate for modeling purposes.

Moreover, although the material examined remains a two-phase system over a very long time during drying since shrinkage is very important in such systems, this saturated state ends in the later stage of drying when the solid phase definitely locks in mechanically. As a result, a three-phase (liquid, solid and gas) zone recedes inwards through the material and porosity is generated. As shown in Fig. 3.3, this final drying period is obviously not predicted by the current mathematical description limited to a fully saturated highly deformable two-phase medium.

The physical configuration considered in order to illustrate the mathematical description is low- or medium-temperature convective drying of sphere of alumina gel. Both the process and the material have been chosen for the following two main reasons:

- Alumina is a viscoelastic particulate gel that has been well identified in the past for temperatures in the range of 0 and 100 °C. The strain-stress relation parameters and all the transport coefficients needed for the simulation are available in (Mercier 1996) and listed in Tab. 3.1. It is assumed here to consist of a solid phase and a water phase, the water phase can be bound to the solid one for low water content.
- For convective drying, closure of the system by boundary conditions can be obtained accurately by using the boundary-layer theory and the classical Lewis analogy.

3.2.2
Averaged Internal Equations

Mass, energy and momentum conservation equations are first written for each phase. Volume conservation (incompressibility) of the liquid phase is a reasonable assumption that is added together with the relevant boundary conditions between the two phases in order to close the phase-scale description. The macroscopic partial differential equations are obtained by volume averaging the microscopic conservation

Tab. 3.1 Physical parameters used in the simulation for alumina gel

Symbol	Description	Value/Expression
$\overline{\rho}_s^s$	Density of the solid phase (kg m^{-3})	2000
$\underline{\underline{D}}$	Transport tensor (m^2 s^{-1})	$\underline{\underline{D}} = \dfrac{D}{\varepsilon_l}\underline{\underline{I}}$ with $D = D_0 \exp(a(X))$ $\begin{cases} a(X) = -1.08 + 2.71X - 2.44X^2 + 1.57X^3 \\ \qquad\quad -0.49X^4 + 0.06X^5 \\ D_0 = 10^{-10}\left(\dfrac{\overline{T}}{273}\right)^{4.04} \end{cases}$
$c_{p,s}$	Heat capacity of the solid phase (J kg^{-1} K^{-1})	1050
$c_{p,l}$	Heat capacity of the liquid phase (J kg^{-1} K^{-1})	4182
$\underline{\underline{\lambda}}_{\text{eff}}$	Thermal conductivity of the two-phase medium (W m^{-1} K^{-1})	$\underline{\underline{\lambda}}_{\text{eff}} = \lambda_{\text{eff}}\underline{\underline{I}}$ with $\lambda_{\text{eff}} = \varepsilon_l\lambda_l + \varepsilon_s\lambda_s$ $\lambda_l = 0.6071,\ \lambda_s = 0.2$
X_{hygr}	Moisture content limit between free and bound water	0.3
a_w	Water activity	1 if $X > X_{\text{hygr}}$, else $X = \dfrac{a_w A}{(1 - Ba_w)(1 + Ca_w)}$ $A = -0.0002\overline{T}^2 - 0.0142\overline{T} + 2.9099$ $B = 0.0022\overline{T} + 0.7226$ $C = -0.0028\overline{T}^2 + 0.0327\overline{T} + 22.629$
$\Delta\overline{h}_D^L$	Desorption enthalpy (J kg^{-1})	0 if $X > X_{\text{hygr}}$, else $\exp(7.0078 - 93.5125X^{2.5})\cdot 10^3$
Rheology	Relaxation functions in simple shear and dilatation (Pa)	$G(t,X) = [2\cdot\exp(t/120) + 1][10^9\cdot\exp(-3.4531X)]$ $K(t,X) = 7.4 G(t,X)$

equations. The hypothesis and the way to obtain the governing equations for heat and mass transports during drying of deformable media have been well established (Whitaker 1986) and are also discussed in Chapter 1. As the reader can find the detailed developments in the previous reference and the main steps in Bogdanis (2001), only the relevant set of the averaged equations is summarized below; all the necessary notations are given in the List of Recommended Notation in the frontmatter of this book.

3.2.2.1 State Equations and Volume Conservation

The change in scale introduces two new local physical quantities: the volume fractions for each phase. The equations obtained by integrating, over the representative volume, the microscopic laws are not sufficient to predict the value of these two quantities. As the medium remains a two-phase system, volume fractions are linked as follows:

$$\varepsilon_l + \varepsilon_s = 1 \tag{3.1}$$

This means that the theory introduces one new unknown. This implies that one new equation besides the set of averaged equations is needed. The missing relation is deduced from the assumption that the volume of the solid phase is conserved, i.e. the solid phase is intrinsically incompressible as is the liquid. The averaged relation is:

$$\bar{\rho}_s^s = \text{constant} \tag{3.2}$$

The averaged expression of the liquid volume conservation gives:

$$\bar{\rho}_l^l = \text{constant} \tag{3.3}$$

3.2.2.2 Mass-conservation Equations

Integration over the representative volume of the microscopic solid and liquid mass-conservation equations leads to:

$$\begin{aligned} \frac{\partial \bar{\rho}_s}{\partial t} + \nabla \cdot (\bar{\rho}_s \bar{v}_s^s) = 0 \\ \frac{\partial \bar{\rho}_l}{\partial t} + \nabla \cdot (\bar{\rho}_l \bar{v}_l^l) = 0 \end{aligned} \tag{3.4}$$

3.2.2.3 Momentum-conservation Equations

After averaging, solid and liquid momentum balances appear in a form that does not permit simulation. They contain integrals of microscopic quantities over the internal solid/liquid interfaces. This means that the aim of removing the microscopic heterogeneity is not yet reached. In order to remedy this, we can add the two equations and replace one of the two, for instance the equation for the solid, by the resulting equation. Due to stress continuity at the interface, integrals vanish and we obtain

$$\nabla \cdot \underline{\underline{\bar{\sigma}}} = 0 \tag{3.5}$$

where $\underline{\underline{\bar{\sigma}}}$ designates the symmetrical second-order stress tensor defined as follows:

$$\begin{aligned} \underline{\underline{\bar{\sigma}}} = \underline{\underline{\bar{\sigma}}}_l + \underline{\underline{\bar{\sigma}}}_s \\ \underline{\underline{\bar{\sigma}}} = -\varepsilon_l \bar{P}_l^l \underline{\underline{I}} + \underline{\underline{\bar{\sigma}}}_s \end{aligned} \tag{3.6}$$

Equation 3.5 is what we call in the Introduction of this chapter the *total momentum-conservation equation* as it depends on the rheological behavior of the combined two-phase system. We must remember also that this relation is obtained under the commonly used assumptions that the flow of each phase is quasisteady and that body forces are negligible compared to the internal phases stresses. Thus, it represents the mechanical equilibrium of the global medium under the load induced by the shrinkage due to solvent removal.

Two kinds of treatment of the total momentum conservation (Eq. 3.5) can be distinguished, depending on the stress-strain relation used. The first method consists in adding an averaged liquid relation and a solid one (Biot, 1941, 1955; Coussy et al. 1998). The rheological behavior of the liquid phase is rightly considered as Newtonian and the average of the viscous stress tensor is always neglected. The solid stress-strain relation has to be determined experimentally:

The idea of neglecting the viscosity effects in the liquid rheological law, but, in the same time, keeping them in the liquid flux expression as Darcy's law, constitutes a contradiction that tempts us to adopt a second method. The rheological behavior of the total two-phase system is considered by introducing a unique law experimentally determined for the humid medium (Jomaa and Puiggali 1991; Mrani et al. 1995; Mercier et al. 1998; Haghighi et al. 1998; Hasatani et al. 1992; Irudayaraj and Haghighi 1993 and Itaya 1997).

Employing the standard rectangular Cartesian coordinates system, the time- and moisture-dependent stress-strain relation for a viscoelastic behavior can be expressed as follows (Tschoegl 1989):

$$\sigma_{ij}(t, \varepsilon_l) = \int_0^t \left[\left(K(t - \tau, \varepsilon_l) - \frac{2}{3} G(t - \tau, \varepsilon_l) \right) \frac{\partial \left(\varepsilon^*_{kk}(\tau) \right)}{\partial \tau} \delta_{ij} + 2G(t - \tau, \varepsilon_l) \frac{\partial \left(\varepsilon^*_{ij}(\tau) \right)}{\partial \tau} \right] d\tau \quad (3.7)$$

In the above expression, ε^*_{ij} are the coefficients of the viscoelastic strain tensor $\underline{\underline{\varepsilon^*}}$ and are not those of the total strain tensor $\underline{\underline{\varepsilon}}$ associated to the total shrinkage. Indeed, most of the total strain is induced by the volume change due to liquid migration towards the surfaces where it is evaporated. By noting that $\underline{\underline{\varepsilon^v}}$ is the tensor that characterizes this volumetric strain, it follows that:

$$\underline{\underline{\varepsilon}} = \underline{\underline{\varepsilon^*}} + \underline{\underline{\varepsilon^v}} \quad (3.8)$$

As is indicated by Eq. 3.7, the volumetric shrinkage would be the real shrinkage value if the material did not build up stresses during drying (case of ideal shrinkage). The supplementary viscoelastic stress influences shrinkage by reducing it to smaller values due to internal forces acting against free deformation. The relaxation functions in simple shear and dilatation, $G(t, \varepsilon_l)$ and $K(t, \varepsilon_l)$, respectively, are equivalent to the shear modulus and bulk modulus encountered in the elastic case.

These parameters depend on time and moisture. They have to be determined experimentally.

At this stage, the question of how to treat rigorously the area integral in the liquid momentum conservation still remains unanswered. In this contribution, we get round this main difficulty, first by replacing the liquid momentum balance by Darcy's law:

$$\bar{\mathbf{v}}_l^l - \bar{\mathbf{v}}_s^s = -\frac{\underline{\underline{\mathbf{k}}}}{\mu_l \varepsilon_l} \cdot \nabla \bar{p}_l^l \tag{3.9}$$

where $\bar{p}_l^l$ is the intrinsic averaged liquid pressure. Using a stress-strain relation for the homogenized medium as Eq. 3.7 rather than the sum of a solid relation and a liquid one as in Eq. 3.6 implies a main inconvenience: how to treat the unknown liquid pressure in Eq. 3.9 so far as it does not appear in the total momentum conservation. As most authors have done in the past (Jomaa and Puiggali 1991; Mrani et al. 1995; Mercier et al. 1998; Kechaou and Roques 1989; Ketelaars 1992), we overcome this difficulty by introducing arbitrarily a phenomenological law that postulates that the pressure depends on the liquid volume fraction:

$$\bar{p}_l^l = f(\varepsilon_l) \tag{3.10}$$

Introducing Eq. 3.10 into Eq. 3.9 leads to the classical relation (Jomaa and Puiggali 1991; Mercier et al. 1998; Kechaou and Roques 1989; Ketelaars 1992),

$$\bar{\mathbf{v}}_l^l - \bar{\mathbf{v}}_s^s = -\underline{\underline{\mathbf{D}}} \cdot \nabla \varepsilon_l \tag{3.11}$$

where the second-order tensor $\underline{\underline{\mathbf{D}}}$ is defined as follows:

$$\underline{\underline{\mathbf{D}}} = \frac{\underline{\underline{\mathbf{k}}}}{\mu_l \varepsilon_l} \frac{\partial \bar{p}_l^l}{\partial \varepsilon_l} \tag{3.12}$$

Equation 3.10 is the additional equation needed to avoid numerical resolution of one of the averaged conservation laws. We chose here to eliminate the mass conservation of the solid phase, Eq. 3.4. However, Eq. 3.10 constitutes an important limitation to the analysis of the transport mechanisms. From a physical point of view, the link between liquid pressure and liquid volume fraction remains without any foundation. From a practical point of view, the equivalent transport coefficient $\underline{\underline{\mathbf{D}}}$ must be identified numerically by matching experimental and predicted data in such a way that further validation becomes negligible. Moreover, whatever the experimental setup, the poor knowledge in the mathematical description of the boundary conditions leads to an internal transport coefficient intimately linked to the apparatus. This dependence is not admissible when the aim is to propose a model in order to describe internal transport that must be valid whatever the solid-liquid separation process examined. Studies on this topic have been undertaken in our group (Mrani et al. 1995; Chausi et al. 2001a; Sfair et al. 2004) and new possibilities are probably to be expected. This point is discussed in Section 3.4.

It is now necessary to evaluate the velocity of the solid phase. After introduction of the well known expression for the total strain-tensor coefficients,

$$\varepsilon_{ij} = \frac{1}{2}(\partial_i \bar{u}_j + \partial_j \bar{u}_i) \tag{3.13}$$

the solid velocity is deduced from the displacement provided by the resolution of the total momentum conservation equation (Eq. 3.5). Let us call $\bar{\mathbf{u}}(x_0, t)$ the displacement vector of material points P of the homogenized medium, where x_0 designates the P coordinates at initial time ($t = 0$). The P coordinates at time t are:

$$x(x_0, t) = x_0 + \bar{\mathbf{u}}(x_0, t) \tag{3.14}$$

By assuming the intrinsic averaged solid-phase velocity $\bar{\mathbf{v}}_s^s$ is equal to the material point velocity, it follows that:

$$\bar{\mathbf{v}}_s^s(x_0, t) = \frac{d}{dt} x(x_0, t) \tag{3.15}$$

The total time derivative can be replaced by the partial derivative because x_0 does not depend on time t:

$$\bar{\mathbf{v}}_s^s(x_0, t) = \frac{\partial}{\partial t} x(x_0, t) \tag{3.16}$$

The value of the material point velocity $\bar{\mathbf{v}}_s^s(x_0, t)$ gives the value of the solid velocity at the geometrical point x in the Eulerian coordinates used in this work: $\bar{\mathbf{v}}_s^s(x, t)$.

3.2.2.4 Energy-conservation Equations

As in momentum balances, microscopic area integrals resulting from application of averaging theorems require the thermal energy equations to be simplified by forming the total thermal energy equation. The solid-liquid system is first considered to be in local equilibrium. A consequence of this reasonable assumption for moderate-temperature convective drying (Quintard and Whitaker 1993) is that the intrinsic phase-averaged temperatures are equal:

$$\bar{T} = \bar{T}_s^s = \bar{T}_l^l \tag{3.17}$$

This equation replaces one of the two initial energy laws, the other being replaced by the sum of the two. In the resulting equation, the area integrals are taken to be proportional to the temperature gradient and incorporated into an effective thermal conductivity, leading to a phenomenological Fourier's law at the local scale:

$$\frac{\partial}{\partial t}(\bar{\rho}_s \bar{h}_s^s + \bar{\rho}_l \bar{h}_l^l) + \nabla \cdot (\bar{\rho}_s \bar{h}_s^s \bar{\mathbf{v}}_s^s + \bar{\rho}_l \bar{h}_l^l \bar{\mathbf{v}}_l^l - \underline{\underline{\lambda}}_{\text{eff}} \cdot \nabla \bar{T}) = 0 \tag{3.18}$$

In this balance equation, the enthalpies per unit mass are defined as follows:

$$\overline{h}_l^l(\overline{T}, X) = C_{p,l}(\overline{T} - T_{\text{ref}}) - \overline{\Delta h}_D^l(\overline{T}, X) \tag{3.18a}$$

$$\overline{h}_s^s(\overline{T}) = C_{p,s}(\overline{T} - T_{\text{ref}}) \tag{3.18b}$$

where $\overline{\Delta h}_D^l$ is the enthalpy of desorption that has to be added to the latent heat of vaporization in order to remove the bound liquid.

3.2.3
Boundary Conditions for Convective Drying

Closure of the system by boundary conditions is obtained by using the boundary-layer theory and the classical heat- and mass-transfer analogy to express the fluxes:

$$(\overline{\rho}_s \overline{h}_s^s (\overline{\mathbf{v}}_s^s - \mathbf{W}) + \overline{\rho}_l \overline{h}_l^l (\overline{\mathbf{v}}_l^l - \mathbf{W}) - \underline{\underline{\lambda}}_{\text{eff}} \cdot \nabla \overline{T}) \cdot \mathbf{n} = h_T(T_{\text{ph}} - T_\infty)$$
$$+ h_m(\rho_{v,\text{ph}} - \rho_{v,\infty}) h_v(T_{\text{ph}}) \tag{3.19}$$

$$\overline{\rho}_l (\overline{\mathbf{v}}_l^l - \mathbf{W}) \cdot \mathbf{n} = h_m(\rho_{v,\text{ph}} - \rho_{v,\infty}) \tag{3.20}$$

Here, h_v is the enthalpy of the vapor,

$$h_v(T_{\text{ph}}) = c_{pv}(T_{\text{ph}} - T_{\text{ref}}) + \Delta h_v(T_{\text{ref}}) \tag{3.20a}$$

and $\mathbf{W}$ is the velocity of the interface between the two-phase medium and the external fluid. This velocity can be easily evaluated by noting that the surface of the gel sample considered here is not permeable to the solid:

$$\overline{\rho}_s (\overline{\mathbf{v}}_s^s - \mathbf{W}) \cdot \mathbf{n} = 0 \tag{3.21}$$

From Eq. 3.21, it follows that:

$$\overline{\mathbf{v}}_s^s \cdot \mathbf{n} = \mathbf{W} \cdot \mathbf{n} \tag{3.22}$$

Hence, Eqs. 3.19 and 3.20 can finally be written in the form:

$$(\overline{\rho}_l \overline{h}_l^l (\overline{\mathbf{v}}_l^l - \overline{\mathbf{v}}_s^s) - \underline{\underline{\lambda}}_{\text{eff}} \nabla \overline{T}) \cdot \mathbf{n} = h_T(T_{\text{ph}} - T_\infty) + h_m(\rho_{v,\text{ph}} - \rho_{v,\infty}) h_v(T_{\text{ph}}) \tag{3.23}$$

$$\overline{\rho}_l (\overline{\mathbf{v}}_l^l - \overline{\mathbf{v}}_s^s) \cdot \mathbf{n} = h_m(\rho_{v,\text{ph}} - \rho_{v,\infty}) \tag{3.24}$$

The last condition is associated with the total momentum conservation (Eq. 3.5). Neglecting the effects of the ambient pressure, stress continuity at the interface allows us to write:

$$\overline{\overline{\sigma}} \cdot \mathbf{n} = 0 \tag{3.25}$$

3.3
Simulation

3.3.1
Numerical Resolution Technique

Numerical resolution is not discussed in detail here. The relevant details are available in Bogdanis (2001). We just want to underline that the scheme presented in Fig. 3.1b of the introduction is applied. Two parts of the modeling procedure according to the mathematical structure of the equations are distinguished:

1. Heat, mass and momentum conservation for each phase except solid momentum conservation consist of the state equations (Eqs. 3.1–3.3), liquid mass conservation (Eq. 3.4), the liquid velocity expression (Eq. 3.11) and the total thermal energy equation (Eq. 3.18). Spatial discretization by a normal finite-element method provides the average liquid density and temperature fields that are introduced in part II.
2. The total momentum conservation (Eq. 3.5) is solved for unknown displacements using the principle of virtual work combined with a finite-element approach. This type of approach has been discussed at length in Zienkiewicz (1989) and thermohydroviscoelasticity has been treated in Haghighi et al. (1988), for instance. Calculation of the load induced by the volume change due to solvent removal is based on the following spherical volumetric strain tensor:

$$\forall i, \varepsilon_{ii}^v = \frac{1}{3}\frac{\Delta\varepsilon_l}{1 - \varepsilon_l} \tag{3.26}$$

In this relation, $\Delta\varepsilon_l$ represents the variation of the liquid fraction between the current and the previous time step of the time discretization. It is given by the solution of part I. The solution of the mechanical equilibrium equation leads to values of displacements, real strain and induced stress. The solid velocity is then evaluated using Eqs. 3.14–3.16; it is then inserted in the mass and thermal balances of part I. With regard to the elastic behavior, viscoelastic equilibrium is not reached instantaneously. At each time step corresponding to a given load, we need to calculate the time-dependent displacements over a very long time period, until only negligible variations of deformation take place. This solution at a given time is stored in order to take it into account in future time steps. The need to predict the future influence of the present event constitutes the main numerical difficulty as it requires large computer storage and many more calculations than in the elastic case.

3.3.2
Comparison between Real Viscoelastic and Assumed Elastic Behavior

Typical simulation results are reported in Figs. 3.4 and 3.5 for drying of a sphere of alumina gel. Rather than the average liquid density, the usual moisture content $X = \bar{\rho}_l / \bar{\rho}_s$ is shown. Moreover, stress evolutions are illustrated by using the Von Mises value defined by:

$$\text{stress} = \sqrt{\sigma_1^2 + \sigma_2^2 - (\sigma_1 \sigma_2) + 3(\sigma_{12}^2)} \tag{3.27}$$

Although spherical coordinates would be practical, two-dimensional Cartesian coordinates are preferred here in order to enable other geometrical configurations to be computed. As previously mentioned, all the necessary transport and physical parameters are listed in Table 3.1. The initial radius, temperature and moisture content are 2 mm, 20 °C and 3.4, respectively, the relative humidity and temperature of the ambient air are 90 °C and 10%, respectively. The air velocity is 2.5 m s^{-1}, yielding an average heat-transfer coefficient h_T of 20 W m^{-2} K^{-1}. Strain, stress, moisture and temperature profiles for various times are presented in Fig. 3.5. The time values have been selected to underline the more interesting period of stress behavior.

During the initial transient period and the constant rate period, stresses are higher near the surface and reach a maximum at the beginning of the falling-rate period. At this stage, the overall stress field decreases. Interestingly, an inversion of profiles from center to surface occurs in such a way that external layers tend towards zero stress faster than at locations within the sphere. Figure 3.4 clearly exhibits this phenomenon by comparing the surface stress to the center one.

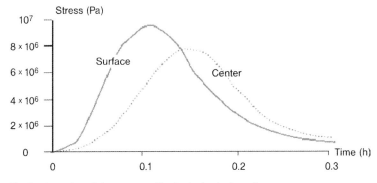

Fig. 3.4 Inversion of the stress profile: At the beginning of drying, stresses are higher at the surface, at the end, stresses become greater inside.

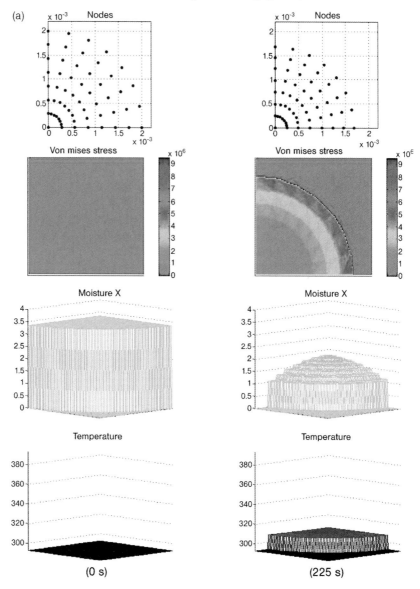

Fig. 3.5 (a) Strain, Von Mises stress, moisture and temperature two-dimensional predicted fields. (0 s) Initial conditions, stress is zero. (225 s) Beginning of the constant rate period, stresses are higher at the surface. (b) Strain, Von Mises stress, moisture and temperature two-dimensional predicted fields. (300 s) Constant rate period, stresses are higher at the surface. (375 s) End of the constant rate period. Maximal stress of the overall process is reached at the surface. (c) Strain, Von Mises stress, moisture and temperature two-dimensional predicted fields. (600 s) Stresses become greater inside, inversion of the stress profile. (850 s) End of drying. Stresses tend toward zero indefinitely.

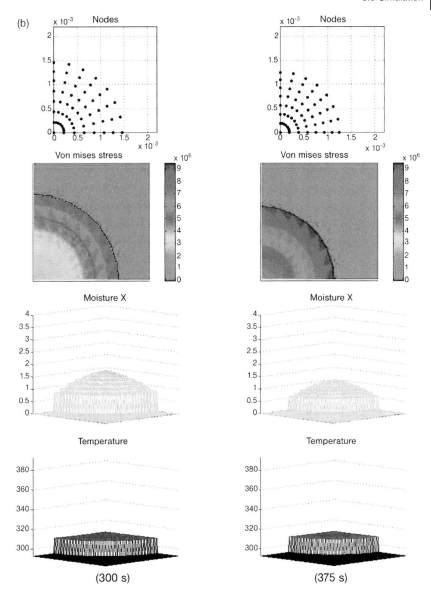

Fig. 3.5 (Continued)

Figures 3.6 and 3.7 emphasize the need to account for the viscoelastic behavior in order to predict realistical stresses. Here, we assume that an elastic stress-strain relation is enough for alumina gel. The shear and bulk modulus values $G(\varepsilon_l)$ and $K(\varepsilon_l)$ are given by the viscoelastic relaxation functions in simple shear and dilatation $G(t, \varepsilon_l)$ and $K(t, \varepsilon_l)$ as time tends toward infinity. Thus,

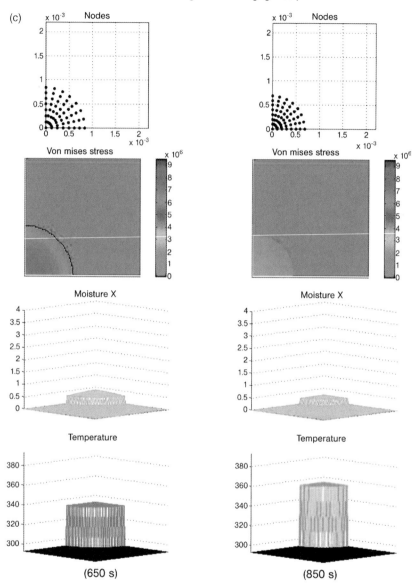

Fig. 3.5 (Continued)

$$\sigma_{ij}(\varepsilon_l) = (K(\varepsilon_l) - \frac{2}{3}G(\varepsilon_l))\varepsilon_{kk}^*\delta_{ij} + 2G(\varepsilon_l)\varepsilon_{ij}^* \quad (3.28)$$

Comparison of assumed elastic and real viscoelastic moisture profiles in the radial direction exhibits clearly the substantial differences in the predicted shrinkage. As the elastic equilibrium induced by water loss is reached instantaneously, the radius

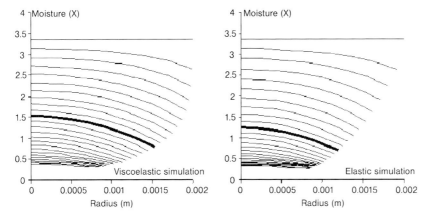

Fig. 3.6 Comparison between the moisture content profiles in the radial direction predicted by using real viscoelatic and assumed elastic strain-stress relations.

decreases quasilinearly. On the other hand, viscoelastic strains induced by the initial loads are time delayed in such a way that most of the shrinkage occurs during the later stages of drying. As a result, viscoelastic stresses are initially lower and become higher as drying proceeds, especially since the rheological parameters G and K increase as the moisture content decreases (Mercier 1996). Moreover, Fig. 3.7 shows clearly that the maximum levels of the overall process are predicted by the viscoelastic simulations. The thick line profiles in Fig. 3.6 illustrate the impact of the rheological behavior on moisture migration. Higher elastic shrinkage induces higher solid

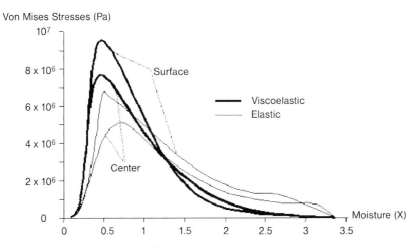

Fig. 3.7 Comparison between von Mises stress versus moisture content predicted using viscoelatic and elastic strain-stress relations.

velocity and then faster drying. In summary, modeling of viscous or viscoelastic materials with elastic models results in considerable underestimation of the stress values ($\approx -30\%$). This can lead to poor prediction of damage due to drying induced stresses.

3.4
Liquid Pressure as Driving Force

A question that comes before all others is how to treat the liquid momentum conservation accurately. We have underscored the fact that the introduction of the phenomenological law (Eq. 3.10) leading to the liquid velocity being proportional to the average liquid density gradient is undoubtedly without any physical foundation. Studies have been undertaken recently in our laboratory (Chausi et al. 2001a and b; Sfair et al. 2004) and a new model has been proposed.

The main novelty (Sfair et al. 2004) is that the arbitrary link between liquid pressure and liquid volume fraction (Eq. 3.10) is not required, although at the same time a global stress-strain relation such as Eq. 3.7 or Eq. 3.28 is used. Thus, the new approach retains mass conservation of the solid phase (Eq. 3.4) that was eliminated

Tab. 3.2 Physical parameters used in the simulation for gelatin gel

$\bar{\rho}_s^s$	Density of the solid phase (kg m^{-3})	1341
K	Permeability (m^2)	$10^{-11} \cdot \varepsilon_l^3$
$c_{p,s}$	Heat capacity of the solid phase (J kg^{-1} K^{-1})	1804
$c_{p,l}$	Heat capacity of the liquid phase (J kg^{-1} K^{-1})	4182
λ_{eff}	Thermal conductivity of the two-phase medium (W m^{-1} K^{-1})	$\lambda_{eff} = \varepsilon_l \lambda_l + \varepsilon_s \lambda_s$ $\lambda_l = 0.6071, \lambda_s = 0.3334$
X_{hygr}	Moisture content limit between free and bound water	0.3
a_w	Water activity	1 if $X > X_{hygr}$, else $X = \dfrac{a_w A}{(1 - Ba_w)(1 + Ca_w)}$ $A = -0.0002\bar{T}^2 - 0.0142\bar{T} + 2.9099$ $B = 0.0022\bar{T} + 0.7226$ $C = -0.0028\bar{T}^2 + 0.0327\bar{T} + 22.629$
$\Delta \bar{h}_D^l$	Desorption enthalpy (J kg^{-1})	0 if $X > X_{hygr}$, else $\exp(7.0078 - 93.5125X^{2.5}) \cdot 10^3$
Rheology	Lamè coefficients (Pa)	$\lambda = -9.92 \times 10^8 + 8.58 \times 10^9 \exp(-\varepsilon_l)$ $\mu = \exp(21.82 - 11.91\varepsilon_l^{1.5})$

in the model presented in Section 3.2 because of the introduction of the additional law (Eq. 3.10). To sum up, the model consists of the following averaged equations: the mass conservation equations for each phase (Eq. 3.4), the volume conservation equations (Eq. 3.2 and Eq. 3.3, also known as the incompressibility relations), the momentum conservation equations (Eq. 3.5 and Eq. 3.9, which is Darcy's law), the energy conservation equations (Eq. 3.17 and Eq. 3.18) and finally the two-phase state equation (Eq. 3.1). Here, liquid pressure is an unknown that has to be calculated in terms of the liquid volume fraction, solid velocity and temperature.

The first step in the numerical resolution of this model has been undertaken in order to simulate one-dimensional convective drying of gelatin. The boundary conditions (Eqs. 3.19–3.21 and Eq. 3.25) remain as before. Although the one-dimensional configuration leads us to reduce the elastic behavior to ideal shrinkage, the numerical results provide some confidence in the proposed mathematical description. As for alumina gel, all the physical coefficients needed for the simulation are given in Tab. 3.2. The initial thickness, temperature, liquid pressure and dry basis moisture content are 0.01 m, 25 °C, 101 326 Pa and 5, respectively, the relative humidity and temperature of the ambient air are 35 °C

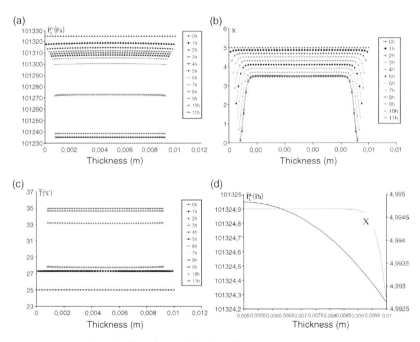

Fig. 3.8 Computed results from the model without the arbitrary link between liquid pressure and liquid volume fraction (Eq. 3.10). Spatial profiles at different times of (a) liquid phase pressure (b) moisture content (c) temperature. Additionally (d) liquid phase pressure and moisture content profiles at $t = 5$ min.

and 55%, respectively. The heat-transfer coefficients h_T is 10 W m^{-2} K^{-1}. The computed liquid pressure, moisture and temperature profiles for various times are presented in Fig. 3.8.

The comparison of the pressure and the water content profiles given in Fig. 3.8d underlines the key difference between the gradients of these two variables. Contrary to the assumptions illustrated by Eqs. 3.10 and 3.11, in the model used in Section 3.2 and commonly encountered in the literature, no link can be observed between the moisture content gradient (and then the liquid volume fraction) and convective transport of the liquid phase. For instance, movement of the liquid phase at the center of the material ($\nabla \overline{P}_l^l \neq 0$) would not exist if the liquid volume fraction gradient was the driving force ($\nabla X = \nabla \varepsilon_l = 0$). This comment can be extended to the whole process.

At this stage, significant numerical efforts are needed to pass from one to two spatial dimensions and then to describe accurately both elastic, and a fortiori, viscoelastic behavior and convective migration under the natural driving force, i.e. the liquid pressure gradient.

3.5
Conclusions

Modeling of heat, mass and momentum transport during drying was carried out for fully saturated viscoelastic media. Numerical resolution of the model equations was performed to account for the coupling effects between the total momentum conservation and the other equations (liquid momentum conservation, mass-conservation equations for the two phases, energy-conservation equations). Finally, no matter what the rheological behavior is, it seems that it can be described accurately at the price of both numerical efforts and experimental characterization. Progress in thermo-hydro-rheology is made necessary since shaping and formulating of a large class of products is of paramount economical importance.

Perhaps, a question that comes before all other is now how to treat rigorously the liquid momentum conservation. We have underlined that the introduction of a phenomenological law proportional to the average liquid density gradient is undoubtedly without any physical foundation. Some recent studies in our laboratory showed that this problem should be overcome by deleting the arbitrary link between liquid pressure and liquid volume fraction, and at the same time by reintroducing the solid mass conservation equation in the model. New possibilities are now imaginable. The main further efforts are essentially numerical in order to pass at two dimension simulations, and then to describe accurately viscoelastic rheology as well as the convective solvent migration under the natural driving force that is the liquid pressure gradient.

The model described in this chapter is based on the homogenization of the phase equations by using a volume-averaging technique. In the following chapter, another technique is used: the theory of mixtures and thermodynamics of irreversible processes.

References

Biot M. A., 1941. General theory of three-dimensional consolidation. *J. Appl. Phys.* **12**: 155–164.

Biot M. A., 1955. Theory of elasticity and consolidation for a porous anisotropic solid. *J. Appl. Phys.* **26**(2): 182–185.

Bogdanis E., 2001. *Modeling of heat and mass transport during drying of an elastic or viscoelastic medium and resolution by the finite-element methods.* PhD Thesis, Université de Pau et des Pays de l'Adour, France.

Chausi B., Couture F., Roques M. A., 2001a. Modeling of mass transport during dewatering – impregnation – soaking. *Proceedings of the 2nd Inter-American Drying Conference (IADC) Veracruz*, Mexique, 08–10/07.

Chausi B., Couture F., Roques M. A., 2001b. Modeling of multicomponent mass transport in deformable media. Application to dewatering impregnation soaking. *Dry. Technol.* **19**(9): 2081–2101.

Christensen R. M., 1971. *Theory of Viscoelasticity – An Introduction.* Academic Press, New York and London.

Coussy O., Dormieux L., Detournay E., 1998. From mixture theory to Biot's approach for porous media. *Int. J. Solids Struct.* **35**: 4619–4635.

Ferry J. D., 1970. *Viscoelastic Properties of Polymers.* 2nd ed., John Wiley and Sons, New York.

Flügge W., 1967. *Viscoelasticity.* Blaisdell Publishing Company, Waltham (Mass.), Toronto, London.

Haghighi K., Segerlind L. J., 1988. Modeling Simultaneous Heat- and Mass-Transfer in an Isotropic Sphere – A Finite-Element Approach. *Trans. ASAE* **31**(2): 629–637.

Hammerle J. R., 1972. Theoretical analysis of failure in a viscoelastic slab subjected to temperature and moisture gradients. *Trans. ASAE* **15**(5): 960–965.

Hammerle J. R., Mohsenin N. N., White R. K., 1971. The rheological properties of corn horny endosperm. *ASAE* **6**(2): 60–72.

Hasatani M., Itaya Y., 1996. Drying-induced strain and stress – a review. *Dry. Technol.* **14**(5): 1011–1040.

Hasatani M., Itaya Y., Hayakawa K., 1992. Fundamental study on shrinkage of formed clay during drying – viscoelastic strain-stress and heat moisture transfer. *Dry. Technol.* **10**(4): 1013–1036.

Irudayaraj J., Haghighi K., 1993. Stress-analysis of viscoelastic materials during drying. 1. theory and finite-element formulation. *Dry. Technol.* **11**(5):901–927.

Itaya Y., 1997. A numerical study of transient deformation and stress behavior of a clay slab during drying. *Dry. Technol.* **15**(1): 1–21.

Jomaa W., Puiggali J. R., 1991. Drying of shrinking materials: Modeling with shrinkage velocity. *Dry. Technol.* **9**(5): 1271–1293.

Kechaou N., Roques M. A., 1989. A variable diffusivity model for drying of highly deformable materials, *Drying 89*, edited by A. S.Mujumdar, Hemisphere Publishing, Washington D.C., 332–338.

Ketelaars A., 1992. Drying Deformable Media – Kinetics shrinkage and stresses. Ph.D Technische Universiteit Eindhoven, Eindhoven, The Netherlands.

Marle C. M., 1967. Ecoulements monophasiques en milieu poreux. *Rev. Inst. Fr. Petrole* **22**: 1471–1509.

Mercier F., 1996. Séchage de gel d'alumine: Maîtrise de la texture de supports de catalyseurs. Thèse de l'Université de Pau et des Pays de l'Adour, France.

Mercier F., Puiggali J. R., Roques M. A., Brunard N., Kolenda F., 1998. Convective and micro-wave drying of alumina beads. Modeling of shrinkage. 3rd World Congress on Particle Technology, 06–09/07, Brighton, UK.

Mrani I., Bénet J. C., Fras G., 1995. Transport of water in a biconstituent elastic medium. *Appl. Mech. Rev.* **48**(10): 717–721.

Quintard M., Whitaker S., 1993. One- and two-equation models for transient diffusion processes in two-phase system. *Adv. Heat Transfer* **23**: 369–464.

Rao S. S., 1989. *The Finite-element Method in Engineering*, Pergamon Press (2), Oxford.

Rao V. N. M., Hamann D. D., Hammerle J. R., 1975. Stress analysis of a viscoelastic sphere subject to temperature and moisture gradients. *J. ASAE* **20**: 283–293.

Sfair A. L., Couture F., Laurent S., Roques M. A., 2004. Modeling of heat and mass transport in two-phase media by considering liquid pressure. *Dry. Technol.* **22**(1&2): 81–90.

Sih G. C., Michopoulos J. G., Chou S. C., 1986. *Hygrothermoelasticity*. Martinus Nijhoff Publishers, Dordrecht, The Netherlands.

Slattery J. C., 1981. *Momentum Energy and Mass Transfer in Continua*. 2nd edn, Robert E. Kreiger Pub. Co. New York.

Tschoegl N. W., 1989. *The Phenomenological Theory of Linear Viscoelastic Behavior, an Introduction*. Springer-Verlag, Berlin.

Whitaker S., 1977. Simultaneous heat, mass and momentum transfer in porous media, a theory of drying. *Adv. Heat Transfer* **13**: 119–203.

Whitaker S., 1986. Flow in porous media III: Deformable media. *Transport Porous Media* **1**: 127–154.

Zienkiewicz O. C., 1989. *Finite-element Method*. McGraw Hill, New York.

4
Continuous Thermohydromechanical Model using the Theory of Mixtures
Stefan J. Kowalski

4.1
Preliminaries

In Chapter 3 a model of drying for fully saturated viscoelastic media based on the homogenization of the phase equations by volume averaging has been presented. This chapter presents the theory of drying that describes fully coupled, multiphase transport in moving, deformable unsaturated porous media by using the theory of mixtures. Such materials may shrink and suffer destruction caused by drying-induced stresses. Drying theory, which includes mechanical effects along with the heat- and mass-transfer phenomena, can be constructed on the basis of the mechanics of continua. The natural tools for development of such a theory are the balances of mass, momentum, moment of momentum, energy, entropy and the principles of irreversible thermodynamics. A version of such a drying theory was presented by Kowalski (2003). In this chapter some modifications are introduced to the former theory, among others, its extension to microwave drying processes.

The present model of drying based on mechanics of continua incorporates the following assumptions:
- The drying material is a mixture of three phases: solid skeleton, liquid and gas in pores. They are represented by mass concentration and volume fraction terms that are continuous functions of space and time due to the averaging procedure applied to *the representative volume element* (RVE, see Chapter 3).
- The skeleton is a deformable body. The deformations can be both reversible (elastic) or irreversible (viscoelastic, plastic).
- The theory excludes thermal shocks and therefore some dynamic terms (accelerations, inertia forces, kinetic energy, etc.) are neglected in the balance equations.

- The stress deviator in moisture is neglected as it is much smaller than the stress deviator of the solid skeleton.
- The skeleton of the material (dry body) has dielectric properties.
- The only effect of microwaves is that they generate volumetric heat sources in the material to be dried.
- There is local thermal equilibrium between moisture and the skeleton, so that the temperatures of these two constituents at a given point are the same.

The following concepts are used in further considerations:
- *Volume fraction* ϕ^α – defines the fraction of the RVE occupied by the constituent α (s – solid, l – liquid, g – gas).
- *Porosity* ϕ – defines the fraction of the RVE occupied by pores. It can be expressed through the volume fraction as: $\phi = \phi^l + \phi^g = 1 - \phi^s$.
- *The volumetric saturation* S – defines the fraction of pore space occupied by liquid, that is: $S = \phi^l/\phi$.
- *Mass concentration* $\rho^\alpha [\text{kg m}^{-3}]$ of constituent α – defines the mass of constituent α contained in a unit of the RVE.
- *Production* $^*\rho^\alpha [\text{kg m}^{-3}\,\text{s}^{-1}]$ of constituent α within the control volume – expresses the rate of mass change due to the phase transitions.
- *Mass content* (dry basis) $X^\alpha = \rho^\alpha/\rho^s$ (Kowalski 2003) of constituent α – is the ratio of constituent mass referred to the mass of dry body.
- *Moisture flux* $J^\alpha [\text{kgm}^{-2}\,\text{s}^{-1}]$ of constituent α with respect to the skeleton (porous body) – defines the amount of moisture in kilograms per unit surface and unit time.
- *Velocity of the skeleton* $v^s [\text{ms}^{-1}]$ – describes the time rate of displacement of the saturated porous body due to shrinkage (or swelling) and the action of drying-induced stresses.
- *Constituent velocity of moisture* $v^\alpha [\text{ms}^{-1}]$ – describes the average motion of constituent ($\alpha \neq s$) being defined as the superposition of convection velocity by skeleton and the relative motion of constituent α with respect to the skeleton, that is, $v^\alpha = v^s + J^\alpha/\rho^\alpha$.

4.2
Global Balance Equations

The balance equations will be developed by applying Euler's description, that is by using the spatial (not material) coordinates $x\{x, y, z\}$. Let us define within the dried body a control volume $V_s(t)$ attributed to the deformable skeleton, being enveloped

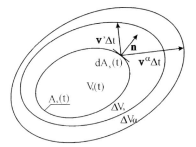

Fig. 4.1 Deformable control volume.

with a smooth control surface $A_s(t)$ oriented spatially with the outward directed unit normal vector $\boldsymbol{n}$, Fig. 4.1.

The particles of the skeleton displace during drying with velocity $\boldsymbol{v}^s$ and the particles of moisture with velocity $\boldsymbol{v}^\alpha$. After time interval Δt the skeleton will occupy the space $V_s(t) + \Delta V_s$ and the moisture the space $V_s(t) + \Delta V_s + \Delta V_\alpha$. The directions of velocities $\boldsymbol{v}^s$ and $\boldsymbol{v}^\alpha$ in general can be taken arbitrary as, for example, in Fig. 4.1.

Let us consider an arbitrary physical quantity $\Psi^\alpha(t)$ of constituent α present in the control volume $V_s(t)$

$$\Psi^\alpha(t) = \int_{V_s(t)} \psi^\alpha(\boldsymbol{x}, t) \mathrm{d}V \tag{4.1}$$

where $\psi^\alpha(\boldsymbol{x}, t)$ is the density of $\Psi^\alpha(t)$.

After time Δt the substance of constituent α occupies the space $V_s(t) + \Delta V_s + \Delta V_\alpha$, therefore

$$\Psi^\alpha(t + \Delta t) = \int_{V_s(t)} \psi^\alpha(\boldsymbol{x}, t + \Delta t) \mathrm{d}V + \int_{\Delta V_s} \psi^\alpha(\boldsymbol{x}, t + \Delta t) \mathrm{d}\Delta V_s \\ + \int_{\Delta V_\alpha} \psi^\alpha(\boldsymbol{x}, t + \Delta t) \mathrm{d}\Delta V_\alpha \tag{4.2}$$

where $\mathrm{d}\Delta V_s = \boldsymbol{v}^s \cdot \boldsymbol{n} \mathrm{d}A_s(t)$, $\mathrm{d}\Delta V_\alpha = (\boldsymbol{v}^\alpha - \boldsymbol{v}^s) \cdot \boldsymbol{n} \mathrm{d}A_s(t + \Delta t)$.

The total time derivative of quantity $\Psi^\alpha(t)$ reads

$$\frac{\mathrm{d}\Psi^\alpha(t)}{\mathrm{d}t} = \lim_{\Delta t \to 0} \frac{\Psi^\alpha(t + \Delta t) - \Psi^\alpha(t)}{\Delta t} = \dot{\Psi}^\alpha(t) + \int_{A_s(t)} \psi^\alpha(\boldsymbol{x}, t)(\boldsymbol{v}^\alpha - \boldsymbol{v}^s) \cdot \boldsymbol{n} \mathrm{d}A_s \tag{4.3}$$

where

$$\dot{\Psi}^\alpha(t) = \int_{V_s(t)} \frac{\partial \psi^\alpha(t)}{\partial t} \mathrm{d}V + \int_{A_s(t)} \psi^\alpha(\boldsymbol{x}, t) \boldsymbol{v}^s \cdot \boldsymbol{n} \mathrm{d}A_s = \int_{V_s(t)} \left(\frac{\partial \psi^\alpha}{\partial t} + \nabla \cdot (\psi^\alpha \boldsymbol{v}^s) \right) \mathrm{d}V \tag{4.4}$$

is the time derivative with convection velocity $\boldsymbol{v}^s$ that describes the variation of quantity Ψ^α inside the control volume $V^s(t)$, which changes with time.

The general structure of the balance equation reads

$$\dot{\Psi}^\alpha = \int_{A_s(t)} [\boldsymbol{f}^\alpha - \psi^\alpha(\boldsymbol{v}^\alpha - \boldsymbol{v}^s)] \cdot \mathbf{n} dA_s + \int_{V_s(t)} b^\alpha dV + \int_{V_s(t)} {}^*p^\alpha dV \qquad (4.5)$$

The integral terms on the right-hand side of Eq. 4.5 describe, respectively: the network flux of Ψ^α through the surface of control volume (including transport with respect to the skeleton); the volumetric supply of Ψ^α; and the production of Ψ^α in the control volume, e.g. the mass production of constituent α due to phase transitions, entropy production, etc.

Applying the Gauss–Ostrogradsky theorem (Nowacki 1969) to the surface integral and recalling the continuity requirements, we can write the local form of balance equation as

$$\frac{\partial \psi^\alpha}{\partial t} + \nabla \cdot (\psi^\alpha \boldsymbol{v}^s) = \nabla \cdot [\boldsymbol{f}^\alpha - \psi^\alpha(\boldsymbol{v}^\alpha - \boldsymbol{v}^s)] + b^\alpha + {}^*p^\alpha \qquad (4.6)$$

Table 4.1 presents definition of Ψ^α for specific balances, where: u^α and s^α denote the specific internal energy and entropy of constituent α_j, $\boldsymbol{\sigma}^\alpha$; $\boldsymbol{q}^\alpha$, r^α, $\boldsymbol{g}$, T denote the partial stress tensor, the heat flux vector, the volumetric heat supply (radiation), gravity acceleration, and the temperature; and ${}^*\rho^\alpha$, ${}^*m^\alpha$, ${}^*e^\alpha$ and ${}^*s^\alpha$ denote the production terms of mass, momentum, energy and entropy, respectively.

We do not consider here the balance of moment of momentum. The only result that follows from this balance is that the total stress tensor is symmetric, and in view of the assumption that the stress tensor in moisture consists of the spherical part only, we conclude that the stress tensor acting on the skeleton is also symmetric.

Masses, momenta and energies of individual constituents do not have to be conserved; however, total mass, momentum and energy of the multicomponent medium as a whole must be conserved. The laws of conservation in our case are:

$$\sum_\alpha {}^*\rho^\alpha = 0, \quad \sum_\alpha {}^*m^\alpha = 0, \quad \sum_\alpha {}^*e^\alpha = 0 \qquad (4.7)$$

The entropy of a multicomponent medium is conserved only in a special case, namely for reversible processes. Drying processes, however, should be considered

Tab. 4.1 Components of the individual balance equations

Balance	Ψ^α	$\boldsymbol{f}^\alpha$	b^α	${}^*\mathbf{p}^\alpha$
Mass	ρ^α	0	0	${}^*\rho^\alpha$
Momentum	$\rho^\alpha \boldsymbol{v}^\alpha$	$\boldsymbol{\sigma}^\alpha$	$\rho^\alpha \boldsymbol{g}$	${}^*m^\alpha$
Energy	$\rho^\alpha u^\alpha$	$\boldsymbol{\sigma}^\alpha \cdot \boldsymbol{v}^\alpha - \boldsymbol{q}^\alpha$	$\rho^\alpha(r^\alpha + \boldsymbol{g} \cdot \boldsymbol{v}^\alpha)$	${}^*e^\alpha$
Entropy	$\rho^\alpha s^\alpha$	$-\boldsymbol{q}^\alpha/T$	$\rho^\alpha r^\alpha/T$	${}^*s^\alpha$

irreversible, so that total entropy change is positive and obeys the second law of thermodynamics, that is

$$\rho^s \dot{s} + \nabla \cdot \left(\frac{q}{T} + \sum_\alpha s^\alpha J^\alpha \right) - \frac{\Re}{T} = \sum_\alpha {}^*s^\alpha \geq 0 \tag{4.8}$$

where $s = s^s + X^l s^l + X^g s^g$ denotes the total entropy of the body per unit mass of the skeleton, q is the total heat flux (sum of constituent fluxes q^α), and $\Re$ is the total volumetric heat supply to the control volume (sum of constituent therms $\rho^\alpha r^\alpha$). The mass balance equation for the solid skeleton (a conserved quantity as ${}^*\rho^s = 0$) and the definition of mass content X^α were used in Eq. 4.8.

Applying the conservation law (Eq. 4.7c) with respect to energy, we obtain

$$\begin{aligned}\rho^s \dot{u} = \boldsymbol{\sigma} \cdot \boldsymbol{d} + \sum_\alpha \left[\rho^s \mu^\alpha \dot{X}^\alpha - J^\alpha \cdot (\nabla \mu^\alpha - \boldsymbol{g}) - \mu^{\alpha *}\rho^\alpha \right] \\ - \nabla \cdot \left(\boldsymbol{q} + T \sum_\alpha s^\alpha \boldsymbol{J}^\alpha \right) + \Re \end{aligned} \tag{4.9}$$

where $u = u^s + X^l u^l + X^g u^g$ denotes the total internal energy of the body per unit mass of the skeleton, $\boldsymbol{\sigma}$ is the total stress tensor (sum of partial stress tensors), $\boldsymbol{d}$ is the strain rate tensor of the solid skeleton, and μ^α is the chemical potential of constituent α. In Eq. 4.9 the equation of total momentum was used, in which the constituent accelerations and the products of velocities were neglected as insignificant in drying processes. This means that the equation of total momentum balance takes the form

$$\nabla \cdot \boldsymbol{\sigma} + \rho \boldsymbol{g} \approx 0 \quad \text{with} \quad \boldsymbol{\sigma} = \sum_\alpha \boldsymbol{\sigma}^\alpha \quad \text{and} \quad \rho = \sum_\alpha \rho^\alpha \tag{4.10}$$

The strain rate tensor and the chemical potential are defined as

$$\boldsymbol{d} = \frac{1}{2}[\nabla \boldsymbol{v}^s + (\nabla \boldsymbol{v}^s)^T] \quad \text{and} \quad \mu^\alpha = u^\alpha - P^\alpha/\rho^\alpha - s^\alpha T \tag{4.11}$$

Note that the chemical potential μ^α appears in Eq. 4.9 in a natural way, while substituting only the spherical part for the stress tensors of liquid and gas constituents, that is, $\boldsymbol{\sigma}^\alpha = P^\alpha \boldsymbol{I}$, where $\boldsymbol{I}$ is the unit tensor and P^α the partial pressure of constituent α.

The mass of skeleton in the control volume is conserved, so that the equation of mass balance reads

$$\dot{\rho}^s + \rho^s \nabla \cdot \boldsymbol{v}^s = 0 \tag{4.12}$$

The local mass balance of moisture is fundamental for developing the differential equation for determination of moisture distribution. Based on Eq. 4.6 and next applying the definition of mass content and Eq. 4.12, we get

$$\rho^s \dot{X}^\alpha = -\nabla \cdot J^\alpha + {}^*\rho^\alpha \quad \text{with} \quad {}^*\rho^l + {}^*\rho^g = 0 \tag{4.13}$$

Inequality 4.8 applies after replacing $\Re$ with the help of that present in the energy equation (4.9) the form

$$-\rho^s(\dot{f} + s\dot{T}) + \boldsymbol{\sigma} \cdot \boldsymbol{d} + \sum_\alpha [\rho^s \mu^\alpha \dot{X}^\alpha - J^\alpha \cdot (\nabla \mu^\alpha - \boldsymbol{g}) - \mu^\alpha {}^*\rho^\alpha]$$
$$-\left(\boldsymbol{q} + T \sum_\alpha s^\alpha J^\alpha\right) \cdot \frac{\nabla T}{T} \geq 0 \tag{4.14}$$

where $f = u - sT$ denotes the free internal energy of the body per unit mass of the skeleton.

The thermodynamic inequality 4.14 is a helpful constraint in developing the rate equations for heat and mass transfer.

4.3
Constitutive Equations in the Skeletal Frame of Reference

Drying processes are irreversible because of heat and mass transfer and also because of irreversible deformation of the skeleton if the drying-induced stresses overcome the yield stress or if the skeleton has a viscoelastic property. Thermodynamic equilibrium in drying means the absence of heat and mass fluxes as well as of phase transitions. Operating in the neighborhood of thermodynamic equilibrium we may use the following Gibbs' equation for description of thermodynamic state in every reversible process

$$-\rho^s(\dot{f} + s\dot{T}) + \boldsymbol{\sigma} \cdot \boldsymbol{d}^{(r)} + \rho^s \sum_\alpha \mu^\alpha \dot{X}^\alpha = 0 \tag{4.15}$$

where $\boldsymbol{d}^{(r)}$ is the strain rate tensor for reversible (elastic) deformations.

According to the literature on mechanics of fluid-saturated porous media (see, e.g. Cairncross et al. 1996; Lade and de Boer 1997) the effective stress tensor $\boldsymbol{\sigma}^{(\text{eff})}$ acting on the porous network is responsible for deformation of the skeleton. It also takes into account the action of pore pressure, and is defined as

$$\boldsymbol{\sigma}^{(\text{eff})} = \boldsymbol{\sigma}^s + p^{\text{por}} \phi^s \boldsymbol{I} \tag{4.16}$$

The pore pressure p^{por} is defined as an average quantity from true pressures of liquid p^{lr} and gas p^{gr} in pore space or from the respective partial pressures $P^l = -p^{\text{lr}} \phi^l$ and $P^g = -p^{\text{gr}} \phi^g$, that is

$$p^{\text{por}} = p^{\text{lr}} S + p^{\text{gr}}(1 - S) = -(P^l + P^g)/\phi \tag{4.17}$$

4.3 Constitutive Equations in the Skeletal Frame of Reference

The effective stress is the one that controls the strain–stress relation and the strength of the porous network independently of the magnitude of pore pressure (Lade and de Boer 1997).

Including the effective stress and Eq. 4.12 with $\boldsymbol{d}^{(r)} \cdot \boldsymbol{I} = \nabla \cdot \boldsymbol{v}^s = -\dot{\rho}^s/\rho^s$ into the Gibbs' equation 4.15, we may write

$$\dot{f} = -s\dot{T} + \frac{1}{\rho^s}\boldsymbol{\sigma}^{(\text{eff})} \cdot \boldsymbol{d}^{(r)} - p^{\text{por}} \dot{V}^s + \sum_\alpha \mu^\alpha \dot{X}^\alpha \qquad (4.18)$$

where $V^s = 1/\rho^s$ is the specific volume of the porous body.

If the body suffers small deformations, the partial mass density ρ^s or the specific volume V^s can be considered as constant, and the third term in Eq. 4.18 is negligibly small. The Gibbs' equation reduces to

$$\dot{f} = -s\dot{T} + \frac{1}{\rho^s}\boldsymbol{\sigma}^{(\text{eff})} \cdot \boldsymbol{d}^{(r)} + \sum_\alpha \mu^\alpha \dot{X}^\alpha \qquad (4.19)$$

Furthermore, it can be assumed $\boldsymbol{d}^{(r)} \approx \dot{\boldsymbol{\varepsilon}}^{(r)}$, where $\boldsymbol{\varepsilon}^{(r)}$ denotes the strain tensor.

Based on the Gibbs' function (Eq. 4.19) we can regard the free energy of the drying body as a function of temperature T, strain tensor $\boldsymbol{\varepsilon}^{(r)}$, and the mass fraction of liquid X^l and gas X^g:

$$f = f(T, \boldsymbol{\varepsilon}^{(r)}, X^l, X^g) \qquad (4.20)$$

Applying the chain rule and setting the time derivative of Eq. 4.20 equal to Eq. 4.19, we get the equations of state

$$s = -\left(\frac{\partial f}{\partial T}\right)_{\boldsymbol{\varepsilon}^{(r)}, X^\alpha} = s(T, \boldsymbol{\varepsilon}^{(r)}, X^l, X^g)$$

$$\boldsymbol{\sigma}^{(\text{eff})} = \rho^s \left(\frac{\partial f}{\partial \boldsymbol{\varepsilon}^{(r)}}\right)_{T, X^\alpha} = \boldsymbol{\sigma}^{(\text{eff})}(T, \boldsymbol{\varepsilon}^{(r)}, X^l, X^g) \qquad (4.21)$$

$$\mu^\alpha = \left(\frac{\partial f}{\partial X^\alpha}\right)_{T, \boldsymbol{\varepsilon}^{(r)}, X^\beta \neq \alpha} = \mu^\alpha(T, \boldsymbol{\varepsilon}^{(r)}, X^l, X^g)$$

Developing the individual equations with respect to the parameters of state and defining the respective derivatives as material coefficients we obtain the physical relations for a drying body.

For example, the strain $\boldsymbol{\varepsilon}^{(r)}$ of an isotropic skeleton can be a superposition of three components: thermal strain $\boldsymbol{\varepsilon}^{(T)} = \kappa^{(T)}(T - T_r)\boldsymbol{I}$, shrinkage strain $\boldsymbol{\varepsilon}^{(X)} = \kappa^{(X)}(X^l - X_r^l)\boldsymbol{I}$, and mechanical strain caused by the effective stress $\boldsymbol{\varepsilon}^{(M)} = 2G'\boldsymbol{\sigma}^{(\text{eff})} + \Gamma'(\boldsymbol{\sigma}^{(\text{eff})} \cdot \boldsymbol{I})\boldsymbol{I}$. Summing these three components, we can find the physical relation between the effective stress and strain. It takes a form similar to that of the

Duhamel–Neuman equation well known in thermoelasticity (see, e.g. Nowacki 1969)

$$\boldsymbol{\sigma}^{(\text{eff})} = 2G\boldsymbol{\varepsilon}^{(r)} + \left[\Gamma\boldsymbol{\varepsilon}^{(r)} - \gamma^{(T)}(T - T_r) - \gamma^{(T)}(X^l - X_r^l)\right]\mathbf{T} \quad (4.22)$$

where G and Γ are the Lame's constants, $\gamma^{(T)} = (2G + 3\Gamma)\kappa^{(T)}$, $\gamma^{(X)} = (2G + 3\Gamma)\kappa^{(X)}$ with $\kappa^{(T)}$ and $\kappa^{(X)}$ being the coefficients of linear thermal and humid expansion; T_r and X_r^l are the reference temperature and liquid content.

Note that chemical potentials μ^α ($\alpha = l$ and g) depend on mass fractions X^l and X^g. As the chemical potentials are intensive parameters, they have to be homogeneous functions of degree zero, that is

$$\mu^\alpha(T, \boldsymbol{\varepsilon}^{(r)}, nX^l, nX^g) = n^0 \mu^\alpha(T, \boldsymbol{\varepsilon}^{(r)}, X^l, X^g)$$

where n is an arbitrary parameter. Differentiating both sides of this equation with respect to n and substituting $n = 1$, we obtain

$$\frac{\partial \mu^\alpha}{\partial X^l} X^l + \frac{\partial \mu^\alpha}{\partial X^g} X^g = 0 \quad \text{for } \alpha = l \text{ and } g$$

Such a system of equations has a nontrivial solution only if the main determinant is equal to zero. This means that potentials μ^l and μ^g are dependent on each other.

The Gibbs' identity for the moisture as a whole, neglecting solid compressibility and capillary pressure, can be expressed as

$$\dot{\mu}^m = -s^m \dot{T} + V^s \dot{p}^{\text{por}} + \mu^l \dot{X}^l + \mu^g \dot{X}^g \quad (4.23)$$

where $\mu^m = f^m + p^{\text{por}} V^s$, $f^m = X^l f^l + X^g f^g$ and $s^m = X^l s^l + X^g s^g$ denote the potential, the free energy, and the entropy for moisture per unit mass of the skeleton.

Substituting the potential as a homogeneous extensive function of degree one, namely $\mu^m = \mu^l X^l + \mu^g X^g$, into Eq. 4.23, we obtain *the Gibbs–Duhem* relationship (Szarawara 1985)

$$\dot{\mu}^l X^l + \dot{\mu}^g X^g = -s^m \dot{T} + V^s \dot{p}^{\text{por}} \quad (4.24)$$

Equation 4.24 relates the chemical potentials μ^l and μ^g. This an important statement for modeling of phase transitions of liquid into vapor inside the drying body.

4.4
Rate Equations for Heat and Mass Transfer

Including Gibbs' equation (Eq. 4.18) and mass conservation (Eq. 4.13b) into the inequality 4.14, we get

$$\boldsymbol{\sigma}^{(\text{eff})} \cdot \dot{\boldsymbol{\varepsilon}}^{(\text{ir})} - \sum_\alpha \mathbf{J}^\alpha \cdot (\nabla \mu^\alpha - \mathbf{g}) - {}^*\rho^l(\mu^l - \mu^g) - \left(\mathbf{q} + T \sum_\alpha s^\alpha \mathbf{J}^\alpha\right) \cdot \frac{\nabla T}{T} \geq 0 \quad (4.25)$$

where $\boldsymbol{\varepsilon}^{(\text{ir})} = \boldsymbol{\varepsilon} - \boldsymbol{\varepsilon}^{(r)}$ is the strain tensor of irreversible deformations.

4.4 Rate Equations for Heat and Mass Transfer

The inequality of Eq. 4.25 expresses the energy that is dissipated during both the heat and mass transfer processes and the irreversible deformations of the skeleton. The dissipated energy contributes to entropy production. Recalling the Curie–Prigogine symmetry principle (see, e.g. Kowalski (2003)), we state that the terms of different tensorial character (scalar, vector, tensor) in inequality 4.25 ought to be positively defined as

$$\boldsymbol{\sigma}^{(\text{eff})} \cdot \dot{\boldsymbol{\varepsilon}}^{(\text{ir})} \geq 0 \tag{4.26}$$

$$-{^*\rho^l}(\mu^l - \mu^g) \geq 0 \tag{4.27}$$

$$-\boldsymbol{J}^\alpha \cdot (\nabla \mu^\alpha - \boldsymbol{g}) \geq 0 \tag{4.28}$$

$$-\left(\boldsymbol{q} \pm T \sum_\alpha s^\alpha \boldsymbol{J}^\alpha\right) \cdot \frac{\nabla T}{T} \geq 0 \tag{4.29}$$

The plus sign in expression 4.29 holds when the heat and mass fluxes have the same directions, otherwise the minus sign is valid.

As we see on the basis of Eq. 4.28, the thermodynamic forces that cause the moisture to flow are the gradient of moisture potential and the gravitational force per unit mass. Expression 4.29 shows that gradient of temperature is the reason for network heat flux, which is the sum (or difference) between the heat flux conducted and heat transported by mass flux.

The constrains from expressions 4.26 to 4.29 imply specific forms for the thermodynamic fluxes, namely

$$\dot{\boldsymbol{\varepsilon}}^{(\text{ir})} = \boldsymbol{C} \cdot \boldsymbol{\sigma}^{(\text{eff})} \quad \text{with} \quad \boldsymbol{C} \cdot \boldsymbol{\sigma}^{(\text{eff})} \cdot \boldsymbol{\sigma}^{(\text{eff})} \geq 0 \tag{4.30}$$

$$^*\rho^l = -{^*\rho^g} = -\varpi(\mu^l - \mu^g) \quad \text{with} \quad \varpi \geq 0 \tag{4.31}$$

$$\boldsymbol{J}^\alpha = -\Lambda^\alpha(\nabla \mu^\alpha - \boldsymbol{g}) \quad \text{with} \quad \Lambda^\alpha \geq 0 \tag{4.32}$$

$$\boldsymbol{q} = -\lambda \nabla T \mp T \sum_\alpha s^\alpha \boldsymbol{J}^\alpha \quad \text{with} \quad \lambda \geq 0 \tag{4.33}$$

The above rate equations constitute sufficient (not necessary) conditions to satisfy the thermodynamic constrains. The coefficients $\boldsymbol{C}, \varpi, \Lambda^\alpha$ and λ are in general functions of state variables.

Equation 4.30 describes viscous or plastic flows of the skeleton, dependent on the interpretation and specification of the material constants $\mathbf{C}$ (see, e.g., Chapter 3 and Perzyna (1966)).

The relation of Eq. 4.31 shows that the rate of mass change from liquid to vapor due to phase transition depends on the difference between chemical potentials μ^l and μ^g. The thermodynamic equilibrium (i.e. absence of phase transition) requires these two potential be equal to each other. In this relation ϖ is the phase-transition coefficient.

The gradient of chemical potential μ^α and the gravity force per unit mass $\mathbf{g} = -\nabla \mu^{\mathrm{grav}}$, where μ^{grav} denotes the gravity potential, constitute the thermodynamic forces for moisture flow. In Eq. 4.32 Λ^α is the moisture transport coefficient (mobility parameter).

Heat flux, which is proportional to the gradient of temperature according to Fourier's law, is, however, decreased or increased by the heat transported with the moisture flux J^α (heat convection); λ is the coefficient of thermal conductivity.

4.5
Differential Equations for Heat and Mass Transfer

4.5.1
Differential Equation for Heat Transfer

The local balance of energy (Eq. 4.9) constitutes the basis for development of differential equation for heat transfer. Substituting Gibbs' equation 4.18 and the rate equations 4.30 to 4.33 into Eq. 4.49, we obtain

$$\rho^s \dot{s} T = \nabla \cdot (\lambda \nabla T) + Q^* + \Re \tag{4.34}$$

where Q^* denotes the internal source of heat from energy dissipated during irreversible processes (e.g. permanent deformations of the skeleton, viscous flow, dispersion of heat during phase transition) (Kowalski 2003). The amount of this heat reads

$$Q^* = \mathbf{C}^{-1} \cdot \dot{\boldsymbol{\varepsilon}}^{(\mathrm{ir})} \cdot \dot{\boldsymbol{\varepsilon}}^{(\mathrm{ir})} + \sum_\alpha \Lambda^\alpha (\nabla \mu^\alpha - \mathbf{g}) \cdot (\nabla \mu^\alpha - \mathbf{g}) + \varpi(\mu^l - \mu^g)^2$$

The value of Q^* is very small in comparison to the heat supplied to the body due to external convective or microwave heating, and will be neglected in further considerations.

The time derivative of entropy reads

$$\dot{s}(T, \varepsilon^{(\mathrm{r})}, X^l, X^g) = \dot{s}^s + \dot{s}^m = \frac{c_v}{T}\dot{T} + \gamma^{(T)}\dot{\varepsilon}^{(\mathrm{r})} + s^l \dot{X}^l + s^g \dot{X}^g \tag{4.35}$$

where $c_v = c_v^s + c_v^l X^l + c_v^g X^g$ is the specific heat of the multicomponent body.

Volumetric strain $\varepsilon = \varepsilon^{(\mathrm{r})} + \varepsilon^{(\mathrm{ir})}$ influences the changes of entropy (and also temperature) insignificantly. The irreversible strain $\varepsilon^{(\mathrm{ir})}$ (e.g. creep strain) is included in the heat production Q^*. The reversible part of strain $\varepsilon^{(r)}$, due to shrinkage or to the drying induced spherical stress, is not an instantaneous strain but one that takes a long time to arise. It influences the changes of entropy much less than the temperature or moisture-content variations. Therefore, we neglect the strain rate term in expression Eq. 4.35 in further considerations.

Substituting expression Eq. 4.35 into Eq. 4.34 and using the equations of mass balance (Eq. 4.13) and the rate equation 4.31, we obtain the differential equation for determination of temperature in the drying body in the form

$$\rho^s c_v \dot{T} = \nabla \cdot (\lambda \nabla T) - \varpi(\mu^l - \mu^g)\Delta h_v + \Re \tag{4.36}$$

where $\Delta h_v = T(s^g - s^l)$ denotes the latent heat of evaporation.

The expression $Ts^l \nabla \cdot J^l + Ts^g \nabla \cdot J^g$ has been neglected in Eq. 4.36. It describes the change of energy due to accumulation of moisture, and is small in comparison to the term containing the latent heat of evaporation.

The term $\Re$ determining the volumetric heat supply due to microwave heating will be developed in the next section.

The initial and boundary conditions necessary for unique solution of the problem will be discussed in Section 4.7 for the special example of convective and microwave drying of a kaolin cylinder.

4.5.2
Determination of the Microwave Heat Source $\Re$

Microwaves are electromagnetic waves in the frequency range from 300 MHz to 300 GHz (of wavelength from 1 m to 1 mm), which can be absorbed by materials containing polar molecules as, e.g. water, proteins, carbohydrates. They can permeate through such materials like glass, teflon, etc., or can be reflected, for example, by metals. Dipole rotations and ion polarizations are the two main mechanisms of conversion of microwave energy into heat.

The fundamental equations describing the electromagnetic field are the Maxwell equations

$$\text{rot } \mathbf{H} = \mathbf{j} + \frac{\partial \mathbf{D}}{\partial t}, \quad \text{rot } \mathbf{E} = -\frac{\partial \mathbf{B}}{\partial t}, \quad \text{div} \mathbf{B} = 0, \quad \text{div} \mathbf{D} = q_e \tag{4.37}$$

where $\mathbf{E}$, $\mathbf{H}$, $\mathbf{D}$ and $\mathbf{B}$ are the vectors of the electric, magnetic, electric induction and magnetic induction fields, q_e is the density of the space electric charge and $\mathbf{j}$ is the vector of electric current.

The Maxwell equations have to be supplemented with the constitutive relations

$$\mathbf{B} = \mu^* \mathbf{H}, \quad \mathbf{D} = \varepsilon^* \mathbf{E}, \quad \mathbf{j} = \sigma^* \mathbf{E} \tag{4.38}$$

where μ^*, ε^* and σ^* denote the magnetic permeability, permittivity and electric conductivity of a material.

The magnetic permeability can be written as $\mu^* = \mu_0 \mu_r$, where μ_0 is the magnetic permeability of free space equal to $4\pi \times 10^{-7}$ H/m, and μ_r is the relative magnetic permeability. For diamagnetic materials $\mu_r < 1$, for paramagnetic materials $\mu_r > 1$, and for ferromagnetic materials $\mu_r \gg 1$.

The permittivity can be written as $\varepsilon^* = \varepsilon_0 \varepsilon_r$, where ε_0 is the permittivity of free space equal to 8.854×10^{-12} Fm^{-1}, and ε_r is the relative permittivity (dielectric constant) that ranges from 1 for gas dielectrics to several thousands for ferroelectrics. For example, ε_r equals 1.00059 for air, 7.00 for glass, 2 to 2.5 for paper, 80.5 for water.

The reverse of electrical conductivity $(1/\sigma^*)$ is the electrical resistance.

Equations 4.37 and 4.38, after elimination of the magnetic field, electric induction and magnetic induction vectors, can be rearranged in the wave equation form

$$\nabla^2 E = \frac{1}{\eta^*} \frac{\partial E}{\partial t} + \frac{1}{c^2} \frac{\partial^2 E}{\partial t^2} + \frac{1}{\varepsilon^*} \nabla q_e \tag{4.39}$$

where c is the velocity of electromagnetic wave and η^* is the magnetic viscosity expressed as

$$c = \frac{1}{\sqrt{\mu^* \varepsilon^*}} = \frac{c_0}{\sqrt{\mu_r \varepsilon_r}}, \quad c_0 = \frac{1}{\sqrt{\mu_0 \varepsilon_0}} \quad \text{and} \quad \eta^* = \frac{1}{\mu^* \sigma^*} \tag{4.40}$$

Here, c_0 denotes the velocity of light in free space. The well-known dependence for the double curl operator $\nabla \times \nabla \times E = \nabla(\nabla \cdot E) - \nabla^2 E$ was used in rearranging Eq. 4.39.

Equation 4.39, expressed by the electric field intensity, shows that microwaves propagate with velocity c, which is a function of magnetic permeability μ_r and permittivity ε_r. The greater the magnetic permeability μ_r and permittivity ε_r, the smaller is the velocity of electromagnetic wave. Also, note that Eq. 4.39 contains a damping term dependent on the magnetic viscosity η^*. This means that the wave amplitude is damped with distance, so that the wave may not arrive at remote areas of the material.

Adapting Eq. 4.39 to drying, we assume that a drying material is composed of a dielectric skeleton and moisture in pores. The moisture is water, which is a strong absorber of microwave energy because of the dipole character of water molecules. Thus, the absorbing properties of materials decrease with decreasing water content in the course of drying. Also, the damping properties decrease with time during drying. As the material becomes drier, it tends to become a dielectric. Dielectrics are characterized by large resistivity ($\sigma^* \to 0$) and thus also by large magnetic viscosity ($\eta^* \to \infty$). Besides, most materials, and in particular dielectrics, have a magnetic permeability close to that in free space, that is $\mu_r \approx 1$. In further considerations we assume that the material is not charged electrically ($q_e = 0$).

The microwaves are cyclic with very high frequency, mostly 2.45 GHz. The solution of Eq. 4.39 for cyclic waves is

$$E = E^* \exp(i\omega t - \gamma x \cdot n) \quad \gamma = \alpha + i\beta = i\omega \sqrt{\varepsilon^* \mu^*} \tag{4.41}$$

where E^* denotes the complex amplitude, ω is the wave frequency, α is the coefficient of damping of amplitude with distance, β is the wave number, $x \cdot n$ is a measure of distance in the direction of wave propagation (x – position vector, n – unit normal vector in the direction of wave propagation), and i denotes the imaginary number.

The dielectric properties of a material undergoing drying depend on the material saturation S and temperature T. The complex parameters of permittivity and magnetic permeability describe both absorption and dissipation of the electromagnetic energy, and can be expressed as

$$\varepsilon^*(S,T) = \varepsilon'(S,T) - i\varepsilon''(S,T), \quad \mu^*(S,T) = \mu'(S,T) - i\mu''(S,T) \quad (4.42)$$

The damping coefficient and the wave number can be expressed with the help of the components of magnetic permeability and permittivity, namely (see Kowalski et al. 2005a)

$$\alpha = \frac{\lambda^*}{4\pi}\omega(\varepsilon'\mu'' + \varepsilon''\mu')$$

$$\beta = \frac{2\pi}{\lambda^*} = \omega\sqrt{\frac{\varepsilon'\mu' - \varepsilon''\mu''}{2}\left(1 + \sqrt{1 + \left(\frac{\varepsilon'\mu'' + \varepsilon''\mu'}{\varepsilon'\mu' - \varepsilon''\mu''}\right)^2}\right)} \quad (4.43)$$

where λ^* is the length of electromagnetic wave.

We can see that a dielectric as a carrier of electromagnetic field is characterized by four independent scalar parameters. In practical applications the problem can be further simplified by assuming that the magnetic permeability is close to that in free space and writing $\mu' = \mu_0$ and $\mu'' = 0$. Thus, the characterization of dielectric properties of homogeneous isotropic materials requires only two parameters.

The real part of permittivity ε' determines accumulation of energy because of polarization, and the imaginary part ε'' determines the energy absorption, that is the energy, which is converted to heat. In most cases this characterization is accomplished by the specific dielectric constant $k' = \varepsilon'/\varepsilon_0$ and the loss parameter δ defined as $\text{tg}\delta = \varepsilon''/\varepsilon'$. Using these simplifications we can rearrange the damping coefficient and the wave number as follows

$$\alpha = \frac{\omega}{c_0}\sqrt{\frac{k'}{2}\left(\sqrt{1 + (\text{tg}\delta)^2} - 1\right)} \quad \text{and} \quad \beta = \frac{\omega}{c_0}\sqrt{\frac{k'}{2}\left(\sqrt{1 + (\text{tg}\delta)^2} + 1\right)} \quad (4.44)$$

We now write the electric field intensity with clearly separated damping term

$$E = E_0 \exp(-\alpha x \cdot n) \quad (4.45)$$

where $E_0 = E^* \exp[i(\omega t - \beta x \cdot n)]$ denotes the microwave energy propagating through the material with velocity $c = \omega/\beta$. Part of this energy is accumulated in the body and the other part is dissipated to heat due to the friction caused by high-frequency polarization of water molecules.

Equation 4.45 shows that the energy of microwaves decreases with distance. The intensity of abatement is determined by parameter α, which depends on the dielectric properties of the medium k' and δ as well as on the microwaves frequency ω.

The ability of a drying material to absorb microwave energy is determined by the electric conductivity $\sigma^* = \omega\varepsilon'' = \omega k'(\text{tg}\,\delta)/\varepsilon_0$, which depends on the absorption parameter ε'' and frequency ω. It has been shown (Di et al. 2000; Feng et al. 2001; Garcia and Bueno 1998; Perre and Turner 1996; Ratanadecho et al. 2001, 2002; Sanga et al. 2002; Zhang and Mujumdar 1992) that microwave power converted into heat per unit volume of drying material is equal to the product of electric conductivity and the square of the electric field intensity, that is

$$\Re = \frac{1}{2}\sigma^* E^2 = \frac{1}{2}\omega\varepsilon'' E_0^2 \exp(-2\alpha\,\mathbf{x}\cdot\mathbf{n}) \tag{4.46}$$

The permittivity ε^* depends on temperature and moisture content, as water is the main absorber of microwave energy. As the drying material consists of three components (solid, liquid and gas), we propose the following expressions for ε' and ε'' (Di et al. 2000)

$$\begin{aligned}\varepsilon' &= \phi[S\varepsilon'^l(T) + (1-S)\varepsilon'^g(T)] + (1-\phi)\varepsilon'^s(T) \\ \varepsilon'' &= \phi[S\varepsilon''^l(T) + (1-S)\varepsilon''^g(T)] + (1-\phi)\varepsilon''^s(T) \approx \phi S\varepsilon''^l(T)\end{aligned} \tag{4.47}$$

where superscripts l, g and s refer to liquid, gas and solid.

The influence of gas and solid is neglected in the final expression for ε'' in Eq. 4.47. Thus, we assume that water is the main absorber of energy. Various expressions for the dielectric constant of water as a function of temperature has been proposed. Following Di et al. (2000) in the temperature range $20 \leq \vartheta \leq 100\,°C$, we have

$$\varepsilon'^l(\vartheta) = 85.2 - 0.3358\vartheta, \quad \varepsilon''^l(\vartheta) = 320.66\vartheta^{-1.03} \tag{4.48}$$

where $\vartheta = T - T_r$, with T_r being the reference temperature.

Summarizing the foregoing considerations, we can state the following:
- The amount of energy absorbed by drying material falls exponentially with distance, as determined by the parameter α, which depends on the dielectric properties of the medium (k' and δ) and on microwave frequency ω.
- The absorption of microwave energy is proportional to the water volume content $\phi^l = \phi S$. In order to make the drying model coherent, we assume here proportionality to the reduced water mass content θ/θ_0, where $\theta = X^l - X_r^l$ denotes excess to the reference liquid content X_r^l and θ_0 is the initial moisture content.

- We introduce the quantity $\mathfrak{R}_0 = \omega E_0^2/2$ describing the power of heating under ideal conditions, which can be determined by measuring the evaporation of pure water in a microwave dryer.

The final expression for the microwave heat source is proposed in the form

$$\mathfrak{R} = \mathfrak{R}_0 \frac{\theta(\mathbf{x},t)}{\theta_0} \varepsilon''(\vartheta)\exp(-2\alpha\,\mathbf{x}\cdot\mathbf{n}) \qquad (4.49)$$

Equation 4.49 shows that the microwave heat source disappears for the dry body, that is, when the excess of water content θ is equal to zero.

4.5.3
Differential Equation for Mass Transfer

The local mass balance (Eq. 4.13) provides the basis for development of differential equations for mass transfer. Substituting into this equation the moisture flux and the rate of mass change due to phase transition given by Eqs. 4.31 and 4.32, we obtain

$$\begin{aligned}\rho^s \dot{X}^l &= \nabla \cdot [\Lambda^l(\nabla\mu^l - \mathbf{g})] - \varpi(\mu^l - \mu^g) \\ \rho^s \dot{X}^g &= \nabla \cdot [\Lambda^g(\nabla\mu^g - \mathbf{g})] + \varpi(\mu^l - \mu^g)\end{aligned} \qquad (4.50)$$

According to Eq. 4.21, the chemical potentials μ^l and μ^g are functions of the state variables $\{T, \varepsilon^{(r)}, X^l, X^g\}$.

The total system consists of differential equations for: temperature (Eq. 4.36); body deformations (to follow); and moisture content (Eq. 4.50). This is sufficient for determination of the thermodynamic parameters of state, and indirectly also stresses. However, such a system of equations would be very difficult to solve. Therefore, a practically justified simplification of the mathematical model of drying is desired.

A first simplifying step follows from the observation that it is important to determine the amount of liquid in drying material, and not the amount of gas. By moderate heating of the drying body, gas flows freely out of the body having no (or insignificant) influence on the mechanical behavior.

Distribution of liquid and its variation in time is responsible for the shrinkage strains and generation of stresses. Thus, for our purpose only the first expression in Eq. 4.50 is necessary; the second expression is redundant. This does not mean that the phase transition of liquid into vapor is not important, as it is evident in the differential equation 4.36 for heat transfer.

Another simplification is obtained by neglecting the dot product of the chemical potential gradient and the gradient of moisture transport coefficient, i.e. $\nabla\Lambda^l \cdot \nabla(\mu^l - \mathbf{g}) \approx 0$. Although Λ^l depends on temperature and liquid content, one can justify this simplification by a continuous distribution of the state variables. It is important to recognize that not all state variables are equally important in the equation for moisture transport. The gradient of volumetric strain $\varepsilon^{(r)}$ affects the moisture flow insignificantly and over a long period of time, mainly due to shrinkage. We also neglect the dependence of liquid chemical potential μ^l on gas content X^g.

Referring to the Gibbs–Duhem equation (Eq. 4.24), we assume the relation between potentials μ^l and μ^g to be of the form

$$\mu^g = \lfloor 1 - a(T, X^l) \rfloor \mu^l \quad (4.51)$$

where $0 \leq a \leq 1$ is the phase transition efficiency; (at thermodynamic equilibrium: $a = 0$).

The rate equation 4.32 and the differential equation 4.50 for liquid transfer can now be rewritten as

$$J^l = -\Lambda^l (c_T \nabla \vartheta + c_X \nabla \theta - g) \quad (4.52)$$

$$\rho^s \dot{\theta} = \Lambda^l \nabla^2 (c_T \vartheta + c_X \theta) - \varpi a (c_T \vartheta + c_X \theta) \quad (4.53)$$

where coefficients c_T and c_X are the derivatives of chemical potential μ^l, the former with respect to temperature at constant liquid content (thermodiffusion) and the latter with respect to liquid content at constant temperature (diffusion).

In drying (see, e.g., Scherer 1990; Kneule 1970; Lykov 1968) we often encounter two main stages: the *constant drying rate period* (first stage) and the *falling-rate period* (second stage). For the purpose of mathematical modeling it is important to note that the two stages of drying are characterized by different mechanisms of moisture transport.

The drying rate during the constant drying rate period (CDRP) is independent of time. In this period the moisture is transported in the form of liquid from the body interior towards the body surface where it evaporates as from an open surface of liquid. In this stage of drying the amount of evaporated liquid inside the body is insignificant in comparison to that evaporated at the body surface. The parameter of phase-transition efficiency can be taken to be $a = 0$. Moreover, the evaporation causes cooling of the body surface and the reduced temperature leads to a lower rate of evaporation. This feedback process equilibrates when the drying surface, and in a short time the whole body, reaches the *wet-bulb temperature*, which remains constant throughout the constant drying rate period. Equation 4.53 becomes, in this case, simply a diffusion equation.

The falling-rate period begins when the body surface dries in some places and the liquid menisci in capillaries start to drive into the body. The local moisture content close the body surface reaches the critical value X_{cr}. Beyond X_{cr} the rate of evaporation decreases and the temperature of the body rises above the wet-bulb temperature tending to equilibrium with the drying air. More and more liquid evaporates inside the body. In order to take into account the changes in mechanisms of moisture transfer between the constant and the falling-rate periods, we postulate the following form for the moisture transport coefficient Λ^l (Kneule 1970; Kowalski and Rybicki 2004)

$$\Lambda^l = \frac{\sigma(T)}{\eta(T)} f(X^l) \text{ with } \frac{f(X^l)}{f_0} = \begin{cases} 1 & \text{for } X^l \geq X_{cr} \\ \left(\dfrac{X^l - X_{eq}}{X_{cr} - X_{eq}}\right)^2 & \text{for } X_{eq} \leq X^l \leq X_{cr} \end{cases} \quad (4.54)$$

where σ and η are the surface tension and the viscosity of moisture, respectively, and f_0 is a structural parameter of the moisture transport coefficient (permeability of porous body).

The phase transitions inside the body are controlled by the following distribution function a

$$\frac{a}{a_0} = \begin{cases} 0 & \text{for } X^l \geq X_{cr} \\ 1 - \left(\dfrac{X^l - X_{eq}}{X_{cr} - X_{eq}}\right)^2 & \text{for } X_{eq} \leq X^l \leq X_{cr} \\ \left(\dfrac{X^l}{X_{eq}}\right)^2 & \text{for } 0 \leq X^l \leq X_{eq} \end{cases} \qquad (4.55)$$

where a_0 is the maximal value of the phase transition efficiency, X_{cr} denotes the critical value of moisture content, and X_{eq} is the moisture content at the body surface, being in equilibrium with the ambient air.

Figures 4.2a and Fig. 4.2b present graphically the distribution functions of Eqs. 4.54 and 4.55. These distribution functions allow us to program a numerical solution of the drying process as a whole. In this numerical model the absence of phase transitions in the constant-drying period is assumed, that is for $X \geq X_{cr}$. The internal phase transition starts to increase and the diffusional moisture transport starts to decrease in the falling drying rate period, i.e. when $X_{eq} \leq X^e \leq X_{cr}$. Since the moment when the moisture content at the surface reaches the equilibrium value with the ambient air X_{eq}, transport of liquid stops totally, and the rest of the liquid inside the body is removed due to evaporation only.

4.6
Thermomechanical Equations for a Drying Body

4.6.1
Physical Relations

One can distinguish several kinds of strain components for materials undergoing drying as, for example, instantaneous, creep, shrinkage and mechanosorptive

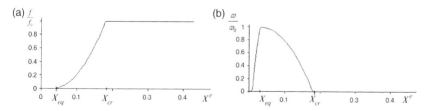

Fig. 4.2 Graphical performance of the distributive functions: (a) the coefficient of structural parameter, (b) the parameter of the phase transition efficiency a.

strains. *Instantaneous strains* (e.g. elastic or plastic strains) are produced immediately after application of mechanical stresses. Under critical conditions, the stresses may destroy the material structure. *Creep* is the time-dependent increment of total deformation due to imposed stress history (see Chapter 3). It is an irreversible strain. The free *shrinkage strain* is assumed to be proportional to the moisture-content change, independent on the state of stress. Some authors introduce an extra *mechanosorptive strain* that results from the interaction between stress and moisture content change. A different deformation of a body occurs when the changes of moisture content and loading are consecutive from that when they are simultaneous (Milota and Quinglin 1994). In our opinion, this effect can be described by making the material coefficients dependent on moisture content.

In our considerations we take into account the shrinkage strain (in many cases the largest strain component by drying), the instantaneous strain in the elastic range caused by the drying-induced stresses, and the creep strain described by the viscoelastic Maxwell model.

The shrinkage (or swelling strain) is proportional to temperature and moisture content, and is expressed as:

$$\varepsilon^{(TX)} = 3(\kappa^{(T)}\vartheta + \kappa^{(X)}\theta) \tag{4.56}$$

where $\kappa^{(T)}$ and $\kappa^{(X)}$ are the respective coefficients of linear thermal and humid expansion.

The physical relation for instantaneous strains is of the form (see also Eq. 4.22):

$$s_{ij}^{(\text{eff})} = 2G e_{ij}^{(r)}, \quad \sigma^{(\text{eff})} = K(\varepsilon^{(r)} - \varepsilon^{(TX)}) \tag{4.57}$$

where $s_{ij}^{(\text{eff})} = \sigma_{ij}^{(\text{eff})} - \sigma^{(\text{eff})}\delta_{ij}$ is the stress deviator, $\sigma^{(\text{eff})} = \sigma_{ii}^{(\text{eff})}/3$ is the spherical stress, $e_{ij}^{(r)} = \varepsilon_{ij}^{(r)} - (\varepsilon_{kk}^{(r)}/3)\delta_{ij}$ is the strain deviator, $\varepsilon^{(r)} = \varepsilon_{ii}^{(r)}$ is the volumetric strain, G and K are the elastic shear and bulk moduli, respectively.

An equivalent physical relation for a viscoelastic material (Maxwell model) reads

$$\dot{s}_{ij}^{(\text{eff})} + \frac{G}{\eta_v} s_{ij}^{(\text{eff})} = 2M\dot{e}_{ij}, \quad \dot{\sigma}^{(\text{eff})} + \frac{K}{\kappa_v}\sigma^{(\text{eff})} = K\left(\dot{\varepsilon} - \dot{\varepsilon}^{(TX)}\right) \tag{4.58}$$

where $e_{ij} = e_{ij}^{(r)} + e_{ij}^{(ir)} = \varepsilon_{ij} - (\varepsilon/3)\delta_{ij}$ and $\varepsilon = \varepsilon^{(r)} + \varepsilon^{(ir)}$ denote the total strain deviator and the total volumetric strain. The superscript (ir) refers here to the creep strain, which is an irreversible one. The new parameters in this relation are η_v and κ_v, which denote the viscoelastic shear and bulk moduli respectively. The appearance of viscoelastic bulk modules assumes the existence of viscoelastic volumetric deformation, which seems to be justified in porous media.

While expressing the physical relations (Eqs. 4.57 and 4.58) in Laplace transforms, we will state a similarity between their forms if the material coefficients are constants. These relations then differ only in the material coefficients. This analogy between the physical relation for elastic and viscoelastic bodies allows us to calculate the stresses for a viscoelastic body on the basis of those for an elastic one. Thus, using Borel's convolution formula, we can write (Kowalski 2003)

$$s_{ij}^{V(\text{eff})}(x,t) = s_{ij}^{E(\text{eff})}(x,t) - \frac{1}{\tau}\int_0^t \exp\left(-\frac{t-\xi}{\tau}\right) s_{ij}^{E(\text{eff})}(x,\xi)d\xi$$

$$\sigma^{V(\text{eff})}(x,t) = \sigma^{E(\text{eff})}(x,t) - \frac{1}{\tau_0}\int_0^t \exp\left(-\frac{t-\xi}{\tau_0}\right) \sigma^{E(\text{eff})}(x,\xi)d\xi$$

(4.59)

where superscript V means viscoelastic and E elastic body, $\tau = \eta_v/G$ and $\tau_0 = \kappa_v/K$.

4.6.2
Differential Equations for Body Deformation

We split the equation of total momentum (Eq. 4.10) into a part expressed by the effective stress tensor and a part of pore pressure. Next, substituting Eq. 4.22 or Eq. 4.57, we obtain the differential equations for the determination of drying-body deformations in the elastic range, and indirectly also the stresses

$$\nabla \cdot \lfloor 2G\varepsilon^{(r)} + (\Lambda\varepsilon^{(r)} - \gamma^{(T)}\vartheta - \gamma^{(X)}\theta)I \rfloor + \rho \mathbf{g} = \nabla p^{\text{por}} \qquad (4.60)$$

where $\vartheta = T - T_r$, $\theta = X^l - X_r^l$ and the strain tensor is defined as

$$\varepsilon^{(r)} = \frac{1}{2}[\nabla \mathbf{u} + (\nabla \mathbf{u})^T] \quad \text{and} \quad \varepsilon^{(r)} = \nabla \cdot \mathbf{u} \qquad (4.61)$$

where $\mathbf{u}$ denotes the displacement vector of the skeleton.

Developing the pore pressure as a function of state variables we can get the physical relation for this quantity, which is rather a complex one. In order to present the pore pressure in a more usable form, we recall here its definition by Eq. 4.17, that is

$$p^{\text{por}} = p^{\text{lr}}S + p^{\text{gr}}(1-S) = (p^{\text{lr}} - p^{\text{gr}})S + p^{\text{gr}} \qquad (4.62)$$

The difference between the liquid and gas pressure is equal to the capillary pressure. We assume here that the gas pressure in closed bubbles of pore space p^{gr} is much greater than the capillary pressure $(p^{\text{lr}} - p^{\text{gr}})S = p^{\text{cap}}S$. So that, we equate the pore pressure with gas pressure. Applying the Clapeyron relation for the gas we can write

$$p^{\text{por}} \approx p^{\text{gr}} = \rho^{\text{gr}} R^{\text{g}} T \qquad (4.63)$$

The gradient of pore pressure can be neglected in many cases, for example in slow convective drying where the moisture transport is of diffusive character. However, intensive internal heating may occur in microwave drying, where the temperature inside the body rises rapidly causing a drastic increase of pore pressure due to phase transition and thermal expansion of gas. It may cause an explosive destruction of the material. We will show such an effect in our numerical example, which has also been confirmed experimentally (Kowalski et al., 2004a, 2005a,b).

We assume here that the gradient of pore pressure in Eq. 4.60 is relevant, if the temperature (or gradient of temperature) in some area overcomes a critical value. This statement can be formulated mathematically as follows

$$\nabla p^{\text{por}} = \mathbf{\Pi} g(T - T_{\text{cr}}) \quad \text{where} \quad g(T - T_{\text{cr}}) = \begin{cases} 0 & \text{for} \quad T \leq T_{\text{cr}} \\ 1 & \text{for} \quad T > T_{\text{cr}} \end{cases} \quad (4.64)$$

where $\mathbf{\Pi}$ denotes the foreseen pore pressure gradient in an area, in which the temperature of the body overcomes the critical value.

Equation 4.60 enables the determination of deformations, and by Eq. 4.22 or Eq. 4.57 of stresses, in the elastic range as functions of temperature and moisture content. Based on Eq. 4.59, one can estimate the stresses in a linear viscoelastic body. Initial and boundary conditions reflect the kind of drying, geometry of dried products, drying conditions, etc. We will illustrate the formulation of such conditions on the example of a cylinder under convective and microwave drying.

4.7
Drying of a Cylindrical Sample made of Kaolin

4.7.1
Convective Drying of a Kaolin Cylinder

The cylinders shown in Fig. 4.3 are the objects of our further considerations. Distribution and time evolution of liquid content and temperature will be determined for the first and second period of drying.

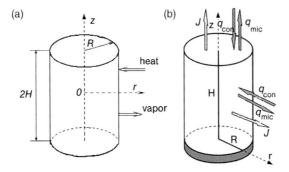

Fig. 4.3 Cylindrical sample under convective drying:
(a) convective drying of the cylinder with all surfaces open,
(b) convective and microwave drying of the cylinder with bottom surface insulated for moisture outflow.

Equations 4.53 and 4.36 take, in cylindrical geometry, the form

$$\rho^s \dot{\theta} = \Lambda^l \left(\frac{\partial^2}{\partial r^2} + \frac{1}{r}\frac{\partial}{\partial r} + \frac{\partial^2}{\partial z^2} \right)(c_T \vartheta + c_X \theta) - \varpi a(c_T \vartheta + c_X \theta) \tag{4.65}$$

$$\dot{\vartheta} = \kappa \left(\frac{\partial^2}{\partial r^2} + \frac{1}{r}\frac{\partial}{\partial r} + \frac{\partial^2}{\partial z^2} \right) \vartheta - \Delta h_v \frac{a}{\rho^s c_v}(c_T \vartheta + c_X \theta) + \frac{\Re}{\rho^s c_v}, \kappa = \frac{\lambda}{\rho^s c_v} \tag{4.66}$$

The boundary conditions of mass and heat transfer for convective drying ($\Re = 0$) of the cylinder with all surfaces open, as in Fig. 4.3a, are

$$\left. \nabla \theta \right|_{\substack{r=0 \\ z=0}} \cdot \mathbf{n} = 0, \; -\Lambda^l \nabla (c_T \vartheta + c_X \theta)|_{\partial B} \cdot \mathbf{n} = k_m(\mu^v|_{\partial B} - \mu_a)$$

$$\left. \nabla \vartheta \right|_{\substack{r=0 \\ z=0}} \cdot \mathbf{n} = 0, \; \lambda \nabla \vartheta|_{\partial B} \cdot \mathbf{n} = h_T(\vartheta_a - \vartheta|_{\partial B}) - \Delta h_v k_m(\mu^v|_{\partial B} - \mu_a) \tag{4.67}$$

In case of a partly insulated cylinder, as in Fig. 4.3b, we write instead

$$\frac{\partial \theta}{\partial r}\Big|_{r=0} = 0, c_X \frac{\partial \theta}{\partial z}\Big|_{z=0} = -c_T \frac{\partial \vartheta}{\partial z}\Big|_{z=0}$$

$$\frac{\partial \vartheta}{\partial r}\Big|_{r=0} = 0, \lambda \frac{\partial \vartheta}{\partial z}\Big|_{z=0} = h_T(\vartheta_a - \vartheta|_{z=0})$$

The initial conditions are

$$\theta(r, z, t)|_{t=0} = \theta_0 = \text{const}, \quad \vartheta(r, z, t)|_{t=0} = \vartheta_0 = \text{const} \tag{4.68}$$

where $\mu^v|_{\partial B}$ and μ_a denote the chemical potentials of vapor at the boundary surface and in the core of drying air, θ_0 and ϑ_0 are the initial moisture content and initial temperature, and k_m and h_T are the coefficients of convective vapor and heat exchange between the drying body and the air, respectively.

The chemical potential for vapor at the evaporation surface changes if the vapor/liquid interface recedes into the body. We introduce the position vector $\mathbf{r}_S$ that measures the distance of the vapor/liquid interface from the cylinder center.

We reformulate the convective term for vapor transfer on the right-hand side of boundary conditions 4.67 to

$$k_m(\mu^v|_{\partial B} - \mu_a) = k_m(\mathbf{r}_S)(\mu_n - \mu_a) \tag{4.69}$$

where $k_m(\mathbf{r}_S)$ is the convective vapor transfer intensity between the body and the surrounding air dependent on the spatial position vector of the vapor/liquid interface $\mathbf{r}_S$ in the cylinder; μ_n and μ_a denote the chemical potentials of vapor in the saturated air and the unsaturated one, respectively.

In our numerical procedure, we shall identify the position of the vapor/liquid interface by checking whether the moisture content at a given point in the cylinder has reached the critical value θ_{cr} or not. In this way, we can determine the displacement of the interface with time.

The chemical potentials of vapor in saturated and unsaturated states of air can be written in the form (see, e.g. Szarawara 1985)

$$\mu_n(p, T_n, x_n) = \mu^v(p, T_n) + R^g T_n \ln x_n$$
$$\mu_a(p, T_a, x_a) = \mu^v(p, T_a) + R^g T_a \ln x_a \tag{4.70}$$

where p is the total pressure, R^g is the gas constant, T_n and T_a are the temperatures, and x_n and x_a the mole fractions of vapor in saturated and unsaturated states of air, respectively.

Expanding the potential μ^v in a Taylor series with respect to the parameters of state and using suitable thermodynamic relations for the derivatives, we can write the driving force for vapor transfer as

$$\mu_n - \mu_a = 0.462 T_a \ln \frac{x_n}{x_a} - (7.36 - 0.462 \ln x_n)(T_n - T_a) \tag{4.71}$$

Such a form of driving force for vapor transfer allows us to control drying processes by suitable choice of drying parameters T_a and x_a. The temperature T_n can be considered as the wet-bulb temperature T_{wb} and x_n as the molar ratio of vapor in a saturated state of air at given drying parameters.

The numerical calculations of the temperature and the moisture content were performed for the following data:

$T_a = 343\,\text{K}, T_n \approx T_{wb} = 323\,\text{K}, H = 40\,\text{mm}, R = 40\,\text{mm}, x_a = 0.075,$

$G = 6.25 \times 10^8\,\text{Pa}, \Gamma = 10^9\,\text{Pa}, \kappa^{(T)} = 3 \times 10^{-8}\,\text{K}^{-1}, \kappa^{(X)} = 3 \times 10^{-5}$

$f_0 = 6 \times 10^{-7}\,\text{kg}\,\text{s}^2\,\text{m}^{-4}, \varpi a_0 = 2.5 \times 10^{-6}\,\text{kg}\,\text{s}\,\text{m}^{-5}, \rho_s = 260\,\text{kg}\,\text{m}^{-3},$

$\Lambda^1 = 6.04 \times 10^{-7}\,\text{kg}\,\text{s}\,\text{m}^{-3}, \kappa = 1.7 \times 10^{-3}\,\text{m}^2\,\text{s}^{-1}, c_T = 3.6\,\text{m}^2\,\text{s}^{-2}\,\text{K}, c_X = 10^{-3}\,\text{m}^2\,\text{s}^{-2},$

$c_v = 1.56 \times 10^5\,\text{J}\,\text{kg}^{-1}\,\text{K}^{-1}, k_m = 8.64 \times 10^{-5}\,\text{kg}\,\text{s}\,\text{m}^{-4}, h_T = 40\,\text{W}\,\text{m}^{-2}\,\text{K}^{-1}$

$\Delta h_v = 2000\,\text{kJ}\,\text{kg}^{-1}, X_A = 5\%, X_0 = 40\%, X_{cr} = 15\%.$

Figure 4.4 illustrates the predicted drying curve and temperature versus time. The temperature of the air was assumed to be $T_a = 343$ K. The wet-bulb temperature at the given drying conditions was $T_{wb} = 323$ K. The body temperature was calculated at 3 points of the cylinder: $(r = 0, z = 0); (r = R, z = 0); (r = R, z = H)$ (see Fig. 4.3a).

We see that the temperatures for individual points $(r = 0, r = R)$ in the middle plane $(z = 0)$ of the cylinder are different only in the heating period and in the

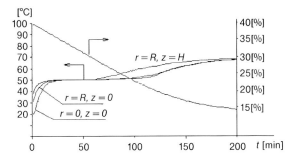

Fig. 4.4 Drying curve and temperature evolution in several points of the cylinder.

transient between the first and second drying periods. The temperature in the corner of the cylinder ($r = R, z = H$) starts to rise much earlier than that in the middle plane. Finally, the body temperature approaches the temperature of the drying air.

Figure 4.5 presents the distribution of temperature (isotherms) in the heating period (Fig. 4.5a), at the end of first period (Fig. 4.5b), and in the second period of drying (Fig. 4.5c). Because of symmetry, the results are presented only in one quarter of the cylinder (Fig. 4.3a).

We see that in the constant drying rate period (Fig. 4.5b) the temperature is constant in the entire cylinder and equal to the wet-bulb temperature. Figure 4.5b shows the end of the first and beginning of the second drying period to illustrate the fact that the temperature starts to increase first in the corner of the cylinder. In the falling-rate period (Fig. 4.5c) the temperature is distributed throughout the cylinder.

Figure 4.6 presents the distribution of moisture content at selected instants of time. The dark area represents the saturated region (above critical moisture content, $X^l > X_{cr}$) and the grey area represents the unsaturated region of the cylinder ($X^l < X_{cr}$). We see that the corner of the cylinder becomes dry first. The distributions of moisture content at 90 min (Fig. 4.6b) and 120 min (Fig. 4.6c) give an image of how

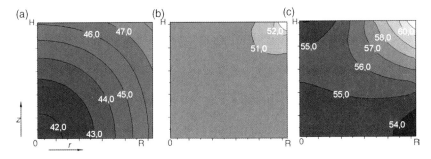

Fig. 4.5 Distribution of temperature [°C] in the cylinder:
(a) heating period, (b) end of first period, (c) second

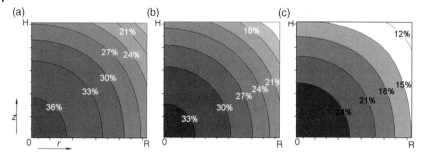

Fig. 4.6 Distribution of moisture content and retreat of the vapor/liquid interface: (a) 60 min, (b) 90 min, (c) 120 min.

the vapor/liquid interface recedes towards the interior of the cylinder in the course of drying. Note that in the present calculations the critical moisture content X_{cr} was assumed to be 15%, and that this value defines the vapor/liquid interface. The moisture content decreases continuously during drying in the whole body that is both in saturated and unsaturated regions.

We neglect the body forces in the numerical calculations of stresses because of the relatively small dimensions of the cylindrical sample and assume that the pore pressure at moderate drying conditions can be considered constant ($p^{(\text{por})} \approx \text{const}$). However, in the case of intensive microwave heating the pore pressure increases rapidly and the pressure gradient cannot be neglected.

In the cylindrical geometry equation 4.60 takes the form

$$G\nabla^2 u_r + \frac{\partial}{\partial r}[(G+\Gamma)\varepsilon - \gamma^{(T)}\vartheta - \gamma^{(X)}\theta] - G\frac{u_r}{r^2} = \frac{\partial p^{(\text{por})}}{\partial r}$$

$$G\nabla^2 u_z + \frac{\partial}{\partial z}[(G+\Gamma)\varepsilon - \gamma^{(T)}\vartheta - \gamma^{(X)}\theta] = \frac{\partial p^{(\text{por})}}{\partial z}, \varepsilon = \frac{\partial u_r}{\partial r} + \frac{u_r}{r} + \frac{\partial u_z}{\partial z}$$

(4.72)

where u_r and u_z are the displacements in the radial and longitudinal directions and ∇^2 denotes the Laplace operator in cylindrical coordinates (see Eq. 4.65 or Eq. 4.66).

The zero-valued stresses at the sample surface $\sigma_{rr}|_{r=R} = 0$, $\sigma_{zz}|_{z=H} = 0$ and the zero-valued displacements $u_r|_{r=0}$, $u_z|_{z=0}$ constitute the boundary conditions.

Figure 4.7 presents spatial distributions of circumferential stresses $\sigma_{\varphi\varphi}$ under the assumption that $\nabla p^{\text{por}} \approx 0$.

The circumferential stresses $\sigma_{\varphi\varphi}$ are tensional close the surface $r = R$, where the material shrinks first. The tensional stresses at the surface compress part of the interior of the cylinder so that the stresses are compressive in the cylinder core. The reasons for the sign change in this case are tensional stresses σ_{zz} on the lateral surface of the cylinder.

Figure 4.8 presents the time evolution of circumferential stresses $\sigma_{\varphi\varphi}$ in the center ($r = 0, z = 0$) and at the surface ($r = R, z = 0$) of the cylinder for elastic and viscoelastic material ($1/\tau = 1/\tau_0 = 10^{-3}\,\text{s}^{-1}$).

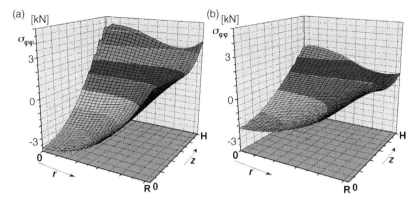

Fig. 4.7 Distribution of circumferential stresses $\sigma_{\varphi\varphi}$ in the cylinder cross section at (a) 60 and (b) 180 min drying time.

It is seen that the circumferential stresses for an elastic material are tensional at the surface and compressive in the core. The shape of the curves for elastic and viscoelastic cylinders is similar; however, the magnitude of the circumferential stresses for a viscoelastic cylinder is significantly smaller and, furthermore, they change their sign after some time of drying. This phenomenon is termed *stress reversal* and can be explained as follows: when the cylinder dries, the drier surfaces attempt to shrink but are restrained by the wet interior. The surfaces are stressed in tension and the interior in compression and large inelastic strains occur. Later, under the surface with reduced shrinkage, the interior dries and attempts to shrink, causing the stress state to reverse (Kowalski 2003; Milota and Quinglin 1994).

The smaller value of circumferential stresses for viscoelastic material follows from the relaxation phenomenon. Similar differences between stresses in elastic and viscoelastic cylinder can be noticed also for other components of the stress tensor.

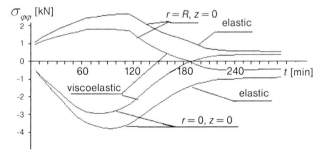

Fig. 4.8 Time evolution of circumferential stresses $\sigma_{\varphi\varphi}$ in the center ($r = 0, z = 0$) and at the surface ($r = R, z = 0$) of the cylinder for elastic and viscoelastic material.

4.7.2
Microwave Drying of a Kaolin Cylinder

For microwave drying we use similar equations as for convective drying with the difference that this time $\Re \neq 0$ and $\nabla p^{\text{por}} \neq 0$ if $T > T_{\text{cr}}$, and the heat flux is pointed from inside to outside. The change of heat flux direction is implemented by multiplying both sides of Eq. 4.67 by minus 1.

Equation 4.49 describes the microwave heat source, in which the distance of microwave propagation is expressed as

$$\boldsymbol{x} \cdot \boldsymbol{n} = \sqrt{[(R-r)\cos \varphi]^2 + [(H-z)\sin \varphi]^2} \tag{4.73}$$

where φ denotes the angle between coordinate r and the direction of microwave radiation.

Figure 4.9 illustrates the distribution of temperature in cylindrical samples after 20 minutes of convective heating or 3 min microwave heating (Kowalski, 2004a, 2005a,b).

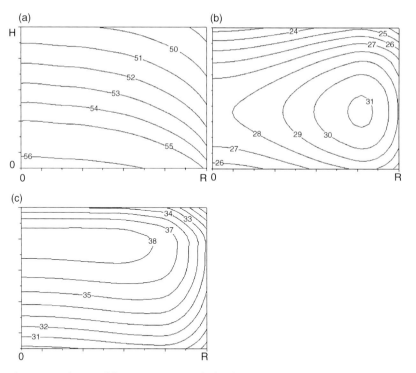

Fig. 4.9 Distribution of the temperature in cylindrical samples: (a) convective heating (20 min), (b) microwave heating (3 min, $\varphi = 0°$), (c) microwave heating (3 min, $\varphi = 45°$).

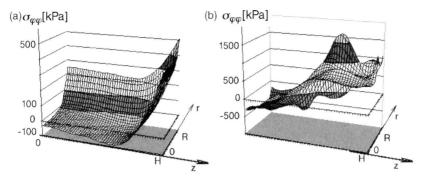

Fig. 4.10 Distribution of circumferential stresses in the cylinder by moderate and intensive microwave heating: (a) when $T < T_{cr}$; (b) when $T > T_{cr}$.

We see that the highest temperature in the cylinder dried convectively occurred in the middle point of the cylinder bottom (Fig. 4.9a). This is because this surface is closed for moisture outflow (see Fig. 4.3b).

It is known that in the case of microwave drying heat is generated inside the cylinder proportionally to the local amount of moisture content; however, due to the attenuation of microwaves with distance, the highest temperature appeared not in the center of the cylinder but somewhere on a circumference, dependent on the direction of microwave radiation. Figure 4.9b presents isotherms for horizontal microwave radiation ($\varphi = 0°$) and Fig. 4.9c for microwave radiation both horizontally and from the top ($\varphi = 45°$).

Figure 4.10a presents the spatial distribution of circumferential stresses in the cylinder by moderate microwave heating and Fig. 4.10b by very intensive microwave heating (Kowalski 2005a,b).

The stresses in Fig. 4.10a were generated when the temperature T is everywhere still below T_{cr}. In Fig. 4.10b, on the other hand, the temperature overcomes in some places T_{cr}. Note that the stresses grow radically when $T > T_{cr}$. Respective locations experience a rapid increase of pore pressure due to intensive phase transition. It is obvious that such a rapid increase of pore pressure may cause sample damage (a kind of explosion), and it actually does. This is seen in results that were carried out in laboratory conditions (Kowalski 2005a,b; Itaya et al. 2004).

Figure 4.11 presents photos of the samples after drastic drying.

We see that the samples in convective and microwave drying were destroyed in different ways. The crack of the sample dried convectively was rather superficial and arose at the top of the cylinder where the tensional stresses are maximal. The sample in microwave drying was damaged volumetrically from the inside as a result of high pore pressure. The crack in this case arose at the area where the temperature of the cylinder had maximal values (see Fig. 4.9c).

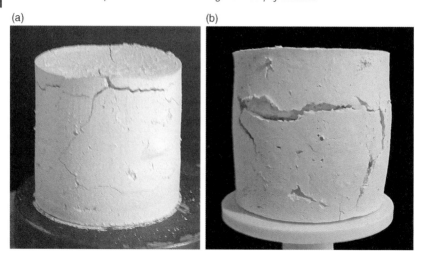

Fig. 4.11 Kaolin samples destroyed in drying: (a) convective drying, (b) microwave drying.

4.8
Final Remarks

An outline of thermohydromechanical theory of convective and microwave drying was developed, in which deformations and stresses along with the heat- and mass-transfer processes were described. The established theory is physically nonlinear, which follows from the dependence of material coefficients on the moisture content and temperature (the variables that change themselves during drying) as well as from the irreversible deformations (see physical relation 4.30). Gibbs' equation 4.18 is also the basis for developing of a finite deformation drying theory (geometrically nonlinear), which can be applied for description of highly shrinking materials, as e.g. sol-gel materials. To this aim, the Couchy–Green strain tensor E should be applied instead of the tensor $\varepsilon^{(r)}$ of small strains. Developing such a theory was beyond the scope of this chapter.

Much effort has been devoted lately to determination of the material coefficients appearing in the theory outlined above and to the experimental validation of the theory for chosen materials, e.g. wood (Kowalski 2004b; Kowalski and Smoczkiewicz 2004; macaroni dough (Kowalski and Mielniczuk 2006); kaolin (Kowalski et al. 2007). The acoustic emission method enabling monitoring *online* the development of mechanical effects in dried materials is very helpful in these studies (Kowalski et al. 2001).

Acknowledgments

This work was carried out as a part of the research project No 3 T09C 030 28, sponsored by the Polish Ministry of Education and Science.

Additional Notation used in Chapter 4

a	phase-transition efficiency [–]
c_T	coefficient of thermodiffusion [m² s⁻² K⁻¹]
c_X	coefficient of diffusion [m² s⁻²]
k_m	coefficient of convective vapor exchange [kg s m⁻⁴]
h_T	coefficient of convective heat exchange [W m⁻² K⁻¹]
e_{ij}	strain deviator [–]
E	electric intensity [–]
f	free energy [J/kg]
$f(X^1)$	moisture transport parameter [kg s² m⁻⁴]
G, Γ	Lame constants [MPa]
$3K = 2M + 3\Gamma$	bulk modulus [MPa]
$\Re$	volumetric heat supply [W m⁻³]
s, s^α	entropy, entropy of α-constituent [J kg⁻¹ K⁻¹]
s_{ij}	stress deviator [Pa]
u	internal energy [J kg⁻¹]
$\mathbf{J}^\alpha$	mass flux of α-constituent [kg m⁻² s⁻¹]
x_a, x_n	mole fractions of vapor in air [–]

Greak Symbols

α	coefficient of microwave damping [1 m⁻¹]
β	microwave number [1 m⁻¹]
δ	loss parameter [–]
$\varepsilon^*(\varepsilon', \varepsilon'')$	magnetic perittivity [F m⁻¹]
ϕ^α, ϕ	volume fraction, porosity [–]
η_v, κ_v	viscous shear and bulk modulus [Pa s]
$\kappa^{(T)}$	coefficient of thermal expansion [1 K⁻¹]
$\kappa^{(X)}$	coefficient of humid expansion [–]
$\gamma^{(T)} = 3K\kappa^{(T)}$	thermal modulus [MPa K⁻¹]
$\gamma^{(X)} = 3K\kappa^{(X)}$	humid expansion modulus [MPa]
σ^*	electric conductivity [Ω⁻¹]
$\hat{\rho}^\alpha$	rate of α-constituent mass change by phase transitions [kg m⁻³ s⁻¹]
μ^α	chemical potential of α-constituent [J kg⁻¹]
$\mu^*(\mu', \mu'')$	magnetic permeability [H m⁻¹]
η, η^*	viscosity [Pa s]
ϖ	phase-transition coefficient [kg s m⁻⁵]
$\vartheta = T - T_r$	relative temperature [°C]
$\theta = X^1 - X^1_r$	relative moisture content [–]
Λ^α	mass transport coefficient of α-constituent [kg s m⁻³]

References

Caircross, R. A., Schunk, P. R., Chen, K. S., Prakash, S. S., Samuel, J., Hurt, A. J., Brinker, C. J., 1996. Drying in deformable partially saturated porous media: sol-gel coatings. Sandia Report, Sand96–2149, UC-905.

Di, P., Chang, D. P. Y., Dwyer, H. A., 2000. Heat and mass transfer during microwave steam treatment of contaminated soils. *J. Environ. Eng.* **12**: 1108–1115.

Feng, H., Tang, J., Cavalieri, R. P., Plumb, O. A., 2001. Heat and mass transport in microwave drying of porous materials in a spouted bed. *AIChE J.* **47**(7): 1499–1512.

Garcia, H., Bueno, J. L., 1998. Improving energy efficiency in combined microwave-convective drying. *Dry. Technol.* **16**(1–2): 123–140.

Itaya, Y., Uchiyama, S., Hatano, S., Mori, S., 2004. Drying enhancement of clay slab by microwave heating. *Proceedings of IDS 2004*, Vol. A: 193–200.

Kneule, F., 1970. Drying. Arkady, Warszawa, p. 363 (in Polish).

Kowalski, S. J., Mielniczuk, B., 2006. Drying induced stresses in macaroni dough. *Dry. Technol.* **24**: 1093–1099.

Kowalski, S. J., Rybicki, A., 2004. Qualitative aspects of convective and microwave drying of saturated porous materials. *Dry. Technol.* **22**(5): 1173–118.

Kowalski, S. J., Smoczkiewicz, A., 2004. Identification of wood destruction during drying, *Maderas: Ciencia Tecnologia*, **6**(2): 133–144.

Kowalski, S. J., 2003. Thermomechanics of Drying Processes. Springer-Verlag, Berlin, Heidelberg, New York, p. 365.

Kowalski, S. J., Banaszak, J., Rajewska, K., 2001. Acoustic emission as a tool for monitoring of fracture processes in dried materials, *Diagnostyka* **27**: 95–100, (in Polish).

Kowalski, S. J., Moliński, W., Musielak, G., 2004b. Identification of fracture in dried wood based on theoretical modeling and acoustic emission. *Wood Sci. Technol.* **38**: 35–52.

Kowalski, S. J., Musielak, G., Banaszak, J., 2007. Experimental validation of the heat and mass transfer model for convective drying, *Dry. Technol.*, **25**(107–121) (special issue on the occasion of 25th anniversary of the Journal).

Kowalski, S. J., Rajewska, K., Rybicki, A., 2004a. Mechanical effects in saturated capillary-porous materials during convective and microwave drying. *Dry. Technol.* **22**(10): 2291–2308.

Kowalski, S. J., Rajewska, K., Rybicki, A., 2005a. Physical Fundamentals of Microwave Drying. Publishers Poznań University of Technology, p. 113 (in Polish).

Kowalski, S. J., Rajewska, K., Rybicki, A., 2005b. Stresses generated during convective and microwave drying. *Dry. Technol.* **23**(9–11): 1875–1893.

Lade, P. V., de Boer, R., 1997. The concept of effective stress for soil, concrete and rock. *Geotechnique* **47**(1): 61–78.

Lykov, A. W., 1968. Theory of Drying, ENERGIA-Moscow, p. 470 (in Russian).

Milota, M. R., Quinglin, W., 1994. Resolution of the stress and strain components during drying of softwood. *Proceedings of the 9th International Drying Symposium*, Gold Coast, Australia.

Nowacki, W. K. (ed.), 1969. Progress in Thermoelasticity. PWN-Polish Scientific Publishers: Warszawa, p. 235.

Perre, P., Turner, I. W., 1996. A complete coupled model of the combined microwave and convective drying of softwood in an oversized waveguide. *Proceedings of IDS 1996*, Vol. A, 183–194.

Perzyna, P., 1966. Theory of Viscoplasticity. PWN-Polish Scientific Publishers: Warszawa, p. 235 (in Polish).

Ratanadecho, P., Aoki, K., Akahori, M., 2001. Experimental and numerical study of microwave drying in unsaturated porous material. *Int. Comm. Heat Mass Transfer* **28**(5): 605–616.

Ratanadecho, P., Aoki, K., Akahori, M., 2002. Influence of irradiation time, particle sizes and initial moisture content during microwave drying of multi-layered capillary porous materials. *J. Heat Transfer* **124**(2): 1–11.

Sanga, E. C. M., Mujumdar, A. S., Raghavan, G. S. V., 2002. Simulation of convection-microwave drying for shrinkage material. *Chem. Eng. Proc.* **41**: 487–499.

Scherer, G. W., 1990. Theory of drying. *J. Am. Ceram. Soc.* **73**(1): 1–14.

Szarawara, J., 1985. Chemical Thermodynamics. WNT-Scientific Technological Publishers: Warszawa, p. 550 (in Polish).

Zhang, D., Mujumdar, A. S., 1992. Deformation and stress analysis of porous capillary bodies during intermittent volumetric thermal drying. *Dry. Technol.* **10**(2): 421–443.

5
CFD in Drying Technology – Spray-Dryer Simulation
Stefan Blei, Martin Sommerfeld

5.1
Introduction

Numerous different types of convective drying processes are used in practice. The most complex of them are those involving multiphase flows such as fluidized bed, pneumatic conveying and spray drying. The design, optimization and scale-up of these multiphase drying processes is rather sophisticated due to the large number of elementary processes involved, such as atomization, particle transport, evaporation of droplets, and interactions between particles and droplets. The presence of a dispersed phase also results in a number of additional nondimensional parameters that complicate lay-out and scale-up. Since a theoretically based design is difficult, pilot-scale experiments are usually performed in a trial and error approach. This is of course very time consuming and costly.

Over the past 15 years, however, the rapid development in computational fluid dynamics (CFD) and the ever-increasing computational power (operational speed and storage size) at decreasing costs make it more and more feasible to numerically calculate complex industrial flow processes with a three-dimensional resolution.

5.1.1
Introduction to CFD

Computational fluid dynamics for turbulent flows may be performed on different levels depending on the degree of turbulence modeling. The following methods are being used:
- direct numerical simulations (DNS),
- large eddy simulations (LES),
- methods based on Reynolds-averaged Navier–Stokes (RANS).

Direct numerical simulations require the solution of the time-dependent three-dimensional Navier–Stokes equations where the numerical grid needs to resolve all scales of the flow down to the dissipation scale of turbulence. This implies

Modern Drying Technology. Edited by Evangelos Tsotsas and Arun S. Mujumdar
Copyright © 2007 WILEY-VCH Verlag GmbH & Co. KGaA. All rights reserved.
ISBN: 978-3-527-31556-7

that the numerical grid must be smaller than the Kolmogorov length scale given by:

$$\eta = \left(\frac{\nu}{\varepsilon}\right)^{1/4} \tag{5.1}$$

wherein ν is the kinematic viscosity of the fluid and ε the dissipation rate.

The ratio of the largest scales, L, in the flow to the Kolmogorov length scale, η, is dependent on the flow Reynolds number in the following way:

$$\frac{L}{\eta} \approx \mathrm{Re}^{3/4} \tag{5.2}$$

Hence, with increasing Reynolds number the Kolmogorov scale decreases, requiring also a decrease of grid size. Therefore, the number of grids increases proportionally to $\mathrm{Re}^{9/4}$. As a result of the limited storage even of advanced computers, the maximal Reynolds number to be considered in DNS is about 5000. Moreover, only simple flows such as homogeneous isotropic turbulence (Shiriani et al. 1981), simple shear flows (Rogers and Moin 1987), and channel flows (Kasagi and Shikazono 1995) can be simulated, where periodic boundary conditions may be applied to limit the computational domain.

Large eddy simulations also require the time-dependent solution of the three-dimensional governing equations, however only the larger scales of the flow are resolved by the numerical grid. Turbulent scales smaller than the grid size are modeled by a so-called subgrid-scale model. Therefore, LES is computationally less expensive than DNS and may be applied to flows with higher Reynolds numbers and in more complex geometries. One of the best known subgrid-scale models is that developed by Smagorinsky (1963) that uses the mixing length approach. For complex and/or high Reynolds number flows the standard Smagorinsky model is not very accurate and better subgrid scale models are needed. A promising approach to subgrid scale modeling is a dynamic procedure, where the smallest scales resolved in the velocity field predicted by LES are used to generate the model. Developments in LES and especially subgrid scale modeling have been summarized by Ferziger (1996). A summary on the capabilities of LES for predicting different flow types is given in the book by Lesieur et al. (2005). Both methods, LES and DNS, have become important tools for fundamental turbulence research and are also used to derive and improve turbulence closure for the Reynolds-averaged equations.

Numerical methods for the calculation of technical turbulent flows are based on the Reynolds-averaged conservation equations in connection with an appropriate turbulence model to close the set of equations. Since the momentum equations are also named Navier–Stokes equations this modeling approach is referred to as the RANS approach. The basic equations (also-called Reynolds equations) are obtained by decomposing the instantaneous properties (e.g. velocities) into a mean value and a fluctuating component:

$$\phi(t) = \overline{\phi} + \phi'(t) \tag{5.3}$$

and time averaging the resulting equations. The set of equations obtained by this procedure has a similar form to the Navier–Stokes equations but contains mean values and additionally averaged values of the fluctuating components, the Reynolds stresses. Hence, all the unsteadiness of the flow resulting from turbulent eddies is removed. In order to close the equations empirical assumptions are needed to calculate the Reynolds stresses. The different approaches for turbulence modeling are:
- algebraic turbulence models,
- one-equation models,
- two-equation models, e.g. k–ε turbulence model,
- Reynolds-stress models.

In all the models it is assumed that turbulence increases the viscosity. Algebraic turbulence models, i.e. mixing-length models, are the simplest way to close the equations and do not require the solution of additional transport equations. The turbulent viscosity is described by algebraic functions in dependence of averaged flow properties and the wall distance. Most common are two-equation turbulence models, such as the well-known k–ε turbulence model (Launder and Spalding 1974). This class of models requires the solution of two additional differential equations that describe the transport of turbulent properties, e.g. the turbulent kinetic energy, k, and the associated dissipation rate, ε. The turbulent viscosity is calculated as a function of k and ε. There exist a number of different versions of two-equation turbulence models that have been proposed throughout the years. A multiscale turbulence model that may account for changes in the shape of the energy spectrum in a flow was for example proposed by Schiestel (1987). For complex flows and highly anisotropic turbulence, Reynolds stress models have been developed to allow for reliable predictions of the fluctuating components (e.g. Gibson and Launder 1978). Such a model requires the solution of six additional transport equations for the Reynolds stresses. The reader is referred to review articles for example by Rodi (1988) and Launder (1990) on the development of turbulence modeling for more details. In recent years turbulence models have been considerably refined and improved by comparisons with direct numerical simulations and large eddy simulations. A description of discretization methods and procedures for solving the fluid dynamic conservation equations together with some applications is given in the book of Ferziger and Peric (2002).

Nowadays, several commercial codes are available and may be applied to a wide range of technical and industrial external as well as internal flow problems. The application of CFD to convective drying processes started in the late 1970s and the early 1980s, initially mostly based on the two-dimensional conservation equations.

In the following years up to the present, several key issues of applying CFD to drying emerged. They are related to either process performance or product properties. The following aspects belong to the first group:
- the scale-up of dryers,
- the evaluation of alternative dryer designs,

- the prediction and avoidance of wall deposits (product and process related).

The second group refers to:
- The prediction of gas-flow patterns and particle trajectories (especially for spray-drying processes) enabling the reconstruction of particle history in a dryer. The main results of such investigations are residence times of particles and the drying history.
- The prediction of particle sizes and drying states of the powder leaving the dryer, including quality aspects such as heat damage and agglomeration.

5.1.2
Introduction to Multiphase Flow Modeling

Over the past 15 years computational fluid dynamics (CFD) has been significantly improved with regard to applications in chemical, process and food engineering. Therefore, CFD is increasingly used in these industries for process analysis, optimization and design. Many of the processes considered involve multiphase flows in complex geometries that may be accompanied by heat and mass transfer as well as chemical reactions. The present chapter focuses on modeling and numerical computations of spray dryers. Two-phase flows occurring in such an apparatus can be regarded as dispersed and dilute, since the spray droplet behavior is governed by aerodynamic transport combined with the evaporation of the solvent from the droplets. Industrial flows of this kind are mostly computed based on the Reynolds-averaged Navier–Stokes equations in connection with an appropriate turbulence model, such as the k–ε turbulence model or full Reynolds-stress models. Dispersed multiphase flows are commonly described by two approaches, namely the two-fluid or Euler–Euler approach or the Euler–Lagrange method. In order to account for the interaction between phases, i.e. momentum exchange and heat and mass transfer, the conservation equations have to be extended by appropriate coupling terms.

In the two-fluid approach (often named the Euler–Euler approach) both phases (i.e. the continuous and the dispersed phase) are considered as interpenetrating continua. Hence, properties such as particle concentration and particle phase velocities are also considered as continuous properties, which are averaged over the control volumes (i.e. computational cells). Additionally, the interfacial transfer terms for momentum, mass, and/or energy require averaging over the computational cells. Especially in turbulent flows the closures of the dispersed phase Reynolds-stresses and the fluid–particle interaction terms are associated with sophisticated modeling approaches (see, e.g., Rizk and Elghobashi (1989) and Simonin et al. 1993). Hence, the resulting Reynolds-averaged equations have similar forms for both phases by accounting for the interfacial transfer. The consideration of a particle-size distribution, however, requires the solution of a set of conservation equations for each size class to be considered. Though the computational effort increases with the number of size classes, this method may be preferable for dense dispersed two-phase flows, as for example found in fluidized beds (Balzer and Simonin 1993; Enwald et al. 1996). A detailed description of the continuum modeling of turbulent reactive particulate flows on the basis of the probability density function kinetic equation is provided by Simonin (2000).

The Euler–Lagrange approach is only applicable to dispersed two-phase flows and accounts for the discrete nature of the individual particles. The dispersed phase is modeled by tracking a large number of particles through the flow field. To this purpose the equations of motion are solved by taking into account the relevant forces acting on the particle (Sommerfeld 2000, 2005) and heat as well as mass transfer, if required (Crowe et al. 1998). Generally, the particles are considered as point particles, i.e. their finite dimension is neglected and the flow around the individual particles is not resolved. Since the number of real particles in a flow system is usually too large to allow tracking of all particles, the trajectories of computational particles (i.e. parcels) that represent a number of real particles with the same properties (i.e. size, velocity and temperature) are calculated. In stationary flows a sequential tracking of the parcels may be adopted, while in unsteady flows all parcels need to be tracked simultaneously on the same time level. Local average properties such as dispersed-phase concentration and velocity are obtained by ensemble averaging. Statistically reliable results for each computational cell require the tracking of typically between 10 000 and 100 000 parcels, depending on the considered flow. The advantage of this method is that physical effects influencing the particle motion, such as particle–turbulence interaction, particle–wall collisions, collisions between particles and agglomeration can be modeled on the basis of physical principles. Moreover, a particle-size distribution may be easily considered by sampling the size of the injected particles from a given distribution function.

Essential for the Euler–Lagrange approach is the coupling between the fluid-flow and particle-phase calculations. The so-called one-way coupling implies that the particles are tracked in the previously calculated flow field by sampling their average properties. Consideration of the influence of the particles on the flow field is denoted by two-way coupling, which was first introduced by Crowe et al. (1977) in the so-called ''particle-source-in-cell'' approach. Here, particles are tracked in the flow field predicted by the flow solver. The influence of the dispersed phase on the gas phase was taken into account by sampling source terms in the computational cells of the domain considered. The source terms are sampled for kinetic interaction as well as for heat and mass transfer across the phase boundaries. A converged solution of the coupled system is obtained by successive computation of flow field and particle phase. Normally an underrelaxation procedure is required for the source terms in order to avoid divergence (Kohnen et al. 1994). The number of coupling iterations depends on the degree of coupling between the phases and the underrelaxation factor. However, convergence problems may still be encountered for very dense two-phase systems. The term four-way coupling is used when, in connection with two-way coupling, interparticle collisions are also taken into account. Hence, the particle-phase properties are not only modified by the changing flow field, but also by modification of the particle-phase velocities through interparticle collisions (see e.g. Lain et al. 2002).

Due to the changing droplet size and size distribution as a result of droplet evaporation and agglomeration the Euler–Lagrange approach is most suitable for spray-dryer computations.

5.1.3
State-of-the-art in Spray-dryer Computations

The scale-up of spray dryers using CFD methods has been investigated by Oakley (1994). He pointed out the limited value of similarity numbers for scale-up purposes due to the very different flow dynamics found in spray dryers of different size. As a result of his considerations he discussed the limits of traditional, empirically based approaches to predict gas-flow patterns as well as the opportunities and limits of the CFD approach. The latter has enormous advantages for the exploration of a greater variety of unknown dryer geometries. However, according to Oakley (1994) one has to know about difficulties in acquiring data for the definition of boundary and inlet conditions (for example for swirling flows and initial droplet velocity and size distributions) and the limitations of the models used for the CFD calculations. The boundary conditions chosen have substantial influence on the process behavior predicted.

Huang et al. (2003) carried out a number of CFD calculations to explore different spray-dryer chamber designs. The purpose of the investigation was to improve the heat- and mass-transfer performance per volume of the dryer. They kept the air inlet and the initial droplet conditions constant for investigating the drying behavior of a cylinder-on-cone, a conical, an hour-glass and a lantern geometry. Huang et al. (2003) faced significant problems related to the definition of boundary conditions for the 2-dimensional CFD calculations and their influence on the result of the calculations. In this case, the authors assumed sticking of all particles colliding with the wall. The authors could not achieve a clear, general solution leading to a recommendation as to which of the geometries should preferably be used. They pointed out the strong correlation between the optimal chamber geometry and the properties of the feed, the atomizer type, the drop-size distribution and other conditions.

Southwell et al. (2001) used the CFX-5 software to explore the possibilities of influencing the gas-flow pattern at the inlet of a dryer by geometrical adjustments. Air-flow patterns at the inlet of spray dryers can be uneven, which can disturb the dryer performance significantly. By comparing experiments and 3-dimensional CFD results they showed that an equalization of the inlet gas-flow patterns is possible.

Langrish and Zbicinski (1994) investigated the wall deposition rates of droplets in a cocurrent spray-dryer chamber. They varied the conditions of inlet gas flow (swirling, nonswirling) and compared the wall deposition rates predicted by the CFD-calculations with that of experiments in a small pilot-scale spray dryer. Ignoring the fact that dry particles bounce from the walls, they noted that turbulence enhancement and a large spray cone angle reduced wall deposition of droplets for the spray dryer considered. The positive influence of turbulence enhancement on the intensity of the drying process was confirmed by Southwell et al. (1999) and Zbicinski et al. (1996). The increased turbulence of the gas flow leads, on average, to increased values of relative velocity between droplets and gas, followed by larger mass and heat transfer rates. At the same time, the residence time of droplets in the swirling gas jet downstream of the tangential gas inlet is increased, compared to a nonswirling jet. Hence, droplets dry faster and the probability that sticky droplets hit the walls is reduced. These trends have been confirmed experimentally and by CFD results.

In spray drying, the final product properties are very important, if not the most important feature to be considered. For a long time the particle movement and the drying history served as criteria for the estimation of drying rate and the state particles leave the dryer (see discussion above). For this reason, the application of CFD on spray dryers was concentrated on the prediction of gas-flow patterns, turbulence intensity and the resulting particle trajectories. Droplet drying models of different complexity have also been applied (Farid 2003, Wijlhuizen et al. 1979, Sano and Keey 1982 and others).

However, from the beginning of CFD calculations applied to spray drying, it has also been recognized that particle collisions and agglomeration play a crucial role for the final powder-size distribution and other important particle properties. Nevertheless, the ability to predict particle collisions, the outcome of collisions and the properties of the particles created by such collisions was (and in some points still remains) limited. The accuracy of predicting particle collisions clearly depends on the accuracy of the CFD solution achieved for the flow field and on which fluid dynamic forces are considered to act on the particles.

Nijdam et al. (2004) and Guo et al. (2003) used the approach of O'Rourke (1981) to predict the occurrence and outcome of droplet collisions in a spray. For this purpose they used efficiencies that summarize the physical effects of impact efficiency (a small particle can be transported by the relative flow field around a larger one), and coalescence efficiency (the probability of two particles coming into contact with each other to coalesce). It seems that Nijdam et al. (2004) and Guo et al. (2003) based their coalescence efficiencies on information reported on collisions of water and hydrocarbons (Qian and Law 1997). The present authors believe this leads to mistakes in predicting particle-size distributions in spray dryers, due to the fact that water and hydrocarbons are fluids dominated by surface-tension forces, but fluids in spray drying are usually more viscous, especially if they reach higher drying states.

Verdurmen et al. (2004) suggested a new overall approach to predict agglomeration in CFD calculations of spray dryers. Here, the impact efficiency and the coalescence or agglomeration are modeled separately. The impact efficiency is based on work by Schuch and Löffler (1978). The result of a collision is modeled in dependence of the drying state and the resulting material properties of the droplet, which are density, dynamic viscosity and surface tension. It is suggested that the collision events are classified into three categories: collisions of surface-tension-dominated droplets, collisions of viscous droplets and collisions of dry particles. More details on the modeling strategies are given in the present work.

It should be pointed out that widely used heat- and mass-balance models, equilibrium-based models and simple rate models are empirical by nature and treat the spray dryer mostly as a "black box". These approaches have advantages if quick engineering approximations are required and will continue to be of value. They quickly reach their limits, though, if details of process performance or product quality are required. To solve such problems, one has to conduct sophisticated experiments or use CFD. This means that the basic processes taking place in the dryer (atomization, particle movement, particle collision, drying of droplets and evaporation of solvent) have to be discretized in space and time in order to enable access to the physical and chemical mechanisms acting on the individual particles.

In the recent past, large advances have been made concerning the prediction of flow patterns, particle trajectories and particle collisions in spray dryers (Langrish and Fletcher 2003; Oakley 2004; Verdurmen et al. 2004). The description of the evaporation process can also be handled in a satisfying way, despite the fact that some assumptions about sphericity of particles and diffusion processes seem to leave room for further improvements. The knowledge of particle interactions in spray dryers is still limited, especially for droplets with a high solid content. Here, further experimental investigation seems to be required, where the difficulty of handling highly viscous products has to be overcome.

At the present stage of development, the atomization process cannot be incorporated satisfactorily into the overall CFD approach of the Euler–Lagrange method. Nowadays, the flow inside the nozzles and in the dense flow region downstream of the nozzle is commonly modeled with the Euler–Euler approach. Volume-of-fluid and direct numerical simulation methods are also in development to analyse the break-up process. For both approaches, the issue of predicting the correct droplet size and velocity distributions arises. In spite of some advances achieved by coupling so-called "maximum-entropy" models with the Euler approach (for example, Platzer and Sommerfeld 2003; Dumouchel and Malot 1999), the problem still remains to be solved. Therefore, the initial conditions needed for Lagrangian particle tracking still have to be gained by measurement.

Nevertheless, the CFD approach offers possibilities for improvement of the models and calculation procedures towards a specific interpretation of the results regarding product properties and process performance. In future, the direct results of a CFD calculation (as particle-size distributions, drying states and related material properties) might be extended by structural information about the morphology of particles and agglomerates. Furthermore, these results could be used to predict the functional behavior of the powder such as wettability, dissolution and flowability. These goals might be reached by investigating the underlying micromechanisms either by experiment or by direct numerical simulation (DNS) methods, and casting the findings into models which are compatible with the Euler–Lagrange approach.

5.2
The Euler–Lagrange Approach: an Extended Model for Spray-dryer Calculations

In the present section the Euler–Lagrange approach, as applied to the calculation of a spray dryer, is introduced. The focus is not related to an improvement of the

5.2.1
Fluid-phase Modeling

The three-dimensional velocity field of the flow inside the spray dryer is calculated by solving the Reynolds-averaged Navier–Stokes equations in combination with a k–ε turbulence model. Here, the standard k–ε turbulence model based on the work of Launder and Spalding (1974) is applied.

The following expression represents the general transport equation used for solving the actual flow problem:

$$\frac{\partial}{\partial x}(\rho u \phi) + \frac{\partial}{\partial y}(\rho v \phi) + \frac{\partial}{\partial z}(\rho w \phi) - \frac{\partial}{\partial x}\left(\Gamma \frac{\partial \phi}{\partial x}\right) - \frac{\partial}{\partial y}\left(\Gamma \frac{\partial \phi}{\partial y}\right)$$
$$- \frac{\partial}{\partial z}\left(\Gamma \frac{\partial \phi}{\partial z}\right) = S_\phi + S_{\phi,p} + S_{\phi,p,ev} \quad (5.4)$$

S_ϕ, $S_{\phi,p}$ and $S_{\phi,p,ev}$ are the source terms of the continuous-phase and the particle-phase source terms due to momentum transfer and droplet evaporation. Γ represents a general transport coefficient, dependent on the variable considered, where μ and μ_t are the molecular and the turbulent dynamic viscosity of the fluid. The source terms and the diffusion coefficient as applied in the present calculations are summarized in Tab. 5.1 and Tab. 5.2. x, y and z are the coordinates of the Cartesian frame of reference on which Eq. 5.4 is based, ρ is the fluid density, u, v, w are the three velocity components, and ϕ is a symbol representing the flow variable

Tab. 5.1 Diffusion term and continuous-phase source terms of the general transport equation, modified for the k–ε turbulence model

ϕ	Γ	S_ϕ
1	0	0
u_i	$\mu + \mu_t$	$\frac{\partial}{\partial x_j}\left(\Gamma_{u_i}\frac{\partial u_j}{\partial x_i}\right) - \frac{\partial P}{\partial x_i} + \rho g_i$
$c_{p,g}T$	$\frac{\mu}{Pr} + \frac{\mu_t}{Pr_t}$	$-2\mu\left(\frac{\partial u_i}{\partial x_i}\right)^2 - \mu\left[\left(\frac{\partial u_1}{\partial x_2}+\frac{\partial u_2}{\partial x_1}\right)^2 + \left(\frac{\partial u_1}{\partial x_3}+\frac{\partial u_3}{\partial x_1}\right)^2 + \left(\frac{\partial u_2}{\partial x_3}+\frac{\partial u_3}{\partial x_2}\right)^2\right]$
Y	$\frac{\mu}{Sc} + \frac{\mu_t}{Sc_t}$	0
k	$\mu + \frac{\mu_t}{\sigma_k}$	$\mu_t\left(\frac{\partial u_i}{\partial x_j}+\frac{\partial u_j}{\partial x_i}\right)\cdot\frac{\partial u_i}{\partial x_j} - \rho\varepsilon$
ε	$\mu + \frac{\mu_t}{\sigma_\varepsilon}$	$\frac{\varepsilon}{k}\cdot\left(C_1\mu_t\left(\frac{\partial u_i}{\partial x_j}+\frac{\partial u_j}{\partial x_i}\right)\cdot\frac{\partial u_i}{\partial x_j}-C_2\rho\varepsilon\right)$

Tab. 5.2 Dispersed phase source terms of the general transport equation, modified for the k–ε turbulence model

ϕ	$S_{\phi,p}$	$S_{\phi,p,ev}$
1	0	$-\sum_p \dfrac{\dot{M}_{p,ev} \cdot N_p}{V_{i,j,k}}$
u_i	$-\sum_p \dfrac{\dot{M}_p \cdot N_p}{V_{i,j,k}} \cdot \left[\left(u_{p,i}^{t+\Delta t} - u_{p,i}^t\right) - g_i \cdot \Delta t\right]$	$\sum_p \dfrac{\dot{M}_{p,ev} \cdot N_p}{V_{i,j,k}} u_{p,i}$
$c_{p,g}T$	$-\sum_p \dfrac{N_p \cdot \dot{M}_p}{V_{i,j,k}} \cdot c_p \cdot \left(T_p^{t+\Delta t} - T_p^t\right)$	$\sum_p \dfrac{\dot{M}_{p,ev} \cdot N_p}{V_{i,j,k}} \cdot \left(c_{P,v} \cdot T_p - \Delta h_{ev}\right)$
Y	0	$\sum_p \dfrac{\dot{M}_{p,ev} \cdot N_p}{V_{i,j,k}}$
k	$\overline{u_i S_{u_i,p}} - \overline{u}_i \cdot \overline{S}_{u_i,p}$	$\overline{u_i s_{u_i,ev}} - \overline{u}_i \cdot \overline{S}_{u_i,ev} + \dfrac{1}{2}(\overline{u_i u_i} \overline{S}_{p,ev} - \overline{u_i u_i S_{p,ev}})$
ε	$c_{\varepsilon,3}^{k-\varepsilon} \cdot \dfrac{\varepsilon}{k} \cdot S_{k,p}$	$c_{\varepsilon,3}^{k-\varepsilon} \cdot \dfrac{\varepsilon}{k} \cdot \overline{S}_{k,p,ev}$

considered (see Tab. 5.1). The reader is referred to well-established standard literature discussing the derivation of the general transport equation from the conservation equations and the rules of mass and energy conservation (Landau and Lifschitz 1991; Rotta 1972).

The particle source terms have to be considered for a spray-dryer calculation. Otherwise highly inaccurate results are obtained for the flow pattern and for particle trajectories. The agglomeration model can not be applied properly and the final product's size distribution can not be predicted without considering the source terms, because they considerably modify the gas flow. Hence, a two-way coupling procedure for momentum, mass and thermal-energy transfer between the continuous and dispersed phase is applied. Also taking into account particle collisions, a so-called 4-way coupling calculation is carried out.

In the following tables, N_p is the number of particles considered, $u_{p,i}$ are the particle's velocity components, $\dot{M}_p$ is the mass flow rate of particles, $\dot{M}_{p,ev}$ the evaporated (or condensed) mass flow rate, $V_{i,j,k}$ the volume of the cell, g_i the gravitation vector, Δt the time step size, c_p the specific heat capacity of the particle material, Δh_{ev} the specific heat of evaporation. The volume-averaged mass transfer caused by evaporation is denoted by $S_{p,ev}$ in Tab. 5.2 (Chrigui and Sadiki 2004). The constant $c_{\varepsilon,3}^{k-\varepsilon}$ is chosen to be 1.1, as suggested by Squires and Eaton (1992).

Further variables and constants in Tab. 5.1 and Tab. 5.2 are: $\mu_t = C_\mu \cdot \rho \cdot k^2/\varepsilon$, $C_\mu = 0.09$, $C_1 = 1.44$, $C_2 = 1.92$, $\sigma_k = 1.0$, $\sigma_\varepsilon = 1.3$.

The general transport equation for a three-dimensional flow problem has to be solved by numerical procedures, due to the fact that an analytical solution is possible only for simple flow problems such as Couette or Hagen–Poiseuille flow. In the following, the numerical solution method is briefly described.

The so-called finite-volume method is applied here. This means that the geometry to be calculated is discretized by a sufficient number of control volumes (here, hexahedral CVs). The partial differential equations have to be fulfilled, on average, for each of these control volumes, but not for every point of the domain considered. The partial differential equations cannot be solved directly. Hence, the system of non-linear differential equations is transformed into a system of algebraic equations that can be solved. The derivation procedure is described in detail in the description of the CFD code used (Fastest-3D, TU-Darmstadt, Schäfer, 2004).

The system of algebraic equations is solved by a combined approach of inner and outer iterations. Inner iteration stands for the solution of the algebraic equations system for one set of estimated coefficients. Here, coefficients means the convective and diffusive fluxes calculated starting from initial estimations that are terms of the equations system. Due to the fact that the equations system represents a thin-filled matrix, the optimized procedure of Stone (1968), called the "strongly implicit procedure" (SIP) is used for this task.

The purpose of the outer iterations is the improvement of the coefficient matrix in a way that the difference between the left-side and right-side terms of the equations system (Eq. 5.4) is minimized. For that, the so-called SIMPLE-algorithm introduced by Patankar and Spalding (1972) (SIMPLE = semi-implicit method for pressure-linked equations) is applied. In the current version of the CFD code, an improved SIMPLE method based on the work of Peric (1985) is implemented.

The SIMPLE algorithm improves the coefficients of the algebraic equations system. For that, a pressure-correction procedure is used that is necessary due to the fact that after an iteration the momentum equation, but not the mass conservation law, is fulfilled. The pressure correction connects the continuity and the momentum equation. The method is described in detail in the CFD code manual (Fastest-3D, TU-Darmstadt, Schäfer, 2004) and in the work of Peric (1985).

During the first iterations of a calculation the differences between the left- and right-hand side terms of the equations usually decrease relatively quickly (the system converges towards a stable solution). However, this process slows down considerably after a few iterations. The reason for this behavior can be found in the progression of the errors of the algebraic equations system across the mesh. The first iterations in particular reduce the short-frequency errors, while long-frequency errors are only very slowly reduced. To improve this characteristic, a multimesh procedure is applied. Here, a second or even more meshes are interpolated from the grid originally selected. These other grids are understandably less dense, and all transport variables are interpolated from the finest grid to the other grid levels. Now, the solution procedures for the general transport equations (SIP and SIMPLE) are alternately applied to the different grid levels and the transport variables are interpolated between the levels in an order to be defined. In this way the different frequency errors are reduced significantly faster than they would be in the case of a single grid method. The multimesh procedure is also described in the manual (Fastest-3D, TU-Darmstadt, Schäfer, 2004).

The time averaging of the system of differential equations based on Eq. 5.4 leads to a loss of information about the fluctuation values of the transport variables. This loss of information has to be compensated by the application of so-called turbulence models. A large amount of standard literature describing turbulence modeling is available (McComb 1990, Pope 2000 and others). Hence, only the standard k–ε turbulence model applied here is briefly introduced.

As mentioned previously, the turbulence model of Launder and Spalding (1974) is used for the present calculations. It is a two-equations model describing the transport of the turbulent kinetic energy k and the rate of turbulent dissipation, ε. The turbulent kinetic energy is defined as:

$$k = \frac{1}{2} \cdot \overline{u'_i u'_i} \tag{5.5}$$

and represents half the sum of the Reynolds-stress tensor. The rate of dissipation ε describes the dissipation of turbulent energy by molecular viscosity. It is a correlation of the fluctuation velocities u':

$$\varepsilon = \frac{\mu}{\rho} \cdot \overline{\frac{\partial u'_i}{\partial x_l} \frac{\partial u'_i}{\partial x_l}} \tag{5.6}$$

Using Prandtl's eddy-viscosity theory (1945), the turbulent stresses of the flow are set proportional to the deformation velocities. That leads to the introduction of the eddy viscosity μ_t and the eddy thermal conductivity λ_t. Both are dependent on the flow structure, but are not identical to the gas-phase properties. Based on dimensional analysis, these two values are defined as:

$$\mu_t = \rho \cdot C_\mu \cdot \frac{k^2}{\varepsilon} \tag{5.7}$$

$$\lambda_t = \rho \cdot c_{P,g} \cdot L_{\mathrm{mom}} \cdot L_{\mathrm{th}} \cdot \left|\frac{\partial u}{\partial y}\right| \tag{5.8}$$

L_{mom} and L_{th} are the hydrodynamic and the thermal mixing length. Both can be related to each other by a turbulent Prandtl number:

$$\mathrm{Pr}_t = \frac{\mu_t \cdot c_{P,g}}{\lambda_t} = \frac{L_{\mathrm{mom}}}{L_{\mathrm{th}}} \tag{5.9}$$

Values of the turbulent Prandtl number are mainly determined experimentally (Eckert and Drake 1987) and are included in the transport equations as empirical constants (see Tab. 5.1 and Tab. 5.2). The transport equations of k and ε are also summarized in Tab. 5.1 and Tab. 5.2.

5.2.2
Fundamentals of Lagrangian Particle Tracking

If particles or droplets are dispersed in a fluid phase, then forces will act on the particles leading to particle motion and acceleration or deceleration. The forces which act on a particle are either fluid dynamic or external-field forces. Fluid-dynamic forces are the drag, Basset, Saffman, Magnus, added mass and pressure force. Field forces are the gravity, magnetic and electrostatic force. It is generally accepted that all forces $\vec{F}_i$ are treated as independent, so that the principle of superposition can be applied to calculate the overall force $\vec{F}_p$ acting on a particle. Hence, Newton's law of motion is given by:

$$M_p \cdot \frac{d\vec{u}_p}{dt} = \sum \vec{F}_i \qquad (5.10)$$

All the forces acting on the particles and the relevant resistance coefficients are summarized by Crowe et al. (1998) and Sommerfeld (2000 and 2005).

5.2.2.1 Drag Force

The drag force is due to the fluid-flow field around the particle and can be divided into two terms. The first term results from the shape of the particle (pressure drag), the second describes the friction that the particle experiences. Within the Stokes-flow regime the shape drag contributes around 1/3 to the overall drag force.

Different regimes of the drag force are usually expressed as functions of the particle Reynolds number. Below a Reynolds number of 0.5 laminar flow prevails around a spherical particle and the corresponding drag force is calculated by integrating the pressure field and the shear stress across the particle surface as:

$$F_{\text{Stokes}} = 3 \cdot \pi \cdot \mu_f \cdot d \cdot \vec{u}_{\text{rel}} \qquad (5.11)$$

For higher Reynolds numbers no analytic solution is available. Hence, experimental data are used to derive interpolation functions of the drag force in dependence of the Reynolds number. Historically, a number of different approximations have been suggested, see the review by Schmidt and Muller (1997). In these approaches, the drag coefficient c_D is introduced, and the drag force is given by:

$$\vec{F}_D = \frac{3}{4} \cdot M_p \cdot \frac{\rho_f}{\rho_p \cdot d} \cdot c_D \cdot (\vec{u}_f - \vec{u}_p) \cdot |\vec{u}_f - \vec{u}_p| \qquad (5.12)$$

Drag coefficients are usually fitted to experimental results. One of the oldest relationships used for spherical particles is that of Schiller and Naumann (1933):

$$c_D = \frac{24}{\text{Re}} \cdot (1 + 0.15 \cdot \text{Re}^{0.687}) \quad \text{for} \quad \text{Re} < 1000 \qquad (5.13)$$

The deviation of this relationship from experimental data is less than 6%. Above a Reynolds number of 1000, the friction loses its importance for the drag force. In this turbulent regime, c_D is usually set to 0.445 in agreement with experimental data. In spray drying, however, both primary particles and agglomerates are mostly of complex, nonspherical, shape. Because drag coefficients are not available for such particles, particle Reynolds numbers and drag coefficients are calculated with the diameter of the sphere of equal volume.

5.2.2.2 Virtual Mass Force

If a particle is accelerated or decelerated or moves in an unsteady flow, a certain mass of the surrounding fluid is accelerated/decelerated as well. This requires additional energy due to the fact that the inertia of the fluid has to be overcome. Crowe et al. (1998) describe the virtual mass force as:

$$\vec{F}_{VM} = M_p \cdot c_{VM} \cdot \frac{\rho_f}{\rho_p} \cdot \left(\frac{d\vec{u}_f}{dt} - \frac{d\vec{u}_p}{dt} \right) \tag{5.14}$$

The coefficient c_{VM} was introduced to extend the applicability of above equation to higher Reynolds numbers. Magnaudet (1997) suggested a value for c_{VM} of 0.5.

5.2.2.3 Basset History Force

The Basset force is caused by the lagging of boundary-layer development on the particle with changing relative velocity (i.e. acceleration or deceleration of particle and flow) and is often referred to as the "history" force. Since this force involves an integral term along the entire particle trajectory it is computationally very time consuming and therefore often neglected.

5.2.2.4 Forces Caused by Pressure Gradients in the Fluid

Pressure gradients in the fluid field create further forces acting on the particle. Under the assumption that the shear across the surface of the particle is constant, the force is expressed as (Crowe et al. 1998):

$$\vec{F}_p = M_p \cdot \frac{\rho_f}{\rho_p} \cdot \frac{d\vec{u}_f}{dt} - M_p \cdot \frac{\rho_f}{\rho_p} \cdot \vec{g} \tag{5.15}$$

Often, the first term of the equation is called the pressure force, the second is the Archimedes buoyancy.

5.2.2.5 Magnus Force

The Magnus force (or slip-rotation lift) is created by rotation of the particle. If a particle rotates, pressure differences can be observed on the surface of the particle, leading to additional forces. Sommerfeld (2005) suggested to calculate the Magnus force as:

$$\vec{F}_M = \frac{\pi}{8} \cdot d^2 \cdot \rho_p \cdot c_M \cdot \frac{|\vec{u}_{rel}|}{|\vec{\omega}_{rel}|} \cdot (\vec{u}_{rel} \times \vec{\omega}_{rel}) \tag{5.16}$$

Here, $\vec{\omega}_{rel}$ is the vector of the relative angular velocity, $\vec{u}_{rel}$ the relative velocity vector between particle and fluid and c_M is the associated resistance coefficient.

5.2.2.6 Saffman Force

The Saffman force (or slip-shear lift) is created by shear gradients in the flow field that lead to pressure gradients across the particle surface. The Saffman force acts in a direction perpendicular to the flow gradient. Mei (1992) suggests the equation

$$\vec{F}_S = 1.615 \cdot d^2 \cdot \sqrt{\rho_f \cdot \mu_f} \cdot c_S \cdot \frac{[\vec{u}_{rel} \times \vec{\omega}_f]}{\sqrt{|\vec{\omega}_f|}} \tag{5.17}$$

for the Saffmann force. Here, $\vec{\omega}_f$ represents the rotational velocity of the fluid and c_S is the resistance coefficient for higher Reynolds numbers.

5.2.2.7 Gravitational Force

The gravitational force

$$\vec{F}_g = M_p \cdot \vec{g} \tag{5.18}$$

is the most important field force for the present computations, all others can be neglected for the tracking of the particles considered here.

The importance of the above-described forces depends on the actual flow situation. For the present case of particles and droplets with sizes from a few and up to some hundred micrometers moving in a spray dryer the drag force is the most important flow force. In addition, the gravity and buoyancy forces are taken into account. For heavy particles in air the pressure force, the added mass and the Basset history force can be neglected. Also, the transverse lift forces and particle rotation are neglected. The wetted air in the spray dryer prevents electrostatic charging of the droplets. Thermophoretic forces are also negligible compared to the large drag forces created by the inhomogeneous gas flow especially in the jet downstream of the gas inlet.

5.2.3 Particle Tracking

Particle tracking is carried out by solving the Basset–Bousinesq–Oseen (BBO) equation for particle movement. Here, the gravity, drag and buoyancy force are taken into account. The differential equations derived from the BBO equation and Newton's law of motion applied in this study read as:

$$\frac{dx_{p,i}}{dt} = u_{p,i} \tag{5.19}$$

$$\frac{du_{p,i}}{dt} = \frac{3}{4} \cdot \frac{\rho}{\rho_p \cdot d_p} \cdot c_D \cdot (u_i - u_{p,i}) \cdot |\vec{u} - \vec{u}_p| - \frac{\rho}{\rho_p} g_i + g_i \tag{5.20}$$

The above equations describe the change of particle location and particle velocity. Since almost no information is available on the rotation of agglomerates, this effect is not considered here.

Despite knowledge of the behavior of agglomerates in turbulent flows, there is still no standard relationship available for drag coefficients of such bodies. Hence, the drag coefficient c_D is calculated according to the already-mentioned correlation for a sphere (Schiller and Naumann 1993):

$$c_D = \frac{24}{Re} \cdot (1 + 0.15 \cdot Re^{0.687}) \quad \text{for} \quad Re < 800 \tag{5.21}$$

$$c_D = 0.445 \quad \text{for} \quad Re \geq 800 \tag{5.22}$$

Equations 5.19 and 5.20 are solved numerically along the particle trajectories until certain termination criteria are fulfilled. These are:
- the particle leaves the computational domain through a dryer outlet
- a particle stops moving for a given number of time steps or does not change its position any longer.

If a particle is involved in an agglomeration or coalescence event, the tracking is continued for the considered particle with adapted particle properties.

The time discretization of the tracking procedure is arranged adaptively. This means, for every parcel tracked the next time step size $\Delta t_{\text{Tracking},i}$ is calculated according to several criteria derived from physical considerations. The criteria are:

1. The parcel should not cross more than a certain part of the actual CV size in one time step. The certain part of the CV is chosen as the minimum of the half lengths of all CV edges ($x_i, i = 1, 2, 3$). Hence,

$$\Delta t_{\text{Grid}} = 0.5 \cdot \min(x_i / |\bar{u}_{pi}|) \tag{5.23}$$

2. The time step should be smaller than the Eulerian time scale t_E:

$$\Delta t_{\text{Euler}} < 0.1 \cdot t_E \quad \text{with} \quad t_E = \frac{0.4 \cdot t_L}{u_p} \cdot \sqrt{\frac{t_L \cdot \varepsilon}{0.24}} \tag{5.24}$$

(The Eulerian time scale t_E is the time a particle needs to cross a turbulent structure.)

3. The time step should be so small that only a binary collision occurs during the step period. Hence, it is inversely proportional to the local particle number density:

$$\Delta t_{\text{coll,max}} = \frac{P_{\text{coll,max}}}{\frac{1}{4}\pi \cdot (d_1(t) + d_2(t))^2 \cdot |\vec{u}_{p,1}(t) - \vec{u}_{p,2}(t)| \cdot n(x,y)} \tag{5.25}$$

Here, $P_{\text{coll,max}}$ is a predefined maximal value of the collision probability to fulfil the criteria of binary collisions, usually set to

0.05; n is the overall particle number density in the current CV, d_1, d_2, $\vec{u}_{p,1}$, $\vec{u}_{p,2}$ are the diameters and velocity vectors of the real particle and its fictitious counterpart at the actual time t.

4. The time step should also be lower than the Lagrangian timescale, defined as

$$t_L = c_t \cdot \frac{u_{RMS,f}^2}{\varepsilon} \tag{5.26}$$

where c_T is a coefficient set to 0.24. The time step $\Delta t_{Lagrange}$ is defined as $\Delta t_{Lagrange} = 0.2 \cdot t_L$. The criterion is only used if the Langrangian length scale of the turbulence is larger than 10% of the particle diameter. Otherwise, it leads to very small time steps; the particle almost does not move within the time steps and gets "artificially stuck". This phenomenon preferably occurs in regions with low fluid fluctuation velocities such as wall boundary layers or others. In this case, $\Delta t_{Lagrange}$ is set to a large value, as is the time step Δt_{Euler} derived from the Euler criterion.

5. The time step should be smaller than the particle response time t_p, including a safety factor. Hence, Δt_p is defined as:

$$\Delta t_p = 0.5 \cdot t_p = 0.5 \cdot \frac{4 \cdot \rho_p \cdot d^2}{3 \cdot \mu_f \cdot Re \cdot c_D} \tag{5.27}$$

6. A predefined minimum time step Δt_{min}, usually set to 10^{-8} s.
7. A predefined maximum time step Δt_{max}, usually set to 1.0 s.

From these criteria, the adaptive Lagrangian time step $\Delta t_{Tracking,i}$ is estimated:

$$\begin{aligned}\Delta t_{Tracking,i} \\ = \max\left(\Delta t_{min}, \min\left(\Delta t_{max}, \Delta t_{Lagrange}, \Delta t_p, \Delta t_{Euler},\right.\right. \\ \left.\left.\Delta t_{coll,max}, \Delta t_{Grid}\right)\right)\end{aligned} \tag{5.28}$$

5.2.4
Particle Turbulent Dispersion Modeling

During the tracking procedure, the instantaneous fluid velocity has to be estimated along the particle trajectory, since this information has been lost as a result of the Reynolds-averaging procedure applied to the conservation equations for the fluid flow. The generation of the instantaneous fluid velocity along the particle trajectory can be based on several stochastic approaches, as for example the simple "eddy-lifetime model" (Gosman and Ioannides 1983), the one-step correlated model based on the Langevin equation or the multistep correlated model (Berlemont et al. 1990).

The principles of these methods were summarized in Chapter 13 of Crowe (2005). Essential for all these schemes is, however, the estimation of the relevant time and length scales of turbulence, which is not a simple task in complex flows. In this study a one-step Langevin equation model is used (Sommerfeld et al. 1993). The mean fluid velocity is interpolated from the neighboring fluid velocity values that are known on the CV corner points. The fluctuation velocity acting on the particle is composed of a random contribution of the local fluid fluctuation velocity and a part that accounts for the correlation of the fluid fluctuation at the actual particle position with that of the particle position one time step earlier:

$$u'_{i,n+1} = R_{p,u_i}(\Delta t, \Delta r) \cdot u'_{i,n} + \sigma_{ui} \cdot \text{GRN}_i \cdot \sqrt{1 - R^2_{p,u_i}(\Delta t, \Delta r)} \tag{5.29}$$

Here, GRN_i is a vector of three independent Gaussian random numbers with a mean value of zero and a standard deviation of unity. With these values, the local fluid root mean square (rms) values σ_{u_i}, and the correlation term $R_{p,u_i}(\Delta t, \Delta r)$ the random contributions for the three velocity directions ($i = 1, 2, 3$) are composed. Furthermore, the correlation term consists of a Lagrangian and a Eulerian part:

$$R_{p,u_i}(\Delta t, \Delta r) = R_L(\Delta t) \cdot R_{E,i}(\Delta r) \tag{5.30}$$

In the present approach, the Lagrangian term is defined using the Lagrangian time scale t_L (see Eq. 5.26) and the actual time step of the tracking procedure:

$$R_L(\Delta t) = \exp\left(-\frac{\Delta t}{t_L}\right) \tag{5.31}$$

The Eulerian part of the correlation term $R_{p,u_i}(\Delta t, \Delta r)$ can be calculated using the longitudinal and transversal correlation coefficients $f(\Delta r)$ and $g(\Delta r)$ (von Karmann and Horwarth 1938):

$$R_{E,i,j}(\Delta r) = \{f(\Delta r) - g(\Delta r)\} \cdot \frac{r_i r_j}{r^2} + g(\Delta r) \cdot \delta_{i,j} \tag{5.32}$$

From the Eulerian correlation tensor $R_{E,i,j}(\Delta r)$ only the three main components are considered, and the correlation coefficients $f(\Delta r)$ and $g(\Delta r)$ are given by

$$f(\Delta r) = \exp\left(-\frac{\Delta r}{L_{E,i}}\right) \quad \text{and} \quad g(\Delta r) = \left(1 - \frac{\Delta r}{2 \cdot L_{E,i}}\right) \cdot \exp\left(-\frac{\Delta r}{L_{E,i}}\right) \tag{5.33}$$

The integral length scales $L_{E,i}$ are defined as (Sommerfeld 2001):

$$L_{E,x} = 3.0 \cdot t_L \cdot \sigma_{\text{fluid}} \quad \text{and} \quad L_{E,y} = L_{E,z} = 0.5 \cdot L_{E,x} \tag{5.34}$$

$L_{E,x}$ is the length scale in the flow direction, the other two are the lateral components.

After tracking a particle for one time step, the drying model is applied to account for the exchange of material and heat. Furthermore, the particle's material properties are updated in accordance with the new drying state. The drying model is described later in detail.

5.2.5
Two-way Coupling Procedure

In order to account for the influence of the particle phase on the fluid flow a consecutive solution of the Eulerian and Lagrangian part is required. The calculation starts with the solution of the fluid flow by not accounting for the source terms of the dispersed phase. After having reached a certain degree of convergence for the single-phase flow, the particle trajectories are calculated; the particle-phase properties (i.e. concentration and particle velocities) and the source terms are sampled for each control volume. When a particle-size distribution is considered, the local particle-size distributions and the size–velocity correlations are also sampled. Thereafter, the continuous phase is recalculated by accounting for the particle phase source terms (Fig. 5.1). In order to avoid convergence problems, an underrelaxation procedure is applied using the equation:

$$S_{\phi,p}^{new} = (1 - \gamma)S_{\phi,p}^{old} + \gamma S_{\phi,p}^{samp} \tag{5.35}$$

where $S_{\phi,p}^{new}$ are the source terms used to calculate the new flow field, $S_{\phi,p}^{old}$ are the source terms used in the previous Eulerian calculation and $S_{\phi,p}^{samp}$ are the new source terms sampled in the Lagrangian calculation. The underrelaxation factor γ depends on the degree of coupling (i.e. the particle concentration and the particle size) and is selected accordingly in the range between zero and one. After a certain number of Eulerian iterations (or after a certain degree of convergence is reached for the fluid flow), the particle tracking is performed again, since the flow field has changed due to the two-way coupling. With the sampled new source terms again the continuous phase is solved, and so forth, until convergence is reached for the coupled system (Fig. 5.1). The overall convergence is decided based on the evolution of a certain reference value (see Kohnen et al. 1994), as for example the fluid velocity at a monitoring location.

5.3
Droplet-drying Models

5.3.1
Introduction

The application of a drying model enables the consideration of mass and heat exchange between the continuous and dispersed phase within the Euler–Lagrange approach. The evaporation of the solvent from suspension droplets furthermore leads to significant changes of material properties and hence of the collision behavior of the drying particles. This again has a considerable effect on the characteristics of

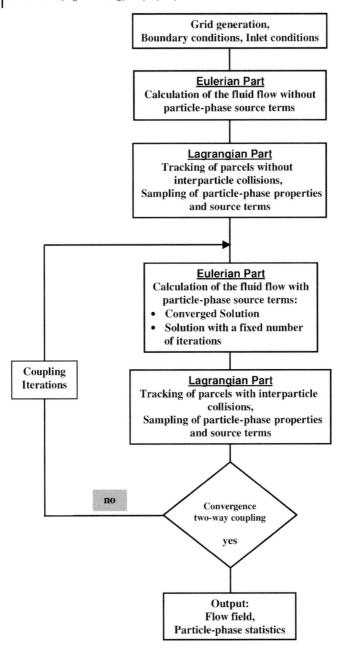

Fig. 5.1 Flow chart of iteration procedure for the Euler–Lagrange approach.

the product obtained by a spray-drying process. Due to this, the proper description of the heat and mass exchange between particles and surrounding gas as well as the connected treatment of the particle's temperature and solvent content and the particle's material properties (viscosity, density, etc.) is crucial for the prediction of product characteristics.

The basic physical phenomena occurring during the evaporation of a liquid droplet can be summarized as the transport of heat and species mass within the droplet, the exchange of heat and mass across the phase boundary droplet–gas and the transport of heat and mass within the gas-side boundary layer. The direction of the transport processes can be both towards the droplet core and away from the droplet (heating/cooling and condensation/evaporation). These processes lead to temperature and, for multicomponent or suspension droplets, concentration gradients within the droplet liquid as long as they have not reached equilibrium. In the case of evaporation, the droplet liquid (or its species) is in equilibrium with the vapor of the surrounding gas at the droplet surface. This can be described with the equation of Clausius–Clapeyron:

$$\ln\left(\frac{p^*(T)}{p^*(T_{\text{ref}})}\right) = \frac{\Delta \tilde{h}_{\text{ev}}}{\tilde{R}} \cdot \left(\frac{1}{T_{\text{ref}}} - \frac{1}{T}\right) \tag{5.36}$$

Here, p^* is the saturation vapor pressure either at a reference temperature T_{ref} or at the temperature T of the droplet surface, $\tilde{R}$ is the universal gas constant, $\Delta \tilde{h}_{\text{ev}}$ the molar evaporation enthalpy of the liquid (latent heat). The vapor is transported away from the droplet in a radial direction by the so-called Stefan flow. This usually is a combined convective and diffusive phenomenon. The main task of models is to describe the mentioned processes in order to obtain the amounts of mass and heat crossing the boundary between gas and liquid with sufficient accuracy. These values are needed for coupling the calculations of the dispersed and continuous phase, as well as for the temporal description of the characteristics of the dispersed phase required for estimating product properties.

5.3.2
Review of Droplet-drying Models

One category of evaporation models assumes spherical symmetry for the droplet geometry as well as for the temperature and concentration fields within the droplet and its surrounding boundary layer. Ranz and Marshall (1952) introduced a model belonging to this category. It assumes an infinite thermal conductivity in the droplet liquid, as does the model of Law (1976). In contrast, Law and Sirignano (1977) developed a model for the case of conduction-controlled heat transport. Abramzon and Sirignano (1989) suggested the use of an effective thermal conductivity for the droplet liquid.

A second category of models assumes an axially symmetric flow of the liquid in the core of the droplet for a stationary process. The models of Prakash and Sirignano (1978, 1980) and Tong and Sirignano (1983) belong to this category. A review of drying and evaporation models can be found in Law (1982), Aggarwal et al. (1984) or Sirignano (1993).

Drying models specially designed for droplets consisting of milk products were reported by Wijlhuizen et al. (1979), Sano and Keey (1982) and Stevenson et al. (1998). All these models assume a uniform temperature distribution throughout the entire droplet and use an effective diffusion coefficient that was determined experimentally and is different from the molecular diffusivity. Cheong et al. (1986) and Farid (2003) suggested models that include a temperature distribution within the droplet due to the fact that for their particular products the Biot and Lewis numbers are very low and, therefore, the drying process is controlled by thermal diffusion.

5.3.3
Exemplary Drying Model for Whey-based Milk Products

The drying model described here was developed in the frame of the EU project "Efficient control and design of agglomeration during spray drying" (EDECAD), where the main work has been done by the group of Prof. C. Tropea (TU Darmstadt, Germany). The products considered within this project are soy- and whey-based liquids used in the food industry. The products may be regarded as suspensions, solutions and emulsions at the same time and, hence, a specially adapted drying model was developed. The model was published by Straatsma et al. (1999) and Verdurmen et al. (2004) and is a derivate of the well-known drying model of Sano and Keey (1982) containing the following fundamental assumptions:

- The drying of skim milk (the products used are considered as such) is assumed to be similar to the drying of a binary mixture with water as the solvent.
- Fickian diffusion in the liquid is modeled with the help of an effective binary diffusion coefficient.
- The rule of Amagat for a liquid–solid mixture (volume additivity of the components) is applied.
- A bubble of moisture vapor arises at the center of the droplet if the equilibrium vapor pressure of the moisture within the droplet becomes larger than the ambient air pressure.
- The overall model describes one cycle of a drying–inflation–drying process, except for the case of the appearance of a bubble within a droplet.
- All phenomena within the droplet are considered as one-dimensional, therefore the governing equations are described in a spherical coordinate system.

The driving forces for heat and mass transfer during the drying of droplets are the temperature difference between the droplet surface (T_{ph}) and the bulk of the surrounding gas (T_∞) and the difference of the respective (solvent) vapor concentrations

$$\Delta C = C_{ph} - C_\infty \tag{5.37}$$

$$\Delta T = T_\infty - T_{ph} \tag{5.38}$$

5.3 Droplet-drying Models

According to Bird et al. (1960), the heat and mass transport across the spherical surface can be described by using transfer coefficients, the contact area A and the driving forces, yielding:

heat flow: $\quad \dot{Q} = \alpha \cdot A \cdot \Delta T = \alpha \cdot \pi \cdot d^2 \cdot \Delta T \quad$ (5.39)

mass flow: $\quad \dot{M} = k \cdot A \cdot \Delta C = k \cdot \pi \cdot d^2 \cdot \Delta C \quad$ (5.40)

Both transfer coefficients are calculated by means of the Ranz–Marshall (1952) relationship:

heat flow: $\quad \alpha = \dfrac{Nu \cdot \lambda_g}{d} = \dfrac{\lambda_g}{d} \cdot (2 + 0.58 \cdot Re^{0.5} \cdot Pr^{0.33}) \quad$ (5.41)

mass flow: $\quad k = \dfrac{Sh \cdot D_g}{d} = \dfrac{D_g}{d} \cdot (2 + 0.58 \cdot Re^{0.5} \cdot Sc^{0.33}) \quad$ (5.42)

In these relations λ_g is the thermal conductivity of the gas phase, D_g the diffusion coefficient of the solvent vapor in the gas, and d the droplet diameter; Nu, Pr, Re, Sh and Sc (the Nusselt, Prandtl, Reynolds, Sherwood and Schmidt numbers, respectively) are the relevant similarity numbers. Alternatives to the classical Ranz–Marshall relation can also be used for the determination of the transfer coefficients. The choice should depend on the actual characteristics of the surrounding flow structure and, hence, on the geometry of the dryer as well as on material properties of the droplet.

The vapor concentration C_{ph} is in thermodynamic equilibrium with the liquid on the wet surface of the droplet and can be expressed as

$$C_{ph} = \dfrac{p_{ph} \cdot \tilde{M}_w}{\tilde{R} \cdot T_{ph}} \quad (5.43)$$

where $\tilde{M}_w$ is the molar mass of water (or another evaporating solvent), p_{ph} is the partial vapor pressure of the liquid on the droplet surface, T_{ph} is the temperature of the droplet surface, and $\tilde{R}$ is the ideal gas constant. C_∞ is also calculated using the equation of state and the vapor pressure in the bulk of the gas.

The partial vapor pressure p_{ph} itself depends upon the saturation vapor pressure of pure liquid $p*$ and the solvent activity (for the present products: water activity):

$$p_{ph} = a_w \cdot p^* \quad (5.44)$$

The water activity a_w can be obtained from sorption isotherms and depends strongly on the products involved in the evaporation processes.

The above relations bridge the gap between phenomena occurring outside and inside the droplet. Within the droplet, diffusion of the solvent takes place, initiated by the driving forces across the phase boundary. Naturally, the laws of mass and heat transfer govern this process as well.

The unsteady diffusion within the droplet can be described (Crank, 1967) by the differential equation:

$$\frac{\partial C}{\partial t} = \frac{1}{r^2} \cdot \frac{\partial}{\partial r}\left(r^2 \cdot D \cdot \frac{\partial C}{\partial r}\right) \tag{5.45}$$

Here, C is the moisture concentration, D is the binary diffusion coefficient and r the radial coordinate of the spherical system. The diffusion coefficient for whey-based products can be described by means of the empirical equation (Wijlhuizen et al. 1979):

$$D = \exp\left(-\frac{B + C \cdot x}{1 + A \cdot x}\right) \cdot \exp\left[-\frac{\Delta H}{R}\left(\frac{1}{T(r)} - \frac{1}{303}\right)\right] \tag{5.46}$$

A, B, C and ΔH are product-dependent parameters, R is the gas constant and $T(r)$ is the local temperature.

5.3.4
Numerical Implementation

The numerical implementation of the droplet-drying model is described in this section. In the framework of the solution algorithm all droplets are discretized into a certain number of shells. To this purpose, a minimum shell diameter d_{min} is predefined as well as a maximum number of shells a droplet can have. In the initial state, all particle shells have more or less equal masses. For this, the number of droplet shells applied is first calculated by estimating the minimum mass $M_{min,shell}$ a shell should not fall below:

$$M_{min,shell} = \frac{\pi}{6} \cdot d_{min}^3 \cdot \bar{\rho}_p \tag{5.47}$$

$\bar{\rho}_p$ is the average droplet density, calculated as:

$$\bar{\rho}_p = \frac{\sum\limits_{N_{shells}} M_i \cdot \rho_i}{M_{p,total}} = \frac{M_{p,total}}{V_{p,total}} \tag{5.48}$$

Here, i is the shell counter and N_{shells} the total number of shells the droplet is discretized in, $V_{p,total}$ is the volume of the entire droplet and $M_{p,total}$ is the entire mass of the droplet. The total number of shells, N_{shells}, used equals the ratio of particle mass and $M_{min,shell}$, rounded to the next higher integer value and, if necessary, corrected by the predefined maximum shell number. Now the preferred masses of the shells are calculated with

$$M_{shell} = \frac{M_{p,total}}{N_{shell}} \tag{5.49}$$

Furthermore, the outer diameters of all shells are calculated and stored as well as their initial temperature and solid content. Once all the values have been determined, the tracking and drying procedures can start.

The drying procedure is carried out for each Lagrangian time step, where the drying algorithm for one droplet is placed after the tracking step, but before the collision algorithm. Very often the actual Lagrangian time step estimated for the tracking of the particle is so large that it could not be applied for the drying step as well. Hence, an adaptive time-step technique is used so the drying process is properly modeled. For this, two criteria are defined, one based on the change of moisture content in the particle, the other based on the change of temperature. Not more than 10% of the moisture of a shell should evaporate within one time step Δt_{ev}, and the temperature ought to increase only moderately:

$$\Delta t_{ev} = \min(\Delta t_m, \Delta t_T, \Delta t_{tracking}) \qquad (5.50)$$

$$\text{with} \quad \Delta t_m = \frac{0.1 \cdot X_i}{dX_i/dt} \qquad (5.51)$$

$$\Delta t_T = \frac{\Delta T_{min}}{dT_p/dt} \qquad (5.52)$$

$$\frac{dT_p}{dt} \approx \frac{\dot{Q}}{M_s \cdot C_s + M_w \cdot C_w} \qquad (5.53)$$

Within these equations, dX_i/dt is the change of moisture of a shell within one time step, X_i is the moisture content [kg$_{water}$/kg$_{solids}$], ΔT_{min} is the predefined maximum temperature change, dT/dt is the temperature change of the particle within a time step and $\dot{Q}$ is the heat flow rate across the particle surface; M and c are the masses and the specific heat capacities of dry solids (subscript "s") and water (subscript "w").

The algorithm for solving the heat- and mass-transfer problem as described above is therefore summarized as follows:

- First, the properties of the gas phase are interpolated for the actual particle position, using values from the computational grid of the flow-field calculation. The properties are solvent vapor concentration, gas temperature, absolute ambient pressure and instantaneous relative velocity between the gas and the particle.
- Second, additional material properties of the gas phase are calculated by using the vapor concentration, the pressure and the gas temperature. These are the gas density, dynamic viscosity, thermal conductivity, the diffusion coefficient of the solvent vapor in the gas phase and the specific heat capacities of the gas and the vapor.
- Following this, the Nusselt and Schmidt numbers are calculated.

- Then a temporal loop is started that lasts until the sum of the smaller drying time steps is equal to the actual Lagrangian time step for a frozen flow field. The loop includes several calculation steps:
  - The actual particle Reynolds number is calculated using the latest particle diameter value.
  - The actual particle properties are calculated for each particle shell (densities, solvent concentrations, diffusion coefficients).
  - The mass-transfer coefficient k (Eq. 5.42), the water activity a_w of the outer particle shell, the saturation pressure of the particle surface p^* and the partial solvent vapor pressure of the particle surface p_{ph} (Eq. 5.44) are calculated.
  - Using these values, the equilibrium solvent concentration in the gas phase C_{ph} (Eq. 5.43) as well as the bulk solvent concentration C_∞ are calculated. Now, the actual solvent flow across the phase boundary can be computed (Eq. 5.40).
  - The heat-transfer coefficient (Eq. 5.41), the latent heat of evaporation at the particle surface temperature and the heat flow rate transferred between the particle and the gas (Eq. 5.39) are calculated.
  - Now, the mass flow rate of the solvent to any shell i can be computed:

$$\dot{M}_{\text{solvent},i} = -4 \cdot \pi \cdot \left[\left(r_{i+1}^2 \cdot D_{i+1} \cdot \left(\frac{dC}{dr}\right)_{i+1} \right) - \left(r_i^2 \cdot D_i \cdot \left(\frac{dC}{dr}\right)_i \right) \right] \tag{5.54}$$

  - The moisture contents and their time derivatives needed for the mass-transfer-related time step are calculated for each particle shell. The smallest time step according to Eq. 5.51 is estimated.
  - The time step related to heat transfer is calculated according to Eqs. 5.52 and 5.53. From all evaluated time steps the drying time step to be used is chosen (Eq. 5.50).
  - Now, new moisture concentrations and new masses are calculated for all particle shells.
  - The new particle temperature is assumed to be constant within the droplet:

$$T_{p,t+\Delta t} = \frac{h_{t+\Delta t}}{c_s + c_{\text{solvent}} \cdot X} + 273.15 \text{K} \tag{5.55}$$

with

$$h_{t+\Delta t} = (T_{p,t} - 273.15 \text{ K}) \cdot (c_s + c_{\text{solvent}} \cdot X) - \frac{\dot{Q} \cdot \Delta t}{M_p \cdot SC_p} \qquad (5.56)$$

Here, h is the enthalpy per unit mass of dry solids, M_p and SC_p are the particle's total mass and solids mass fractions, X is the dry-based solvent content; c_s and c_{solvent} are the specific heat capacities of the dry solid and the solvent.
– Finally, the new outer diameter d_i of each particle shell is calculated using the updated material density of each shell:

$$d_i = \sqrt[3]{\frac{6 \cdot M_i}{\pi \cdot \rho_i}} \quad \text{for the inner shell} \qquad (5.57)$$

$$d_i = \sqrt[3]{\frac{6 \cdot M_i}{\pi \cdot \rho_i} + d_{i-1}^3} \quad \text{for all other shells} \qquad (5.58)$$

This results in a new particle diameter.

A reduced shrinkage of the droplet as can be observed for many dried products is not considered in the present approach, but is of course of interest for the modeling of spray-drying processes. The quantity of shrinkage to be predicted by the model is usually incorporated empirically, depending on the product behavior observed.

If particle drying is finished for one Lagrangian time step, the average particle density is calculated and new dynamic viscosity values are interpolated for each particle shell. The viscosity values are needed for the agglomeration model and may be determined by separate measurements for the liquids of interest in dependence on temperature and solvent content.

5.4
Collisions of Particles

5.4.1
Introduction

In the framework of the Euler–Lagrange approach, collision models are required to predict whether two particles moving in a gaseous or liquid surrounding will collide with each other or not. This prediction has major relevance for industrial processes in which phenomena such as agglomeration, coalescence or breakage of particles strongly influence product properties (e.g., particle-size distribution, porosity) or process performance (for example by catalyst demobilization caused by abrasion).

The basic mechanisms leading to particle collisions are the mean relative drift between two particles, laminar and turbulent shear flows the particles are moving in, and fluctuations of the particle velocity induced by turbulent fluctuations of the surrounding fluid flow (Crowe et al. 1998). Particle collisions can also be driven by Brownian motion in the case of high Knudsen numbers (Nowakowski and Sitarski 1981).

These mechanisms do not have the same degree of influence on all particles and their probability to collide. This strongly depends on the properties of the particle in relation to its environment, namely the ability of the particle to react to fluctuations of the surrounding continuous phase and the size ratio of the potentially colliding particles. The first is generally expressed in terms of a so-called Stokes number representing the ratio of the particle relaxation time t_p to the representative time scale of turbulence, t_L, of the flow field (Sommerfeld 2001):

$$St = \frac{t_p}{t_t} = \frac{\rho_p \cdot d^2}{18 \cdot \mu_f t_L} \tag{5.59}$$

Low Stokes numbers (St $\ll$ 1) indicate a strong influence of the fluid flow fluctuations on particle movement, high Stokes numbers a weak influence.

Sommerfeld (2001) reviewed investigations dealing with Stokes-number-dependent collision rates. This review pointed out remarkable differences of expected collision rates for inertial and noninertial particles due to the different importance of the described physical mechanisms.

In the Euler–Lagrange approach, the occurrence of a collision, the outcome of the collision and the properties of the particles involved after the collision event have to be predicted.

5.4.2
Extended Stochastic Collision Model

A novel stochastic interparticle collision model for the calculation of two-phase flows was introduced by Sommerfeld (2001). Up to now this model has mainly been applied to systems in which the outcome of particle–particle collisions is governed by one mechanism only in the entire computational domain, namely the elastic-plastic bouncing of solid particles (Lain et al. 2002b), the collision and coalescence of water droplets (Rüger et al. 2000) or the collision of air bubbles in water (Lain et al. 2002a). In contrast to these investigations the dispersed system in a spray dryer shows different collision phenomena. The particles, either surface-tension-dominated droplets just after atomization, viscous droplets in a later stage of the drying process, or dry particles after complete drying or fine particles returned back into the process, interact with each other according to very different physical mechanisms. At the same time, the dispersed phase undergoes time-dependent changes of its material and geometrical properties due to the drying process. These characteristics require an extension of the stochastic collision model of Sommerfeld (2001) to enable the consideration of variable material properties, temperatures and the different mechanisms of particle interaction.

The main characteristic of the stochastic collision model of Sommerfeld (2001) is the creation of a fictitious particle from statistical data locally sampled for the dispersed phase. This fictitious particle is individually created every Lagrangian time step, for every tracked parcel in the computational domain. The statistics required for the creation of the fictitious collision partners are sampled every Lagrangian time step. These are, for each control volume:
- the relative number cumulative size distribution (based on diameter classes);
- the number fraction of dry (or wet) particles of each size class;
- the mean and rms solid content of the wet particles in each size class;
- the mean and rms temperature of all particles;
- the mean and rms values of all velocity components for each individual size class.

The sizes of the fictitious particles are calculated using the local cumulative size distributions, applying a random process. At first, a uniform random number between 0 and 1 is created. Then the size class is found for which the random number is just smaller than the value of the normalized relative number cumulative distribution Q_0, starting with the smallest size class (Fig. 5.2). After this has been accomplished, the diameter is calculated with the help of another uniform random number between 0 and 1:

$$d_{\text{fict}} = d_{\min,i} + RN \cdot (d_{\max,i} - d_{\min,i}) \tag{5.60}$$

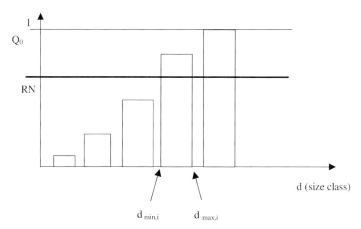

Fig. 5.2 Sampling of the fictitious particle diameter with help of a cumulative size distribution.

with $d_{min,i}$ - minimum diameter of size class i;
        $d_{max,i}$ - maximum diameter of size class i;
        RN   - uniform random number between zero and one (two of them are created).

The mean and fluctuating components of the fictitious particle's velocity components are sampled using a procedure described by Sommerfeld (2001). Here, the fictitious particle's fluctuation velocity components $u'_{p,fict,i}$ are correlated to those of the real one ($u'_{p,real,i}$) using a correlation function based on the particle Stokes number:

$$u'_{p,fict,i} = R(St) \cdot u'_{p,real,i} + \sigma_{p,i} \cdot GRN \cdot \sqrt{1 - R(St)^2} \quad (5.61)$$

$\sigma_{p,i}$ is the local rms value of the particle's velocity component i and GRN represents a Gaussian random number with zero mean and a standard deviation of unity. According to an investigation of Sommerfeld (2001) and the comparison with large-eddy simulations (LES), the term $R(St)$ is modeled as:

$$R(St) = \exp(-0.55 \cdot St^{0.4}) \quad (5.62)$$

Recently, Berlemont et al. (2001) reviewed Lagrangian models (including the previously described one) and the impact of collision modeling on the properties of a two-phase flow. Based on comparisons with LES data they pointed out that the fluid–particle correlation, and here especially the correlation during particle collisions, is of major importance for the quality of the multiphase-flow modeling. Equation 5.62 for $R(St)$ is an improvement compared to uncorrelated modeling, but still remains an empirical correlation evaluated for homogeneous isotropic turbulence.

To enable collision calculations for a spray-drying process, the fictitious particle needs further properties such as temperature and material properties like particle density, viscosity and surface tension.

The first step towards the prediction of these properties is the decision whether the fictitious particle is dry or a droplet. For this, the fraction of dry particles in the local computational cell i has to be calculated. If a uniform random number in the range [0...1] is smaller than this fraction,

$$RN < \frac{N_{solid,i}}{N_i} \quad (5.63)$$

the fictitious particle is a dry solid particle. Otherwise, it is considered to be a droplet. The temperature of the fictitious particle is sampled with the help of the local mean and rms value of the real particle temperatures:

$$T_{fict} = \overline{T}_i + GRN \cdot \sigma_{T,i} \quad (5.64)$$

Here, $\overline{T}_i$ is the mean temperature of the actual size class i and $\sigma_{T,i}$ represents the temperature rms value of this size class. GRN is, again, a Gaussian random number sampled from a normal distribution that has a mean value of zero and a standard deviation of 1.

During spray drying the drying state of a droplet is usually expressed in terms of the solid mass fraction SC. The solid mass fraction is the ratio of the mass of dissolved or suspended materials to the total droplet mass:

$$\text{SC} = \frac{M_{\text{solid}}}{M_{\text{droplet}}} = 1 - \frac{M_{\text{solvent}}}{M_{\text{droplet}}} \quad (5.65)$$

Material properties such as density, viscosity and surface tension are generally expressed as functions of the solid mass fraction and the temperature. Hence, the fictitious particle is given a solid mass fraction value in a similar way as used for the temperature by accounting for the droplet size distribution:

$$\text{SC}_{\text{fict}} = \overline{\text{SC}}_i + \text{GRN} \cdot \sigma_{\text{SC},i} \quad (5.66)$$

All other material properties are then calculated by either known or measurable functions of the type $f(T, \text{SC})$.

It should be mentioned that for systems having a bi- or multimodal temperature or solid mass fraction distribution Eqs. 5.64 and 5.66 lead to errors in the temperature and solid mass fraction of the fictitious particle. In such cases respective values should be sampled by using classes (temperature classes, solid mass fraction classes), as already done for the particle size. Consequently, the statistical description of the dispersed phase becomes much more challenging. The distributions of temperature and solid mass fraction are considered to be monomodal in the present work.

Based on the kinetic theory of gases the collision probability of the tracked parcel with its fictitious counterpart is calculated. A further random process and the consideration of the impact efficiency (described by Ho and Sommerfeld 2002) decide finally about the occurrence of a collision event:

$$P_{\text{coll}} = f_{\text{coll}} \cdot \Delta t = \frac{\pi}{4} \cdot (d_{\text{real}} + d_{\text{fict}})^2 \cdot |\vec{u}_{\text{p,real}} - \vec{u}_{\text{p,fict}}| \cdot n_i \cdot \Delta t \quad (5.67)$$

Here, f_{coll} (1/s) is the collision frequency, n_i (1/m³) the total particle number density in control volume i and Δt the actual time step. The properties of the real, respectively the fictitious particle are indicated with the indices "real" and "fict".

In order to allow for the determination of the impact point on the surface of the collector particle the collision problem is calculated in a coordinate system where the collector particle is stationary. It should be noted that the collector particle can be the real or the fictitious particle depending on the size of the collision partners. The larger or low-viscosity particle is always the collector particle. In this situation, the point of impact on the surface of the collector particle can only be located on the hemisphere facing the oncoming particle (Fig. 5.3). Hence, a collision cylinder is defined as the domain where the center of the collector particle must be located if a collision takes place and the relative velocity vector is aligned with the axis of the collision cylinder.

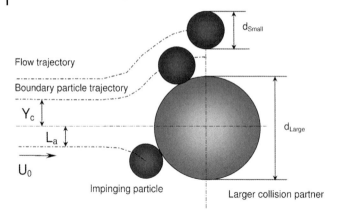

Fig. 5.3 Scheme for modeling impact efficiency for quasilaminar flow conditions.

The collision calculation becomes rather simple, since a central oblique collision is considered (Sommerfeld 2001).

If a large and a small particle are interacting with each other, the collision probability P_{coll} will be reduced by aerodynamic interaction of both partners (Pinsky et al. 1999). The small particle will partly follow the relative flow around the large one (Fig. 5.3). In order to account for this phenomenon, an inertia parameter Ψ_I and the so-called collector (or impact) efficiency η_p suggested by Schuch and Löffler (1978) are calculated:

$$\Psi_I = \frac{\rho_{p,small} \cdot |\vec{u}_{p,small} - \vec{u}_{p,large}| \cdot d_{small}^2}{18 \mu_f \cdot d_{large}} \quad \eta_p = \left(\frac{2 Y_C}{d_{large} + d_{small}}\right)^2 = \left(\frac{\Psi_I}{\Psi_I + a}\right)^b \quad (5.68)$$

Here $\rho_{p,small}$, $\vec{u}_{p,small}$ and d_{small} are the density, velocity and diameter of the small collision partner, $\vec{u}_{p,large}$ and d_{large} are the velocity and the diameter of the large collision partner, and a and b are Reynolds-number-dependent parameters described by Löffler (1988). The quantity Y_C is the radial distance of the small particle from the axis of the relative flow, for which the small particle just can hit the large one (boundary particle trajectory, Fig. 5.3).

The impact efficiency is taken into account for all types of collisions occurring in a spray dryer. Hence, a collision will only take place if two conditions are fulfilled:

1. A random number RN, again created out of a uniform distribution [0...1], must be smaller than the collision probability.
2. The lateral displacement L_a, representing the distance between the smaller particle trajectory and the axis of the collision cylinder (Fig. 5.3), has to be smaller than Y_C:

$$RN < P_{coll} \quad \text{and} \quad L_a < Y_C \quad (5.69)$$

The lateral nondimensional displacement is defined as:

$$L = \frac{2 \cdot L_a}{d_{\text{large}} + d_{\text{small}}} \tag{5.70}$$

The point of impact on the surface of the fictitious particle is sampled randomly using the procedure described by Sommerfeld (2001). By generating two uniform random numbers XX and YY in the range [0, 1], the lateral nondimensional displacement, L, and the angle ϕ are obtained:

$$L = \sqrt{XX^2 + YY^2}, \quad L < 1 \tag{5.71}$$

$$\phi = \arcsin(L) \tag{5.72}$$

An additional random number with uniform distribution in the range 0 to 2π is necessary to determine the angular position of the collision point.

Depending on the drying states of the colliding partners, the appropriate collision, coalescence and/or agglomeration model is chosen to calculate the postcollision properties of the considered droplet or particle. Thereafter, the velocities of the real particle in the laboratory frame of reference are obtained by retransformation. The velocities of the fictitious particle are not of interest any longer.

5.4.3
Modeling of Particle Collisions: Coalescence and Agglomeration

Within a spray-drying process, different types of particles exist having distinguished collision behavior. Just after atomization, the liquid droplet's behavior is usually dominated by surface-tension forces, while during the drying process, either a thickening (increase of viscosity) or crystallization reduce the surface-tension force. Especially for amorphous products like lactose-based formulations for the food industry, the viscosity of the droplet liquid increases dramatically with decreasing content of solvent. This leads to different mechanisms and results of particle collisions that require a classification for collision modeling.

5.4.3.1 Surface-tension Dominated Droplets (STD Droplets)
The collision behavior of droplets belonging to this class is dominated by the interfacial tension forces. They are distinguished from droplets dominated by viscous forces (i.e. having a higher solids content) in terms of the Ohnesorge number:

$$\text{Oh}^2 = \frac{\mu_{\text{droplet}}^2}{d_{\text{droplet}} \cdot \rho_{\text{droplet}} \cdot \sigma_{\text{droplet}}} < 1 \tag{5.73}$$

Viscosity, surface tension, density and diameter are functions of the drying state and therefore time-dependent properties. All STD droplets are considered to be spherical and nonoscillating. The result of collisions of this type of particles is modeled either as coalescence (creation of a new spherical droplet) or separation after temporary coalescence. Tracking of the considered droplet or the coalesced one is continued with adapted velocity components. The decision as to which type of collision occurs is made on the basis of a correlation proposed by Podvysotsky and Shraiber (1984). Details of this model are provided by Blei and Sommerfeld (2004).

5.4.3.2 Droplets Dominated by Viscous Forces (VD Droplets)

Two boundaries are to be defined for droplets dominated by viscous forces: one to distinguish them from STD droplets and one for the boundary towards dry particles. The first boundary is again defined by means of the Ohnesorge number. If the square of this number is larger than unity, the droplet is considered to be dominated by viscous forces:

$$Oh^2 > 1 \tag{5.74}$$

In investigating drying processes, it was observed that the surface tension of a liquid droplet changes only slightly. Such changes can be related to changes in liquid density. The dynamic viscosity is the decisive property varying strongly during the drying process. A plot showing viscosities that correspond to the boundary between the two mentioned regions is depicted in Fig. 5.4.

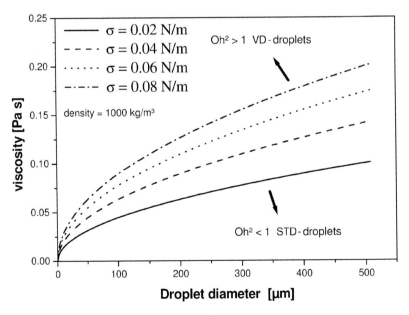

Fig. 5.4 Boundary between surface tension and viscosity dominated droplets for different surface tension values.

In the case of amorphous materials the boundary between a viscous and a dry particle can be defined using the glass-transition temperature. The average temperature of viscous droplets must exceed the actual glass-transition temperature. The glass-transition concept in combination with CFD calculations is briefly described by Verdurmen et al. (2004). If the thickness of the outer, viscous shell of a nucleus exceeds one third of the droplet radius, the droplet is considered to be viscosity dominated. The shape of VD nuclei is considered to be spherical.

5.4.3.3 Dry Particles

The definition of a dry particle depends on the material it consists of. If the origin of the particle is a suspension, the boundary between droplet and dry particle can be determined by the solvent content. Below a certain solvent content the particle is considered to be dry. Some fluids that are solutions with high solvent contents change their state to a nonequilibrium amorphous structure. Here the glass-transition temperature is a useful criterion to distinguish between viscous droplets consisting of supercooled liquid and dry particles whose material is in the amorphous glassy state (Roos 2002; Verdurmen et al. 2004).

Dry particles are considered to be spherical and might have a residual water content. They represent a primary particle (smallest nondivisible element of an agglomerate), and have the characteristic collision behavior of elastic-plastic bodies (Sommerfeld 2001).

The classification of particle types described above leads to different collision scenarios that have to be modeled in a different way:

STD Droplet–Droplet Collisions. These collisions lead either to coalescence or to separation of temporarily coalesced droplets or to a destruction of droplets.

STD Droplet with VD Droplets or Dry Particles. Such collisions are associated either with separation or with merging to form a larger particle. In the second case a change of solid content has to be accounted for. The body created is still considered as a spherical particle.

Collisions between a Dry Particle and a VD Droplet or between two VD Droplets. Here, colliding particles might be partially penetrated to form a porous agglomerate with a certain solids content. This underlines the outstanding importance of this collision mode for agglomerate growth during spray drying and the resulting product properties. In such a case kinetic energy is dissipated during the collision process. Separation or complete penetration might also occur.

In the following, the different models used for the description of different collision types are only briefly introduced. All of them are explained in detail in Blei (2006).

5.4.4
Collisions of Surface-tension Dominated Droplets (STD–STD)

This type of collision can be modeled using the formulations of Podvysotsky and Shraiber (1984). They characterized the droplet interaction in terms of a probability of mass exchange between two colliding droplets

$$P\left(\frac{dM_1/dt}{M_2}\right) = \phi_{1,2} \tag{5.75}$$

dependent on the Reynolds and Laplace numbers and the size ratio of the colliding droplets.

Podvysotsky and Shraiber (1984) estimated the parameter $\phi_{1,2}$ as

$$\phi_{1,2} = 1 - 0.246 \cdot Re_{1,2}^{0.407} \cdot Lp_1^{-0.096} \cdot \left(\frac{d_1}{d_2}\right)^{-0.278} \tag{5.76}$$

for quiescent air environment, droplets of the same liquid and head-on collisions on the basis of numerous experiments. Further details and efforts to validate the model with data gained from droplet-collision experiments using whey-based liquids are explained in Blei (2006).

5.4.5
Collisions of Viscous Droplets

During the drying process droplets that are typical for spray drying (milk products, flavors, cheese whey, etc.) undergo a change of viscosity in the range from a few mPa s up to 10^7 Pa s. The latter is a generally accepted value for the occurrence of stickiness and agglomeration related to polymer-like fluids in the glassy or rubbery state (Roos 2002, Sperling 1992). The influence of surface tension on the collision process of droplets can clearly be neglected for such fluids. The collision of viscous droplets with each other or with dry particles is the major mechanism that leads to structured agglomerates during spray drying. Colliding viscous droplets roughly keep their shape and only partly penetrate each other. This is different to collisions of surface-tension-dominated droplets, where mostly new spherical droplets are generated in the case of coalescence.

For the modeling of viscous droplet collisions several assumptions are made:
- The primary drops keep their size during the collision process (no mass exchange between bouncing or sticking VD droplets will occur).
- The VD droplets are seen as deformable, fully plastic spheres without elastic restitution.
- Newtonian behavior of the liquid (which is clearly a strong simplification).

If two viscous droplets collide, kinetic energy is dissipated and the droplets may either coalesce by building up a new, roughly spherical body, agglomerate or separate. In the case of two colliding viscous droplets the forces resisting the movement of one of the droplets are created by the interaction with the fluid of the other droplet. If the whole situation is seen in a spherical frame of reference, these forces can be distinguished as drag forces acting in the radial direction and shear forces acting in the tangential directions. Hence, a set of differential equations can be formulated describing the movement of one droplet in the fluid of the other. In the present approach it is assumed that the droplet with the higher viscosity value moves in the fluid of the droplet with the lower viscosity. The radial force resisting the droplet movement is assumed to be the Stokesian drag force:

$$M_{high} \cdot \frac{du_r}{dt} = -3 \cdot \mu_{low} \cdot d_{contact} \cdot u_r \quad \text{with} \quad dr = u_r \cdot dt \tag{5.77}$$

The relative tangential movement of the droplet couple is resisted by shear forces of the liquid placed between the centers of mass. With the help of the definition of the shear force between two plates (here: shearing of liquid across the area of contact) the following equations are formulated:

$$M_{high} \cdot \frac{du_\theta}{dt} = -\mu_{low} \cdot d_{contact} \cdot u_\theta \quad \text{with} \quad d\theta = \frac{u_\theta \cdot dt}{r} \tag{5.78}$$

$$M_{high} \cdot \frac{du_\varphi}{dt} = -\mu_{low} \cdot d_{contact} \cdot u_\varphi \quad \text{with} \quad d\varphi = \frac{u_\varphi \cdot dt}{r} \tag{5.79}$$

This set can be solved if a collision event of two viscous particles is predicted by the stochastic collision model. The solution procedure is described in detail in Blei (2006). The model itself still has to be validated for different types of fluids.

5.4.6
Collisions of Dry Particles

In the case of dry-particle collision, only van der Waals forces are considered. In the present approach the dry particles are assumed to be spherical. Hiller (1981) suggested a critical velocity u_{crit} to decide whether an agglomeration between spheres takes place or not. This critical velocity can be extracted from an energy balance as:

$$u_{crit} = \frac{1}{d} \cdot \frac{(1-e^2)^{0.5}}{e^2} \cdot \frac{H}{\pi \cdot s^2 \cdot \sqrt{6 \cdot P_l \cdot \rho_{solid}}} \tag{5.80}$$

Here, H is the Hamaker constant, s the contact distance, e the energetic restitution coefficient and P_l the material limiting contact pressure.

According to Ho and Sommerfeld (2002) agglomeration takes place if the normal relative velocity between particles 1 and 2 is less than the critical velocity given by Eq. 5.80:

$$u_{crit} \geq |\vec{u}_{p,1} - \vec{u}_{p,2}| \cdot \cos\varphi \tag{5.81}$$

The velocity components of the two particles after a collision ending with separation can be calculated using an approach based on an impulse balance, considering elastic–plastic rebound (Ho and Sommerfeld 2002; Blei 2006).

5.5
Example of a Spray-dryer Calculation

In this section an exemplary calculation of a spray dryer, applying the Euler–Lagrange approach, is discussed. The calculation is carried out by solving the Reynolds-averaged conservation equations in combination with a standard k–ε turbulence model for the gas phase. The dispersed phase is modeled by sequential tracking of particles through the flow field, considering particle collisions, agglomeration, coalescence and drying (evaporation of the solvent). For particle tracking the gravitational and the drag forces are considered. Particle collisions are predicted by the extended stochastic interparticle collision model on the basis of impacts between surface-tension-dominated droplets, viscous droplets and dry particles, as described before. The evaporation model is adapted to the drying process of an amorphous skim milk product, which has also been discussed in the previous sections.

The calculations are carried out at different levels of modeling depth, a so-called 1-way coupling (only the gas phase acts on the dispersed phase, but not vice versa), 2-way coupling (the influence of the dispersed phase on the gas phase is considered by particle source terms, but particle collisions are only sampled at the end of the calculation procedure), and 4-way coupling (particle collisions are considered during each iteration loop between gas and dispersed phase calculation). For details on the calculation procedures, see Blei (2006) or Decker (2006).

5.5.1
Geometry and Spatial Discretization of the Spray Dryer

The spray dryer discussed in this work is a fairly large pilot plant spray dryer with the main air inlet centrally at the top of the drying chamber. The upper part of the dryer is typically cylindrical, the lower part having a conical shape. The cylindrical part has a diameter of 4.3 m, the total dryer height is 7.3 m and the total dryer volume is approximately 78.6 m^3. The air inlet has an area of 0.4074 m^2 (Fig. 5.5). At the bottom of the dryer a fluid bed for collecting the agglomerated product is located, hence, a small amount of air flows into the dryer from that direction as well. The bottom inlet diameter is 1 m.

The spray dryer has two air outlets, located at the top of the dryer. They have a diameter of 0.37 m, so that an outflow area of 2 $\times$ 0.11 m^2 is available. The nozzle is located in the center of the main air inlet at the top of the dryer. The calculation of the air flow, species and temperature field requires the discretization of the geometry in a grid structure on which the transport equations can be solved. For that, a nonorthogonal, block-structured hexahedral grid type is chosen. The final grid consists of 102 blocks with altogether 692224 computational volumes (CVs).

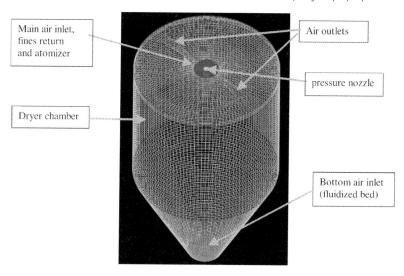

Fig. 5.5 View of the mesh-hull defining the dryer geometry.

The bulk gas velocity at the inlet is set to 25 ms^{-1}, whereby the gas flow has an inlet temperature of 400 K. The liquid to be dried is atomized by a pressure nozzle. The resulting spray is simulated using data gained from measurements with milk fluids undertaken at the University of Bremen (Menn 2005). Details of the droplet spray produced are shown in Fig. 5.6.

5.5.2
Results for the Fluid Phase

The large-scale flow-field pattern of the spray dryer investigated agrees with the expected behavior. A strong jet of inflowing gas is directed downwards into the center of the drying chamber, keeping a relatively high velocity down to the lower third of the dryer (Fig. 5.7). There, the gas flow direction is turned by 180 degrees and the gas flows up towards the gas outlets with a low velocity of around 1–5 ms^{-1}. In this region, bounded by the dryer walls and the central gas jet ranging from the top to the bottom of the dryer, large vortex structures are observed (Fig. 5.9). Some of the upstreaming gas is re-entrained into the central jet, other portions of gas move in horizontal vortices before reaching the dryer outlets. This flow structure increases particle residence time in the drying chamber. A very high gas velocity is observed only near the outlets (Fig. 5.7), without significant influence on a larger region of the drying chamber.

For the present case, the atomization has only a minor influence on the gas-flow field due to the fact that a pressure nozzle is used. Figures 5.7 and 5.8 compare the gas-flow patterns of the atomization zone with and without consideration of the particle source terms for momentum, turbulent kinetic energy and dissipation. Only minor differences, which are not of importance for the drying chamber gas-flow field,

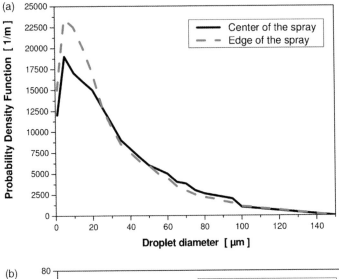
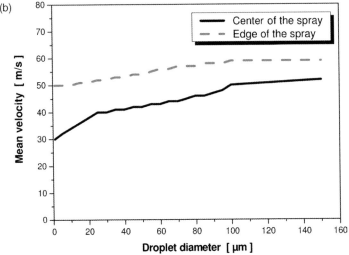

Fig. 5.6 Number droplet size distribution (a) and velocity-size correlation (b) as used for the injection at a level 10 mm below the nozzle. The radius of the spray core is 6.4 mm, the radius of the outer spray edge is 6.9 mm.

can be observed. However, the coupling procedure is very important for the velocity and turbulence of the dispersed phase, as will be discussed later.

The particle source terms for vapor concentration and temperature are very important for both the flow field and the dispersed phase. Evaporation of solvent and the initially low liquid temperature strongly affect the gas-phase properties. The gas temperature is lowered, and the solvent vapor concentration is increased. This in turn affects the evaporation of the droplets and the location in the dryer where it takes

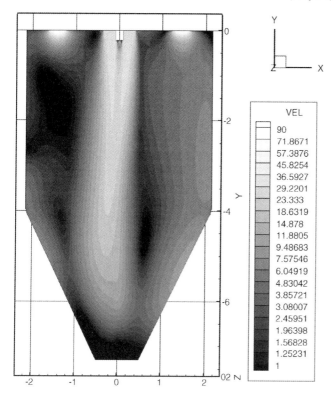

Fig. 5.7 Contour of the gas velocity [m/s], drawn for a vertical plane through the center of the dryer, calculation without particle source terms.

place. The evaporation of solvent is reduced, and certain levels of droplet drying are reached later.

5.5.3
Results of the Dispersed Phase

The movement of droplets and particles in the spray dryer is strongly affected by the gas flow. Hence, the droplets created by the atomization move downwards first and follow the main central gas jet (Fig. 5.8). Near the bottom of the dryer, the gas flow is reversed and large vortices are created. If the particle Stokes number is small enough, the particles follow the vortex structures. For the present case however, the majority of particles moves at the outer edges of the vortices. Hence, a strongly increased particle number density is observed just near the walls in the conical dryer section (Fig. 5.10). In this region, many interparticle collisions will occur. If the calculations are carried out 2-way coupled, the locally very high number concentrations disappear and fewer particle collisions are observed, leading to less intense agglomeration (Fig. 5.10). Hence, for the present case a 1-way coupled

196 | 5 CFD in Drying Technology – Spray-Dryer Simulation

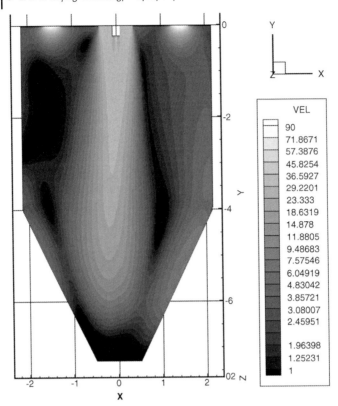

Fig. 5.8 Contour of the gas velocity [m/s], drawn for a vertical plane through the center of the dryer, calculation with particle source terms.

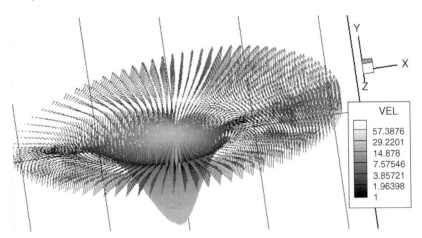

Fig. 5.9 Velocity vectors in a cut plane through the dryer located 3 m downstream of the gas inlet. The colors refer to the absolute velocity values [m/s].

5.5 Example of a Spray-dryer Calculation

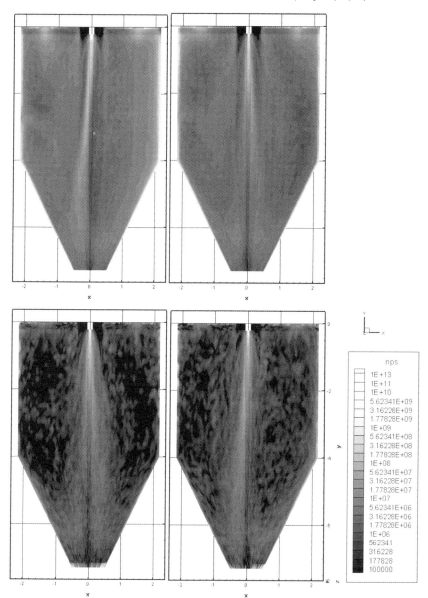

Fig. 5.10 Particle number concentration [1/m³] in the dryer. The data are taken from calculations without fines return. From the left to the right, results of a 1-way coupled calculation without collisions, a 2-way coupled calculation without collisions, 2-way coupled calculation with collisions and 4-way coupled calculation are shown.

calculation procedure leads to results that look reasonable at a first view, but are not correct in general.

The source terms for the momentum exchange between dispersed and fluid phase are of major importance especially if particle collisions are to be computed. As already discussed, the effect of the dispersed phase on the gas-flow field is not very large in the main drying chamber (Fig. 5.7). However, in the atomization zone close to the nozzle exit 2-way coupling effects are very strong. For the present case, the gas velocity near the nozzle is relatively low if the dispersed phase is neglected (Fig. 5.7). Introducing the source terms for the u, v, w velocity components and those for k and ε, the fluid is accelerated. Therefore, without these source terms, the residence time of the particles tracked in the atomization zone is too large and the time-averaged particle number concentration close to the nozzle is overpredicted (Fig. 5.10). This is associated with a large increase of the number of interparticle collisions observed in this region, leading to artifically large agglomerates (Fig. 5.13).

The number of collisions predicted for a particle moving through the spray dryer strongly depends on the calculation method used (Fig. 5.11). A 1-way coupled computation with collisions strongly overpredicts the number of collisions per particle. For some particles, up to 300 000 collisions have been observed, which is far from being realistic. This leads to agglomerates with a much too large size (Fig. 5.13) that contain too many primary particles (Fig. 5.12). In the 2- and 4-way coupled calculations much more particles that experience only up to about 100 collisions are sampled. However, the 2-way coupled calculation still shows a larger number of particles with collision numbers between 1000 and 20 000. The fraction of

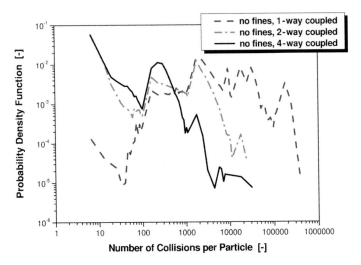

Fig. 5.11 Distribution of the number of collisions sampled for a population of particles in the spray dryer using different calculation methods. A return of fines is not considered.

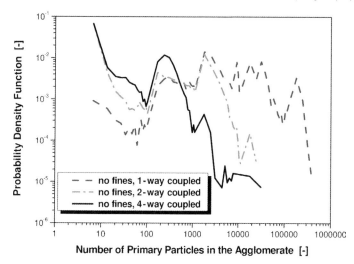

Fig. 5.12 Distribution of the number of primary particles in the agglomerates, sampled for a population of particles in the spray dryer using different calculation methods.

particles in this range is further reduced by applying the 4-way coupling calculation procedure, where the coupling iteration is performed including interparticle collisions and the particle properties (e.g. particle-size distribution) are updated in every iteration.

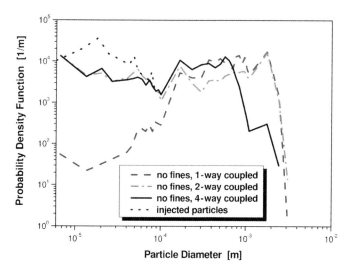

Fig. 5.13 Particle-size distribution, sampled for a population of particles in the spray dryer using different calculation methods.

The size of the agglomerates created during collisions is, of course, strongly related to the number of collisions a particle experiences. As shown in Fig. 5.12, the distributions of the number of primary particles found in the agglomerates after their tracking is completed have similar structures as those of the number of collisions. Applying the 1-way coupling calculation procedure, agglomerates occur consisting of up to 300 000 primary particles. This is not realistic for the present case.

The particle-size distributions of the powder produced in the dryer are shown in Fig. 5.13 for the different calculation methods. All methods lead to a reduction of the number of small particles in the powder population, which can be seen by comparing them with the distribution of the injected particles. The 1-way coupling leads to a very strong decrease of the fraction of small particles and a strong increase of the number of large particles (>200 μm). For the 2-way coupled calculation, the decrease of the number of small particles is not nearly as strong as for the 1-way coupled result, but the number of large particles, especially with diameters above 1 mm, is almost identical. The number of particles in the range 200–1000 μm is reduced compared to the 1-way coupled calculation. The size distribution resulting from the 4-way coupled calculation shows an almost equal number of particles in this middle size range (i.e. between 200 and 500 μm), but almost no agglomerates with sizes above 1 mm. Compared to experimental observations this result looks more reasonable. However, more experimental results from pilot or industrial-scale spray dryers are needed for further validation.

5.6
Prediction of Product Properties

Drying processes can create a huge variety of product properties desired for different application purposes. For powders produced by spray drying, such properties can be the particle-size distribution, particle stability, porosity, bulk solids density, dissolution behavior or wettability. The discussion of this wide topic is concentrated on some approaches and models that have been introduced and used in the frame of CFD during the past few years.

5.6.1
Particle-size Distribution

The prediction of particle-size distributions requires a proper description of both the primary particle or droplet sizes and the changes of size caused by the process (collision followed by agglomeration or coalescence, drop break-up, shrinkage). The topic of size enlargement by agglomeration of colliding particles has already been discussed in detail. The most influencial factors are:
- The collision frequency of particles, which depends on the overall particle number concentration, the average particle Stokes number describing the impact of gas-phase turbulence on the particle fluctuation velocity, and the difference in particle sizes.

- Liquid material properties as surface tension and viscosity, which are important for the outcome of the collision process.

The prediction of these parameters by modeling within the Euler–Lagrange approach has been discussed in the previous sections.

5.6.2
Heat Damage

Heat damage represents a serious loss of product quality, usually to be avoided during a drying process. The combination of gas-flow prediction, particle tracking and droplet-drying models within a CFD simulation presents the opportunity to draw conclusions on the heat damage a particle may face in a spray dryer. This requires knowledge of the thermal sensitivity of the product material. Hence, it has to be estimated what temperature a particle must exceed for how long before the chemical composition of the particle is detrimentally affected. Schwartzbach et al. (2001) suggested a respective approach that can be incorporated in an Euler–Lagrange CFD method for spray drying. They called this approach the "heat-damage index number". Here, a range of critical temperatures is defined above which the product is damaged. For every tracked parcel, the time during which this temperature band is exceeded is sampled during the Lagrangian tracking using a step function H_T, which has the value 1 if overheating occurs and zero if not. After the tracking has been stopped, the sum of the overheating periods is divided by the overall tracking time for the parcel. This results in a time-weighted value of the heat damage. Finally, the sum over all tracked parcels can be calculated, giving an expression for the heat-damage index number:

$$\text{HDIN} = \frac{\sum_i \frac{\sum H_T \cdot \Delta t_n}{\sum \Delta t_n} \cdot \rho_{i,\text{end}} \cdot d_{i,\text{end}}^3}{\sum_i \rho_{i,\text{end}} \cdot d_{i,\text{end}}^3} \tag{5.82}$$

Here, H_T is the value of the step function, and ρ and d are the density and the diameter of the droplets at the end of the tracking. The approach of Schwartzbach et al. (2001) is promising regarding the estimation of heat damage. It has the potential for extensions that may relate the critical temperature values to the chemical changes inside the particles and the times required for these changes.

5.6.3
Particle Morphology

The morphology of particles is another important property of powder produced by spray-drying processes. Rigid solid particles can be formed as well as hollow spheres. Layering of shells consisting of different materials can also occur. These phenomena are governed by the interaction of the evaporation of solvent, the diffusion of solvent in the interior of the particle and the heating of the particle.

During the first period after atomization, the heat transferred from the gas phase to the particle is rate controlling and the droplet is cooled by enhanced evaporation. Inside the droplet, the evaporation creates concentration fields, so that diffusion processes start. Here it should be noticed that the evaporation of solvent is usually followed by a decreased concentration of solvent in the outer zone of the particle, leading to following tendencies:

- The stability of the outer particle shell rises so that the shrinkage of the particle can be hindered. The magnitude of the force resisting the shrinkage depends on the ability of solidifying material to build stable structures, which means on the stability and mobility of the shell created.
- The increased concentration of solidifying material (either as suspension, crystals or amorphous network) leads to backdiffusion of this material towards the center of the particle. The velocity of this process depends on the diffusion coefficient of the dissolved material (or crystals/suspended particles) and the strength of the convective forces caused by the solvent flowing towards the droplet surface.
- If the outer shell is strong enough to hinder the diffusion of solvent significantly, the first drying period is finished. This leads to a reduction of the evaporation rate and, hence, to an increase of droplet temperature.
- Hollow spherical particles are created if the concentration profile of the solid components in the droplet has not become uniform before the appearance of a stable outer solid shell. Additionally, any remaining liquid in the interior of the particle can heat up and start to boil, leading to a gaseous phase inside the droplet. The pressure inside the droplet increases and may lead to growth of particle size in the case of a flexible hull, or to cracking/destruction of the particle if the outer shell is not strong enough to resist the evolving overpressure.

To assess these phenomena, knowledge about the diffusion coefficient in dependence of the solvent content is necessary, as well as knowledge about the flexibility and elasticity of the solidifying material again in dependence of the solvent content. The following general conclusions can be drawn:

The higher the diffusion coefficient of the solvent in the solidifying material, the faster are evaporation and heating of the droplet, and the less likely is the occurrence of hollow spheres. Also, a higher initial solid content leads to a reduced number of hollow spheres. Manipulating the phenomenon, properties such as bulk density and particle stability can be influenced.

Similar observations apply to the creation of shells in particles produced during spray drying of multicomponent liquids (two or more components solidify). If the two solidifying components do not tend to remain well mixed and have very different diffusion velocities in the solvent, then the creation of shells of different materials inside the particle is likely to occur. Here, knowledge about the diffusion coefficients, or the relation of the diffusion velocities of the two components in the solvent is again necessary for modeling in the frame of the Euler–Lagrange approach to become feasible.

5.7 Summary

Computational fluid dynamics is nowadays a tool widely used, also in convective drying technology. The Euler–Lagrange approach, intensively discussed here, is especially suited to describe the two-phase flows occurring in equipment such as spray dryers. To ensure a good quality in modeling of industrial-scale drying processes, the degree of physical detail (elementary processes occurring at the particle scale) that has to be described and incorporated into the model framework is rather high. Though the necessary computational effort becomes very large, it is still manageable on modern computational facilities. The most important elementary processes to be modeled in a spray dryer are flow turbulence, particle transport and interaction with turbulence, droplet drying (evaporation of the solvent from a suspension droplet) and the interaction between the dispersed phase particles (i.e. interparticle collisions). Here, the coupling procedure between the phases and the estimation of fluctuation velocities are of special importance for the prediction of the dispersed-phase concentration field and collision probabilities. Moreover, the gas-phase dynamics are strongly linked with the dispersed phase properties eminently affected by interparticle collisions (four-way coupling). The accuracy with which the properties of particles can be predicted relies on the accuracy of the phase coupling and, hence, on the accuracy of the gas-phase flow-field prediction.

Looking at the dispersed phase, one can definitely state that without considering the change of suspension droplet properties throughout the dryer due to evaporation and accounting for this by considering collisions between different types of particles, a prediction of final product properties would not be a realistic aim. Detailed knowledge about the links between material properties that change in the course of drying and, especially, the collision characteristics is indispensable. A basic task at this point is the proper description of the evaporation process and the estimation of parameters like diffusion coefficients, whereby the drying progress has to be connected with material properties like viscosity, density of particle material and particle morphology.

Nowadays a detailed description of the physics and elementary processes at the microscale is possible, even if still gaps exist, for example about the collision behavior of very viscous droplets dispersed in turbulent flows. The present contribution demonstrated that it is possible to build up a hybrid computational modeling frame, which incorporates all relevant physics down to a certain depth for the prediction of macroscale processes like spray drying. However, further validation of the individual models on the basis of detailed experiments and validation of the overall computational model by measured dried powder characteristics are urgently necessary.

Additional Notation used in Chapter 5

C	mass concentration	kg m^{-3}
c	resistance coefficient	–
F	force	N
f	frequency	s^{-1}
GRN	Gaussian random number	–
k	turbulent kinetic energy	m^2 s^{-2}
P	probability	–
R	correlation coefficient	–
RN	random number	–
S	source term	various
SC	solids mass fraction	–
u	velocity in x-direction	m s^{-1}
v	velocity in y-direction	m s^{-1}
w	velocity in z-direction	m s^{-1}

Greek letters

Γ	general transport coefficient	kg m^{-1} s^{-1}
ε	turbulent dissipation rate	m^2 s^{-3}
η	Kolmogorov velocity	m s^{-1}
σ	root mean square value	various
ϕ	general field variable	various
ω	angular velocity	s^{-1}

Subscripts

E	Eulerian
ev	evaporation
f	fluid
L	Lagrangian
m	referring to moisture
mom	momentum
RMS	roots mean square
T	referring to temperature
t	turbulent
t	referring to time
th	thermal

Superscripts

$'$	temporal deviation variable

References

Abramzon, B., Sirignano, W. A., 1989. Droplet vaporization model for spray combustion calculations. *Int. J. Heat Mass Transfer* **32**: 1605–1618.

Aggarwal, S. K., Tong, A. Y., Sirignano, W. A., 1984. Comparison of vaporization models in spray calculations. *AIAA J.* **22** (10): 1448–1457.

Balzer, G., Simonin, O., 1993. Extension of Eulerian Gas-Solid Flow Modeling to Dense Fluidized Bed Prediction: Refined Flow Modeling and Turbulence Measurements. *Proc. of the 5th Int. Symp., Presses Pont et Chaussees, Paris*, pp. 417–424.

Berlemont, A., Desjonqueres, P., Gouesbet, G., 1990. Particle Lagrangian simulation in turbulent flow. *Int. J. Multiphase Flow* **16**: 19–18.

Berlemont, A., Achim, P., Chang, Z., 2001. Lagrangian approaches for particle collisions: The colliding particle velocity correlation in the multiple particles tracking method and in the stochastic approach. *Physics of Fluids* **13**: 2946–2956.

Bird, R. B., Stewart, W. E., Lightfoot, E. N., 1960. *Transport phenomena*. Wiley, New York.

Blei, S., 2006. *On the Interaction of non-Uniform Particles during the Spray Drying Process: Experiments and Modeling with the Euler-Lagrange Approach*. Dissertation, Shaker Verlag, Aachen, Germany.

Blei, S., Sommerfeld, M., 2004. Computation of agglomeration for non-uniform dispersed phase properties - an extended stochastic collision model. 5th International Conference on Multiphase Flow, ICMF '04, Yokohama, Japan, Paper No. 438.

Cheong, H. W., Jeffreys, G. V., Mumford, C. J., 1986. A receding interface model for the drying of slurry droplets. *AIChE Journal* **32** (8): 1334–1346.

Chrigui, M., Sadiki, A., 2004. Prediction performance of a thermodynamically consistent turbulence modulation for multiphase flows. 3rd International Symposium on Two-Phase Flow Modeling and Experimentation, Pisa, September 2004.

Crank, J., 1967. *Mathematics of diffusion*. University Press, Oxford.

Crowe, C. T., Sharma, M. P., Stock, D. E., 1977. The Particle-Source-in-Cell (PSI-cell) method for gas-droplet flows. *Trans. ASME - J. Fluids Eng*. **99**: 325–332.

Crowe, C. T., Sommerfeld, M., TsujiY., 1998. Multiphase flows with droplets and particles. CRC Press LLC, Boca Raton, USA.

Crowe, C. T., 2005. *Multiphase Flow Handbook*. CRC Press, Taylor & Francis Group, Boca Raton, USA

Decker, S., 2006. *Zur Berechnung gerührter Suspensionen mit dem Euler-Lagrange-Verfahren*. Diss. Martin-Luther-Universität, Halle-Wittenberg, Germany.

Dumouchel, C., Malot, H., 1999. Development of a three-parameter volume-based drop-size distribution through the application of the Maximum Entropy Formalism. Particle and Particle System Characterization **16**: 220–228.

Eckert, E., Drake, R., 1987. *Analysis of heat and mass transfer*. McGrawHills, Education; 2Rev. Edition (December 1959).

Enwald, H., Peirano, E., Almstedt, A.-E., 1996. Eulerian two-phase flow theory applied to fluidization. *Int. J. Multiphase Flow* **22**, Suppl.: 21–66.

Farid, M., 2003. A new approach to modeling of single droplet drying. *Chemical Engineering Science* **58**: 2985–2993.

Ferziger, J. H., 1996. Recent advances in large eddy simulation. In: *Engineering Turbulence Modeling and Experiments 3*, ed. by W. Rodi and G. Bergeles, Elsevier Science, pp. 163–175.

Ferziger, J. H., Peric, M., 2002. *Computational Methods for Fluid Dynamics*. Springer-Verlag, Berlin, 2002.

Gibson, M. M., Launder, B. E., 1978. Ground effects of pressure fluctuations in atmospheric boundary layers. *J. Fluid Mech*. **86**: 491–511.

Gosman, A. D., Ioannides, I. E., 1983. Aspects of computer simulation of liquid-fuelled combustors. *J. Energy* **7**: 482–490.

Guo, B., Fletcher, D. F., Langrish, T. A. G., 2003. Simulation of the agglomeration in a spray using Lagrangian particle tracking.

Applied Mathematical Modeling **28**: 273–290.

Hiller, R. B., 1981. Der Einfluss von Partikelstoß und Partikelhaftung auf die Abscheidung in Faserfiltern. *Diss. Universität Karlsruhe*, VDI-Verlag GmbH Düsseldorf, Germany.

Ho, C. A., Sommerfeld, M., 2002. Modeling of micro-particle agglomeration in turbulent flows. *Chem. Eng. Sci.* **57**: 3073–3084.

Huang, L., Kumar, K., Mujumdar, A. S., 2003. Use of computational fluid dynamics to evaluate alternative spray-dryer chamber configurations. *Drying Technology* **21**(3): 385–412.

Kasagi, N., Shikazono, N., 1995. Contribution of direct numerical simulation to understanding and modeling turbulent transport. *Proc. R. Soc. Lond.* A **451**: 257–292.

Kohnen, G., Rüger, M., Sommerfeld, M., 1994. Convergence behavior for numerical calculations by the Euler-Lagrange method for strongly coupled phases. In: *Numerical Methods for Multiphase Flows*, ed. by C. T. Crowe, R.Johnson, A.Prosperetti, M.Sommerfeld and Y. Tsuji, ASME Fluids Engineering Division Summer Meeting, Lake Tahoe, U.S.A., June 1994, *ASME FED* **185**: 191–202.

Laín, S., Bröder, D., Sommerfeld, M., Götz, M. F., 2002a. Modeling hydrodynamics and turbulence in a bubble column using the Euler-Lagrange procedure. *Intern. J. Multiphase Flow* **28**: 1381–1407.

Laín, S., Sommerfeld, M., Kussin, J., 2002b. Experimental studies and modeling of four-way coupling in particle-laden horizontal channel flow. *Int. Journal of Heat and Fluid Flow* **23**: 647–656.

Landau, L. D., Lifschitz, E. M., 1991. *Hydrodynamik*. Akademie Verlag, Berlin, Germany.

Langrish, T. A. G., Fletcher, D. F., 2003. Prospects for the modeling and design of spray dryers in the 21st century. *Drying Technology* **21**(2): 197–215.

Langrish, T. A. G., Zbicinski, I., 1994. The effects of air inlet geometry and spray cone angle on the wall deposition rate in spray dryers. *Trans I. Chem. E* . **72**, part A: 420–430.

Law, C. K., 1976. Unsteady droplet vaporization with droplet heating. *Combustion and Flame* **26**: 17–22.

Law, C. K., Sirignano, W. A., 1977. Unsteady droplet combustion with droplet heating - II: conduction limit. *Combustion and Flame* **29**: 175–186.

Law, C. K., 1982: Recent advances in droplet vaporization and combustion. Prog. Energy Combustion Sci. **8**: 171–201.

Lesieur, M., Metais, O., Comte, P., 2005. *Large-Eddy Simulations of Turbulence*. Cambridge University Press, Cambridge, U.K.

Löffler, F., 1988. *Staubabscheiden*. Thieme VerlagStuttgart, Germany.

Launder, B. E., Spalding, D. B., 1974. The numerical computation of turbulent flows. *Comp. Meth. Appl. Mech. and Eng.* **3**: 269–289.

Launder, B. E., 1990. Phenomenological modeling: Present and future. *Proc. Whither Turbulence Workshop*, ed. by J.Lumley, Lecture Notes in Physics 357, Springer-Verlag, Berlin, Germany.

Magnaudet, J. J. M., 1997. The forces acting on bubbles and rigid particles. In: *ASME Fluids Engineering Division Summer Meeting, Vancouver, Canada*, Paper No. FEDSM97–3522 (1997).

McComb, W. D., 1990. *The physics of fluid turbulence*. Clarendon Press, Oxford University Press, Oxford, U.K.

Mei, R., 1992. An approximate expression for the shear lift force on a spherical particle at finite Reynolds number. *Int. J. Multiphase Flow* **18**: 145–147.

Menn, P., 2005. *Charakterisierung von Sprays - Experimentelle PDA-Messergebnisse und ihre mathematische Nachbearbeitung*. Diss., Universität Bremen, Germany.

Nijdam, J. J., Guo, B., Fletcher, D. F., Langrish, T. A. G., 2004. Challenges of simulating droplet coalescence within a spray. *Drying Technology* **22**(6): 1463–1488.

Nowakowski, B., Sitarski, M., 1981. Brownian coagulation of aerosol particles by Monte Carlo Simulation. *Journal of Colloid and Interface Science* **83**: 606–613.

Oakley, D. E., 1994. Scale-up of spray dryers with the aid of computational fluid dynamics. *Drying Technology* **12**(1&2): 217–233.

Oakley, D. E., 2004. Spray dryer modeling in theory and practice. *Drying Technology* 22(6): 1371–1402.

O'Rourke, P. J., 1981. Collective drop effects on vaporizing liquid sprays. PhD-thesis, Princeton University, Princeton, USA.

Patankar, S. V., Spalding, D. B., 1972. A calculation procedure for heat, mass and momentum transfer in three-dimensional parabolic flows. *Int. J. Heat and Mass Transfer* 110: 1787–1806.

Peric, M., 1985. A finite volume method for the prediction of three-dimensional fluid flow in complex ducts. PhD- thesis, University of London, London, U.K.

Pinsky, M., Khain, A., Shapiro, M., 1999. Collisions of small drops in a turbulent flow. Part I: Collision efficiency. Problem formulation and preliminary results. *J. of the Atmospheric Science:* 56: 2585–2600.

Platzer, E., Sommerfeld, M., 2003. Modeling of turbulent atomisation with an Euler/Euler approach including the drop size prediction. Paper 2-3, ICLASS 2003, Sorrento, Italy.

Podvysotsky, A. M., Shraiber, A. A., 1984. Coalescence and break-up of drops in two-phase flows. *Int. J. Multiphase Flow* . 10: 195–209.

Pope, S. P., 2000. *Turbulent Flows.* Cambridge University Press, Cambridge, U.K.

Prakash, S., Sirignano, W. A., 1978. Liquid fuel droplet heating with internal circulation. *Int. J. Heat Mass Transfer* 21: 885–895.

Prakash, S., Sirignano, W. A., 1980. Theory of convective droplet vaporization with unsteady heat transfer in the circulating liquid phase. *Int. J. Heat Mass Transfer* 23: 253–268.

Prandtl, L., 1945. Über ein neues Formelsystem für die ausgebildete Turbulenz. *Nachr. Akad. Wiss., Göttingen, Math.-Phys* . K1, 6–19.

Qian, J., Law, C. K., 1997. Regimes of coalescence and separation in droplet collision. *J. Fluid Mech.* 331: 59–80.

Ranz, W. E., Marshall Jr., W. R., 1952. Evaporation from drops, part I. *Chem. Eng. Prog* . 48(3): 141–146, 173–180.

Rizk, M. A., Elghobashi, S. E., 1989. A two equation turbulence model for dispersed dilute confined two-phase flow. *Intern. J. Multiphase Flow* 15: 119–133.

Rodi, W., 1988. Recent developments in turbulence modeling. *Proc. 3rd Int. Symp. on Refined Flow Modeling and Turbulence Measurements*, Tokyo, Japan.

Rogers, M. M., Moin, P., 1987. The structure of the vorticity field in homogeneous turbulent flows. *J. Fluid Mech* . 176: 33–66.

Roos, Y. H., 2002. Importance of glass transition and water activity to spray drying and stability of dairy products. *Lait* 82: 475–484.

Rotta, J. C., 1972. *Turbulente Strömungen.* B. G. Teubner, Stuttgart, Germany.

Rüger, M., Hohmann, S., Sommerfeld, M., Kohnen, G., 2000. Euler-Lagrange calculations of turbulent sprays: the effect of droplet collisions and coalescence. *Atomization and Sprays* 10: 47–81.

Sano, Y., Keey, R. B., 1982. The drying of a spherical particle containing colloidal material into a hollow sphere. *Chem. Eng. Science* 37(6): 881–889.

Schiestel, R., 1987. Multiple-scale modeling of turbulent flows in one-point closures. *Phys. Fluids* 30: 722–731.

Schiller, L., Naumann, A., 1933. Über die grundlegenden Berechnungen bei der Schwerkraftaufbereitung. *Ver. Deut. Ing.* 77: 318–320.

Schmidt, E., Müller, O., 1997. Strömungskräfte auf Partikeln in Gasen. VDI-Verlag GmbH, Düsseldorf, Germany.

Schuch, G., Löffler, F., 1978. Über die Abscheidewahrscheinlichkeit von Feststoffpartikeln an Tropfen in einer Gasströmung durch Trägheitseffekte. *Verfahrenstechnik* 12: 302–306.

Schwartzbach, C., Nikas, K., Bergeles, G., 2001. A heat damage index number (HDIN) as an indicator of spray dryer suitability. In: *Proc. Spray Drying '01 and Related Processes* ed. by P.Walzel, Dortmund, Germany.

Shiriani, E., Ferziger, J. H., Reynolds, W. C., 1981. Mixing of passive scalar in isotropic and sheared homogeneous turbulence. Stanford University Report, No. TF-15.

Simonin, O., Deutsch, E., Boivin, M., 1993. Comparision of large eddy simulation and second-moment closure of particle fluctuating motion in two-phase turbulent shear flows. *Proceedings, 9th Symposium on Turbulent Shear Flows*, Vol. 2, Session 11–22, Paper 15–2, Japan.

Simonin, O., 2000. Statistical and continuum modeling of turbulent reactive particulate flows; Part I: Theoretical derivation of dispersed phase Eulerian modeling from probability density function kinetic equation. Lecture Series 2000–06, Theoretical and Experimental Modeling of Particulate Flow, ed. by J.-M. Buchlin, von Karman Institute for Fluid Dynamics, Belgium.

Sirignano, W. A., 1993. Fluid dynamics of sprays – 1992 Freeman Scholar Lecture. *J. Fluids Eng*. **115**: 345–378.

Smagorinsky, J., 1963. General circulation experiments with the primitive equations, Part 1: The basic experiment. *Mon. Wea. Rev.* **91**: 99.

Sommerfeld, M., Kohnen, G., Rüger, M., 1993. Some open questions and inconsistencies of Lagrangian particle dispersion models. In: *Proceedings of the Ninth Symposium on Turbulent Shear Flows*, paper no. 15–1, Kyoto, Japan.

Sommerfeld, M., 2000. Theoretical and experimental modeling of particulate flow: Overview and fundamentals. Von Karman Institute for Fluid Mechanics, Lecture Series No. 2000–6, Rhode Saint Genèse, Belgium.

Sommerfeld, M., 2001. Validation of a stochastic Lagrangian modeling approach for inter-particle collisions in homogenious isotropic turbulence. *Int. J. Multiphase Flow* **27**: 1829–1858.

Sommerfeld, M., 2005. Bewegung fester Partikel in Gasen und Flüssigkeiten. In: *VDI-Wärmeatlas, 9. Auflage*, Springer-Verlag, Berlin/Heidelberg, Germany.

Southwell, D. B., Langrish, T. A. G., Fletcher, D. F., 1999. Process intensification in spray dryers by turbulence enhancement. *Trans I. Chem. E*. **77**, part A: 189–205.

Southwell, D. B., Langrish, T. A. G., Fletcher, D. F., 2001. Use of computational fluid dynamics techniques to assess design alternatives for the plenum chamber of a small spray dryer. *Drying Technology* **19**(2): 257–268.

Sperling L. H. (1992): *Introduction to physical polymer science*. John Wiley and Sons New York USA.

Squires, K. D., Eaton, J. K., 1992. On the modeling of particle-laden turbulent flows. *Proc. 6th Workshop on two-phase flow predictions*, ed. by M.Sommerfeld, Erlangen, pp. 220–229.

Stevenson, M. J., Chen, X. D., Fletcher, A., 1998. Modeling the drying of milk. Chemeca 98, Queensland, Australia, paper 177.1.

Stone, H. L., 1968. Iterative solution of implicit approximation of multidimensional partial differential equations. *J. Num. Anal* . **5**(3): 530–557.

vanStraatsma, J., Houwelingen, G., Steenbergen, A. E., De Jong, P., 1999. Spray drying of food products: 1. Simulation models. *J. Food Engineering* **42**: 67–72.

Tong, A. Y., Sirignano, W. A., 1983. Analysis of vaporizing droplet with slip, internal circulation and unsteady liquid phase heat transfer. *ASME/JSME Thermal Engineering Joint Conference Proceedings*, Vol. 2, pp. 481–487.

Verdurmen, R. E. M., Menn, P., Ritzert, J., Blei, S., Nhumaio, G. C. S., Sonne, S. T., Gunsing, M., Straatsma, J., Verschueren, M., Sibeijn, M., Schulte, G., Fritsching, U., Bauckhage, K., Tropea, C., Sommerfeld, M., Watkins, A. P., Yule, A.J., Schönfeldt, H., 2004. Simulation of agglomeration in spray drying installations: the EDECAD project. *Drying Technology* **22**(6): 1403–1461.

vonKarman, T., Horwarth, L., 1938. On the statistical theory of isotropic turbulence. *Proc. Royal Society London A164*: 192–215.

Wijlhuizen, A. E., Kerkhof, P. A. J. M., Bruin, S., 1979. Theoretical study of the inactivation of phosphatase during spray drying of skim-milk. *Chem. Eng. Science* **34**: 651–660.

Zbicinski, I., Grad, J., Strumillo, C., 1996. Effect of turbulence on heat and mass transfer in the atomisation zone. *Drying-Technology* **14**(2): 231–244.

6
Numerical Methods on Population Balances
Jitendra Kumar, Mirko Peglow, Gerald Warnecke, Stefan Heinrich, Evangelos Tsotsas, Lothar Mörl, Mike Hounslow, Gavin Reynolds

6.1
Introduction

Most commercial processes in drying technology are concerned with discrete or particulate materials. In these processes individual particles will vary in properties and, as a consequence, also vary in drying behavior. The most obvious example is that the particles may vary in size and, as a consequence, dry at different rates. The purpose of this chapter is to explain how problems of this type can be posed and solved. The first task is to identify a method of representation for processes where discrete particles are present and each particle can have a variety of properties. The method we adopt is to describe the distribution of particle properties by a population density function. So if particle size, x, is the distributed property, then the distribution of particle sizes would be given by $n(t, x)$, where the number of particles per unit volume (or some other basis) with sizes in the range $(x, x + dx)$ is $n(t, x) \, dx$. It follows that the total number of particles smaller than some size x is

$$N(t,x) = \int_0^x n(t, \varepsilon) d\varepsilon \tag{6.1}$$

As Hulburt and Katz (1964) have shown, this method may readily be extended to situations where the particles have multiple properties – termed by them multiple internal coordinates. So, for example, each particle might be characterized by its size and its moisture content. The choice of how many internal coordinates might be considered is essentially a practical one. If a system can be described by one coordinate only, then there is no need to add further coordinates. If, however, the drying behavior is a complex function of material properties and particles states, then a longer list of coordinates might be desirable. In this case, practical computational limits usually mean that two or at most three coordinates are considered.

It is also possible that the distribution of properties will vary in space. In this way the population density will be a function of spatial – or in Hulburt and Katz's language external coordinates. If these are given by a vector $\mathbf{x_e}$, and the internal coordinates by a vector $\mathbf{x_i}$, the total coordinate space is given by $\mathbf{x} = (\mathbf{x_i}, \mathbf{x_e})$.

The next step is to recognize that it is possible to write a conservation statement – or population balance equation (PBE) – that describes how the number density varies with time and in space. Various derivations of such an equation can be found, most famously by Hulburt and Katz (1964), Randolph and Larson (1978) and Ramkrishna (2000). In essence, for a small volume in the space of internal and external coordinates the rate of change of the number density must match the divergence in the flux of particles and source and sink terms (called birth and death terms, B and D). The flux is, of course, the local product of velocity

$$\mathbf{V} = \frac{dx}{dt} \qquad (6.2)$$

and number density. The velocity components related to the external coordinates give the conventional speeds, while those related to the internal coordinates describe, for example, the rate of change of size, or moisture content. The population balance equation is then

$$\frac{\partial n}{\partial t} + \nabla \cdot (\mathbf{V}n) = B(t, \mathbf{x}) - D(t, \mathbf{x}) \qquad (6.3)$$

In order to use the PBE it is necessary that the velocities are known as functions of position in space and that rate laws can be found for the birth and death terms. Finally, numerical techniques are required for their solution.

In order to illustrate the significance of population balances in drying, two applications are considered, namely drying of particles in a continuous fluidized bed and simultaneous particle size enlargement and drying.

Application 1: The first example is a continuous fluidised bed dryer. The wet solid material can be conveyed into the dryer by adequate equipment such as a votary value. At the same time the dried product is discharged so that the total mass of solid in the apparatus remains constant (Fig. 6.1). A traditional approach for modeling of this process considers the disperse solids as a phase with average properties like particle size, moisture content and enthalpy. Such a model will always predict a uniform moisture content for the solids in the dryer. However, Kettner et al. (2006) demonstrated that uniform properties do not occur in practice. They have measured the moisture content of single particles in a lab-scale dryer using a combination of NMR and coulometric techniques. Some typical results of these measurements are presented in Fig. 6.2. It can be seen that the moisture of dried solids is not uniform, but widely distributed. Kettner et al. (2006) claim that different residence times, or in other words the age of the particles, cause such broad distributions. One might also argue that spatial distributions may have an impact, which is certainly the case for large-scale fluidized bed dryers with horizontal product transport. This does not apply to lab scale equipment. By tracer experiments, Burgschweiger and Tsotsas (2002) proved that such small fluidized bed dryers can be treated as a well-mixed

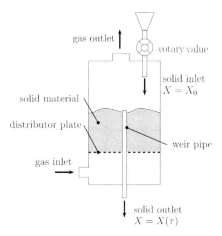

Fig. 6.1 Scheme of a continuous fluidized-bed dryer.

system where spatial distributions of the solid phase can be neglected. Thus, we concentrate on the residence time or, respectively, the age of particles as the only additional particle property. As soon as the age is introduced as a new property of solids, the moisture and enthalpy need to be considered as distributed properties, because they depend directly on the age of particles. Thus, a number of internal coordinates have been identified that are required to be incorporated in a more precise drying model. They are the moisture of solids l, the enthalpy h and the age of particles τ.

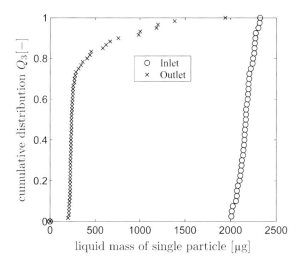

Fig. 6.2 Measured moisture distribution of single particles at the solid inlet and outlet of a continuous fluidized-bed dryer.

The population balance approach allows now to combine the distributed properties of the solid phase in one equation. For a well-mixed system the temporal change of number density distribution of the solid phase can be derived to

$$\frac{\partial f}{\partial t} + \frac{\partial G_\tau \cdot f(\tau,l,h)}{\partial \tau} + \frac{\partial G_l \cdot f(\tau,l,h)}{\partial l} + \frac{\partial G_h \cdot f(\tau,l,h)}{\partial h} = \dot{f}_{in} - \dot{f}_{out} \quad (6.4)$$

In this equation, the parameter f indicates the number density distribution of particles. The second term on the left-hand side represents the aging of particles with $G_\tau = 1$, the third term symbolizes the drying, where the variable G_l indicates the drying rate of particles in kg s^{-1}. The fourth expression corresponds to the change of enthalpy due to heat and mass transfer, whereby G_h denotes the rate of change of enthalpy. The two variables on the right-hand side are a source and a sink for particles being conveyed and discharged, respectively. Burgschweiger and Tsotsas (2002) have shown that an extended model based on population balances predicts the performance of a fluid-bed dryer more precisely than traditional models.

Application 2: The second example concerns the process of particle formulation in fluidized beds. Generally there are three main mechanisms; growth, agglomeration and breakage, that influence the size of particles. For simplicity only agglomeration will be considered in this example. The size enlargement by agglomeration transforms fine primary particles into an instant, free-flowing and dustless product. The fluidized bed technology offers the possibility to combine agglomeration and drying in a single apparatus. In the literature, many attempts can be found to describe particle formation in fluidized beds in terms of population balances, but usually particle size or particle volume has been considered as the only significant internal coordinate. The influence of operating conditions on the evolution of particle size has been investigated by various authors (Adetayo et al. 1995; Watano et al. 1996; Schaafsma et al. 1998). Watano et al. (1996) point out that moisture content is one of the most important particle properties to control the agglomeration process. It is obvious that such a significant particle property needs to be incorporated in a more complex model of the solid phase.

Figure 6.3 depicts exemplarily, how the mechanisms of agglomeration, drying and wetting are coupled. Let us consider two primary particles with a dry solid volume u and $v - u$. Both particles contain a certain amount of liquid denoted by γ and $l - \gamma$, respectively. It is clear that the solid volume and the amount of liquid of a newborn particle are v and l, respectively. Thus, the agglomeration influences not only the particle size but also the moisture distribution. During the agglomeration process the particles are wetted with a liquid binder solution. Thus, the particle moisture will be increased. In addition, the particles will be dried due to contact with the hot fluidization gas. Therefore, the amount of liquid within a particle is influenced by drying and wetting but not by the particle size. However, drying and wetting rates may depend on the size of particles. The consequence of a size-dependent drying rate is a nonuniform distribution of moisture within the

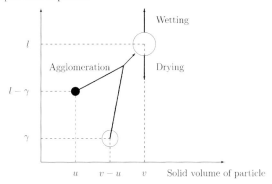

Fig. 6.3 Coupled mechanisms of agglomeration, drying and wetting.

disperse phase. Similarly to the moisture content, the enthalpy will not only change by the agglomeration of particles, but also by the size-dependent heat and enthalpy transfer rates. The population-balance approach can be used to model the combined processes of agglomeration and drying as

$$\frac{\partial f(t,v,l,h)}{\partial t} + \frac{\partial G_l \cdot f(t,v,l,h)}{\partial h} + \frac{\partial G_h \cdot f(t,v,l,h)}{\partial h}$$
$$= \frac{1}{2}\int_0^v\int_0^l\int_0^h \beta \cdot f(t,v-u,l-\gamma,h-\epsilon) \cdot f(t,u,\gamma,\epsilon)\, d\epsilon\, d\gamma\, du \qquad (6.5)$$
$$- \int_0^\infty\int_0^\infty\int_0^\infty \beta \cdot f(t,v,l,h) \cdot f(t,u,\gamma,\epsilon)\, d\epsilon\, d\gamma\, du$$

Again, f denotes the number density distribution. The advection terms on the left-hand side of the equation represent the change of mass of liquid and enthalpy, respectively. The parameter G_l indicates the rate of change of liquid mass. Positive and negative values of G_l correspond to wetting and drying. The expression on the right-hand side represents the birth and death of particles due to agglomeration. The triple integral explains that three independent particle properties are considered. The above equation provides the possibility to incorporate additional particle properties in the agglomeration kinetics β. Thus, the function β might not only depend on particle size but also on moisture content, which is certainly the case in practice.

These and other applications will be analyzed in detail in the other parts of the Modern Drying Technology series. In the present chapter, we will try to provide the background necessary for this analysis by systematically discussing numerical methods for the solution of population balance equations.

6.2
Pure Breakage

6.2.1
Population-balance Equation

Population balances for breakage are widely known in high-shear granulation, comminution, crystallization and atmospheric science. The general form of population-balance equation (PBE) for breakage is

$$\frac{\partial n(t, x)}{\partial t} = \int_x^\infty b(t, x, \varepsilon) S(t, \varepsilon) n(t, \varepsilon) \mathrm{d}\varepsilon - S(t, x) n(t, x) \tag{6.6}$$

The breakage function $b(t, x, \varepsilon)$ is the probability density function for the formation of particles of size x from particles of size ε. The selection function $S(t, x)$ describes the rate at which particles of size x are selected to break. Equation 6.6 can only be solved analytically for very simple forms of the breakage and the selection functions (Ziff 1991; Ziff et al. 1985; Dubovskii et al. 1992). Therefore, numerical methods are needed and fall into several categories: stochastic (Lee and Matsoukas 2000; Mishra 2000), finite-element (Everson et al. 1997), sectional (Kumar and Ramkrishna 1996a, 1996b), and moment methods (Kostoglou and Karabelas 2002, 2004).

The stochastic methods (Monte Carlo) are very efficient for solving multidimensional population balance equations, since other techniques become computationally very expensive in such cases. A wide variety of finite-element, weighted-residual, orthogonal-collocation, and Galerkin methods are also used for solving population-balance equations. In these methods, the solution is approximated as a linear combination of basis functions over a finite number of subdomains. In recent times, the sectional methods have become computationally very attractive (Vanni 2000). In moment methods, the fragmentation equation is transformed into a system of ODEs describing the evolution of the moments of the particle-size distribution. These only give severely limited information on the distribution function.

6.2.2
Numerical Methods

Sectional methods are the most important alternative for solving PBEs since they are simple to implement and predict particle properties accurately. Several sectional methods have been recently proposed by Hill and Ng (1995), Kumar and Ramkrishna (1996), as well as Vanni (1999). Our objective is to first briefly discuss previous discretized methods and then propose a new approach that is more accurate and general.

Hounslow et al. (1988) proposed for aggregation problems the first discretized method that preserves the first two moments. Following this, Hill and Ng (1995) developed a discretized method for the general breakage PBE. They used two correction factors in the discretized equation and imposed two conditions in order to calculate them: the correct evaluation of total mass (the first moment) and the correct evaluation of total number (the zeroth moment). Since the calculation of the correction factors was not possible for the general case, they considered three different forms of the breakage function and a special form of the selection function $S(x) = S_0 x^\alpha$. Afterwards, Vanni (1991) modified the method of Hill and Ng (1995) to make it more general. Keeping the entire formulation the same, he changed the second condition to calculate the correction factor. He considered the correct evaluation of the death term, while Hill and Ng (1995) imposed the correct prediction of the total number of particles. The method becomes more general but leads to less accurate solutions.

Following the approach of Eyre et al. (1988), Hounslow et al. (2001) investigated a new discretization for breakage PBE. They calculated the selection and the breakage functions by imposing the condition of correct evaluation of the total number and the movement of granule volume from one interval to another. The formulation predicts the total number and the total mass correctly, but the selection and the breakage functions require computation of many single and double integrals. Later, Tan et al. (2004), used this discretization on fluidized bed melt granulation.

Kumar and Ramkrishna (1996a) developed the so-called fixed-pivot method that is more general than previously existing techniques. This method divides the entire range into small sections of variable size, each section having its representative size. It relies on assigning the particles that do not appear at the representative size of any section to the neighboring representatives. The technique not only preserves the number and mass of the particles, but it can also be generalized for the preservation of many desired properties of the population.

The main advantage of the fixed-pivot technique, besides simplicity, is its generality and consistency with selected moments. Moreover, it can easily be formulated for consistency of any two desired properties of the population. Kumar and Ramkrishna considered that particles with number concentrations N_i, $i = 1, 2, 3\ldots, I$ are sitting at sizes x_i, $i = 1, 2, 3\ldots, I$, respectively. Mathematically, the number density function can be represented in terms of a Dirac-delta distribution as

$$n(t, x) = \sum_{i=1}^{I} N_i \delta(x - x_i) \tag{6.7}$$

Their final set of discretized equations for the exact preservation of number and mass take the form

$$\frac{dN_i(t)}{dt} = \sum_{k=i}^{I} \eta_{i,k} S_k(t) N_k(t) - S_i(t) N_i(t) \tag{6.8}$$

where

$$\eta_{i,k} = \int_{x_i}^{x_{i+1}} \frac{x_{i+1} - x}{x_{i+1} - x_i} b(x, x_k) dx + \int_{x_{i-1}}^{x_i} \frac{x - x_{i-1}}{x_i - x_{i-1}} b(x, x_k) dx \tag{6.9}$$

The fundamental concept behind the fixed-pivot technique can be summarized as follows. Suppose a new particle of a size that is not a representative of any cell appears due to breakage of larger particles. The particle has to be divided between neighboring representatives in such a way that number and mass are conserved. In this process, numerical diffusion is possible due to the assignment of particles to representatives to whom they do not really belong. In the following we present a new approach to reduce the numerical diffusion based on taking average of all newborn particles in a cell. If the average value does not belong to the representative of the cell, then the particles have to be divided into neighboring cells in order to have a scheme consistent with moments. We will see that the fixed-pivot method is a first-order method, whereas our new method is of second order.

We will present two new approaches for solving breakage PBE. One is based on the cell-average technique recently proposed by Kumar et al. (2006a) for aggregation problems. The other is based on a very general finite-volume scheme. This approach was originally applied by Filbet and Laurenot (2004) for aggregation problems by transforming the number density PBE to a mass-based conservation law. Then, the idea of finite-volume schemes can easily be implemented.

6.2.2.1 The Cell-average Technique

The fixed-pivot technique can be improved by taking the volume average of all the particles coming from a breakage event within a cell. Kumar et al. (2006a) proposed the cell-average technique for aggregation problems. It was shown that instead of assigning each particle immediately to neighboring nodes, it is better to take first the average of all particles appearing within the cell and then assign them to the nodes neighboring the average value. The mathematical formulation of the cell-average technique for breakage problems can be established in the following way. The total birth rate of particles in an interval is calculated as

$$B_{\text{break},i} = \int_{x_{i-1/2}}^{x_{i+1/2}} \int_{x}^{\infty} b(t, x, \epsilon) S(t, \epsilon) n(t, \epsilon) d\epsilon dx \qquad (6.10)$$

Substitution of $n(t, x)$ from Eq. (6.7) gives

$$\hat{B}_{\text{break},i} = \sum_{k \geq i} N_k(t) S_k(t) \int_{x_{i-1/2}}^{p_k^i} b(t, x, x_k) dx \qquad (6.11)$$

where $p_k^i = x_i$ for $k = i$ and $p = x_{i+\frac{1}{2}}$ elsewhere. Similarly, the volume flux into this cell is

$$\hat{M}_{\text{break},i} = \sum_{k \geq i} N_k(t) S_k(t) \int_{x_{i-1/2}}^{p_k^i} x b(t, x, x_k) dx \qquad (6.12)$$

Now, the volume average of all the particles can be evaluated as

$$\bar{a}_i = \frac{\hat{M}_{\text{break},i}}{\hat{B}_{\text{break},i}} \tag{6.13}$$

We assume that the newborn particles $B_{\text{break},i}$ are assigned temporarily at $\bar{a}_i$. These particles have to be divided depending upon the value of $\bar{a}_i$ between neighboring nodes in such a way that the formulation is consistent with the total number and mass. Suppose that the average value $\bar{a}_i$ is greater than x_i and the fractions a and b are the birth contributions assigned to x_i and x_{i+1}, respectively. For the consistency of the total number and mass, these fractions must satisfy

$$\begin{aligned} a + b &= \hat{B}_{\text{break},i} \\ ax_i + bx_{i+1} &= \hat{B}_{\text{break},i}\bar{a}_i \end{aligned} \tag{6.14}$$

Solving the above equations, we get

$$\begin{aligned} a &= \hat{B}_{\text{break},i}\lambda_i^+(\bar{a}_i) \\ b &= \hat{B}_{\text{break},i}\lambda_{i+1}^-(\bar{a}_i) \end{aligned} \tag{6.15}$$

where the functions λ_i^+ and λ_i^- are defined as

$$\lambda_i^\pm(x) = \frac{x - x_{i\pm 1}}{x_i - x_{i\pm 1}} \tag{6.16}$$

The birth contribution at x_i from all possible cells is shown in Fig. 6.4. Collecting all contributions, the birth term can be modified as

$$\begin{aligned} \hat{B}_{\text{break},i}^{\text{CA}} &= \hat{B}_{\text{break},i-1}\lambda_i^-(\bar{a}_{i-1})H(\bar{a}_{i-1} - x_{i-1}) + \hat{B}_{\text{break},i}\lambda_i^-(\bar{a}_i)H(x_i - \bar{a}_i) \\ &\quad + \hat{B}_{\text{break},i}\lambda_i^+(\bar{a}_i)H(\bar{a}_i - x_i) + \hat{B}_{\text{break},i+1}\lambda_i^+(\bar{a}_{i+1})H(x_{i+1} - \bar{a}_{i+1}) \end{aligned} \tag{6.17}$$

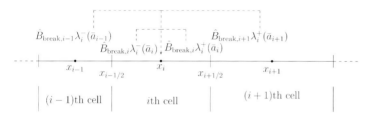

Fig. 6.4 Assignment of particles at x_i from all possible cells.

The discontinuous Heaviside function or *unit step function* is here defined by

$$H(x) = \begin{cases} 1, & x > 0 \\ \frac{1}{2}, & x = 0 \\ 0, & x < 0 \end{cases} \quad (6.18)$$

The death term $S(x)\,n(t, x)$ can be discretized by substituting $n(t, x)$ from Eq. 6.7, to give

$$D_{\text{break},i}^{\text{CA}} = S_i N_i \quad (6.19)$$

Now, the final set of discrete equations can be written as

$$\frac{dN_i}{dt} = \hat{B}_{\text{break},i-1}\lambda_i^-(\bar{a}_{i-1})H(\bar{a}_{i-1} - x_{i-1}) + \hat{B}_{\text{break},i}\lambda_i^-(\bar{a}_i)H(x_i - \bar{a}_i)$$

$$+ \hat{B}_{\text{break},i}\lambda_i^+(\bar{a}_i)H(\bar{a}_i - x_i) + \hat{B}_{\text{break},i+1}\lambda_i^+(\bar{a}_{i+1})H(x_{i+1} - \bar{a}_{i+1}) - S_i N_i,$$

$$i = 1, 2, \ldots, I \quad (6.20)$$

The formulation 6.20 is a mathematical model for the assignment of particles at the ith representative size from neighboring cells. The first and the fourth terms describe the particle birth coming from the $(i - 1)$th and the $(i + 1)$th cells, respectively. The second or the third term is the fraction of the particle birth in the ith cell.

Discussion: Consider the example of a uniform binary breakage function with monodisperse particles of size unity as initial condition. Now we compute the rate of change of particles in a cell $j < I$ at time $t = 0$ using the fixed-pivot technique (FP) and the cell-average technique (CA) calculating all the integrals appearing in the schemes exactly. Note that I is the total number of cells and $x_I = 1$. The rate of change of particles in the jth cell according to the fixed-pivot technique is given by

$$\left(\frac{dN_j}{dt}\right)_{\text{FP}} = N_I S_I(x_{i+1} - x_{i-1}) \quad (6.21)$$

Since the cell averages lie exactly at the center in this case, i.e. $\bar{a}_j = x_j$ for $j = 1, 2, \ldots I - 1$, the cell average technique gives

$$\left(\frac{dN_j}{dt}\right)_{\text{CA}} = 2N_I S_I(x_{i+1/2} - x_{i-1/2}) \quad (6.22)$$

The analytical birth rate at $t = 0$, that is

$$\left(\frac{dN_j}{dt}\right)_{\text{ana}} = 2N_I S_I(x_{i+1/2} - x_{i-1/2}) \quad (6.23)$$

and the numerical solutions by the cell-average technique are, in this case, the same. This is because the cell averages lie exactly on the pivots, so that no division is made in

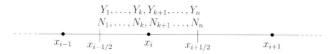

Fig. 6.5 Birth of particles in a cell.

order to be consistent with the first two moments. Though it seems logical not to divide particles that are already concentrated at the representative size of a cell, the fixed-pivot technique always distributes particles to the neighboring cells, independently of available local information.

Consider now a more general example shown in Fig. 6.5. As a result of aggregation, breakage or any other event, particles $N_1, N_2, \ldots, N_n$ appear at the positions $Y_1, Y_2, \ldots, Y_n$ in cell i, with $Y_k < x_i$ and $Y_{k+1} > x_i$. In order to assign them to the pivot in such a way that particle number and mass remain conserved, we use the fixed-pivot and the cell-average mechanisms. The particles assigned to x_i by the fixed-pivot mechanism are

$$N_{\text{FP}} = \sum_{j=1}^{k} \left(\frac{Y_j - x_{i-1}}{x_i - x_{i-1}} \right) N_j + \sum_{j=k+1}^{n} \left(\frac{x_{i+1} - Y_j}{x_{i+1} - x_i} \right) N_j \qquad (6.24)$$

According to the cell-average mechanism, the total number of particles assigned to x_i is calculated by either

$$N_{\text{CA}} = \left(\frac{\bar{a} - x_{i-1}}{x_i - x_{i-1}} \right) \sum_{j=1}^{n} N_j, \quad \text{if } \bar{a} \leq x_i \qquad (6.25)$$

or

$$N_{\text{CA}} = \left(\frac{x_{i+1} - \bar{a}}{x_{i-1} - x_i} \right) \sum_{j=1}^{n} N_j, \quad \text{if } \bar{a} \geq x_i \qquad (6.26)$$

depending upon the position of $\bar{a}$, which is defined as

$$\bar{a} = \frac{\sum_{j=1}^{n} Y_j N_j}{\sum_{j=1}^{n} N_j} \qquad (6.27)$$

It is easy to verify that $N_{\text{CA}} \geq N_{\text{FP}}$. Equality holds if all particles appear at the same side of the representative x_i, i.e. $N_j = 0$ for all $j \geq k+1$ or $N_j = 0$ for all $j \leq k$.

From the preceding discussion it is evident that the cell-average technique keeps more information about the original particles that belong to the cell during the consistency process. When the average of particles is equal to the representative size no distribution to neighboring cells takes place. Assignment of particles to neighboring cells causes numerical diffusion. Since the cell-average technique maintains consistency with less particle distribution to neighboring cells, we expect

Tab. 6.1 Binary breakage, CA and $S(v) = v$, $t = 1000$

Grid points, I	Error, L_1	EOC
61	32.8591	–
122	11.6129	1.50
244	3.4487	1.75
488	0.9378	1.88

the technique to be more accurate with respect to local moments (particle-size distribution) and less diffusive. In the following we will compare cell average with fixed pivot for two test cases where analytical results are available.

Numerical results: The following problems have been considered for the comparison

- binary breakage, $b(u, v) = 2/v$ with selection function, $S(v) = v$, and
- binary breakage, $b(u, v) = 2/v$ with selection function, $S(v) = v^2$.

In both problems, monodisperse particles with size unity are taken as initial data. The analytical solutions can be found in Ziff and McGrady (1985). For numerical computation, the volume domain has been divided by the rule $x_{i+1/2} = 2^{1/q} x_{i-1/2}$.

In order to check the efficiency of the schemes we first calculate the experimental order of convergence (EOC). Here, EOC measures the numerical order of convergence by comparing computations on two meshes and is set to:

$$\text{EOC} = \ln(E_{r_I}/E_{r_{2I}})/\ln(2)$$

where E_{r_I} and $E_{r_{2I}}$ are the errors defined by the L_1 norm $|N^{\text{ana}} - N^{\text{num}}|$. The symbols I and $2I$ correspond to the degrees of freedom. For the case $2I$, each cell of case I was divided into two equal parts, doubling the degrees of freedom. The variable N describes the number distribution of particles. Tables 6.1 and 6.2 contain EOC tests by both the schemes for the first problem. The EOC for the second problem are summarized in Tables 6.3 and 6.4 by the cell-average and the fixed-pivot technique, respectively. This shows clearly that the cell-average technique is of second order, while the fixed-pivot technique is only first-order accurate.

Tab. 6.2 Binary breakage, FP and $S(v) = v$, $t = 1000$

Grid points, I	Error, L_1	EOC
61	173.4532	–
122	89.7167	0.95
244	45.2410	0.99
488	22.6329	0.99

Tab. 6.3 Binary breakage, CA and $S(v) = v^2$, $t = 2000$

Grid points, I	Error, L_1	EOC
61	2.5326	–
122	0.7733	1.71
244	0.2226	1.80
488	0.05954	1.90

Tab. 6.4 Binary breakage, FP and $S(v) = v^2$, $t = 2000$

Grid Points, I	Error, L_1	EOC
61	13.8570	–
122	7.0647	0.97
244	3.5574	0.98
488	1.7826	0.99

Since the numerical results for the first two moments by both techniques are the same, we now compare the complete particle-size distribution. Similar behavior is observed for both problems so that we show results only for the first problem in Fig. 6.6. Figures 6.6(a) and 6.6(b) have been plotted on semi-log and log-log scale, respectively. A small overprediction is observed by the fixed-pivot technique. On the other hand, excellent agreement between the numerical results obtained by the cell-average technique and the analytical results can be seen in the figure. For very fine grids the numerical results by both techniques come close to each other.

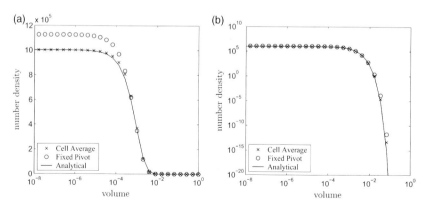

Fig. 6.6 Comparison of particle size distribution in binary breakage, $S(v) = v$, grid points $= 31$.

6.2.2.2 The Finite-volume Scheme

Finite-volume schemes are a class of discretizations that are highly successful in approximating the solution of a wide variety of conservation law systems. A typical conservation law in one dimension takes the form

$$\frac{\partial}{\partial t} u(t, x) + \frac{\partial}{\partial x} f(u(x, t)) = 0 \qquad (6.28)$$

Here, conservation means that the quantity $\int_{-\infty}^{\infty} u(t, x)dx$ is constant with respect to t. Conservation laws are involved in many practical problems in science and engineering. The finite-volume schemes are popular for solving the PDEs of this class since they automatically conserve the quantity $\int_{-\infty}^{\infty} u(t, x)dx$. To build a general setup of the finite-volume scheme for Eq. 6.28, let us first fix a finite domain $[0, R]$ for the computation. We discretize space into I cells $\Lambda_i = [x_{i-1/2}, x_{i+1/2}]$ and time in discrete level t_n. Integrating the conservation law 6.28 on a cell in space-time $\Lambda_i \times [t_n, t_{n+1}]$ we obtain

$$\int_{x_{i-1/2}}^{x_{i+1/2}} u(t_{n+1}, x)dx = \int_{x_{i-1/2}}^{x_{i+1/2}} u(t_n, x)dx - \int_{t_n}^{t_{n+1}} (f(t, x_{i+1/2}) - f(t, x_{i-1/2}))dt \qquad (6.29)$$

The preceding equation can be approximated as

$$u_i^{n+1} = u_i^n - \frac{\Delta t}{\Delta x_i}(f_{i+1/2}^n - f_{i-1/2}^n) \qquad (6.30)$$

where u_i^n denotes an approximation of the cell average of $u(t^n, x)$ on cell i at time t_n, and $f_{i+1/2}^n$ approximates the flux at the boundary of the cell. It is the so-called numerical flux. The accuracy of the finite-volume scheme is determined by the accuracy with which u is reconstructed in each cell and the flux function at the boundary is evaluated.

Breakage in the form of a conservation law: In a batch system the breakage process can be defined as the flow of mass from bigger to smaller particles and can be modelled by the mass-conservation law

$$\frac{\partial xn(t, x)}{\partial t} + \frac{\partial F(t, x)}{\partial x} = 0 \qquad (6.31)$$

where $n(t, x)$ is the number density and $F(t, x)$ is the mass flux across mass x. Now the objective is to model the flux function F and to solve the preceding equation using the idea of finite-volume schemes.

As mentioned earlier, the breakage function $b(t, u, \epsilon)$ is the probability density function for the formation of particles of size u from particle of size ϵ. The selection function $S(t, \epsilon)$ gives the death rate of particles of size ϵ as $n(t, \epsilon) S(t, \epsilon)$. Clearly this death results in flow of mass towards smaller particles. So the total mass flux at x for

$x < \epsilon$, resulting from this death is

$$-\int_0^x ub(t, u, \epsilon) S(t, \epsilon) n(t, \epsilon) du \tag{6.32}$$

The negative sign appears due to the negative direction of mass flow. Considering all the breakage events that produce a flux at x, we obtain the mass flux as

$$F(t, x) = -\int_x^\infty \int_0^x ub(t, u, \epsilon) S(t, \epsilon) n(t, \epsilon) du d\epsilon \tag{6.33}$$

Putting the value of $F(t, x)$ in Eq. (6.31), we get

$$\frac{\partial x n(t, x)}{\partial t} = \frac{\partial}{\partial x} \int_x^\infty \int_0^x ub(t, u, \epsilon) S(t, \epsilon) n(t, \epsilon) du d\epsilon \tag{6.34}$$

This equation can easily be transformed to the standard form of PBE (6.6) using the Leibnitz integration rule. Note that here we get an unusual form of the conservation law (6.31) due to the fact that the flux function is nonlocal and given by an integral operator.

Discretization: Noting g_i^n as an approximation of the cell average of the mass $g(t^n, x) = x n(t^n, x)$ on cell i at time t^n, the finite-volume formulation 6.30 can be written as

$$g_i^{n+1} = g_i^n - \frac{\Delta t}{\Delta x_i} (F_{i+1/2}^n - F_{i-1/2}^n) \tag{6.35}$$

where $F_{i+1/2}^n$ is the numerical flux that approximates the flux F at the boundary. The numerical flux may be approximated as

$$\begin{aligned}
F_{i+1/2}^n &= -\int_{x_{i+1/2}}^R \int_0^{x_{i+1/2}} ub(t, u, \epsilon) S(t, \epsilon) n(t, \epsilon) du d\epsilon \\
&= -\sum_{k=i+1}^I \int_{\Lambda_k} S(t, \epsilon) n(t, \epsilon) \int_0^{x_{i+1/2}} ub(t, u, \epsilon) du d\epsilon \\
&\approx -\sum_{k=i+1}^I g_k^n \int_{\Lambda_k} \frac{S(t, \epsilon)}{\epsilon} d\epsilon \int_0^{x_{i+1/2}} ub(t, u, x_k) du
\end{aligned} \tag{6.36}$$

The integrals coming in the computation of numerical flux can explicitly be evaluated by any second- or higher-order quadrature formula. The formulation 6.35 is easy to implement. It can also be rewritten in semidiscrete form as

$$\frac{dg_i}{dt} = \frac{1}{\Delta x_i} (F_{i-1/2} - F_{i+1/2}) \tag{6.37}$$

Tab. 6.5 Binary breakage with the finite-volume scheme and $S(v) = v$, $t = 1000$

Grid points, I	Error, L_1	OC
61	33.8559	–
122	8.8548	1.93
244	2.2363	1.98
488	0.5612	1.99

where

$$F_{i+1/2} = \sum_{k=i+1}^{I} g_k \int_{\Lambda_k} \frac{S(t, \epsilon)}{\epsilon} d\epsilon \int_0^{x_{i+1/2}} ub(t, x_k, u) du \qquad (6.38)$$

Any second- or higher-order ODE solver can be used for the time discretization. This formulation is second-order accurate.

Numerical results: We tested the scheme on our test problems stated above. The ode45 MATLAB solver has been used for the time integration. Geometric grids of the type $x_{i+1/2} = 2^{1/q} x_{i-1/2}$ are used for the spatial discretization. Tables 6.5 and 6.6 clearly show that the scheme is second-order accurate.

We now compare the numerical results obtained by the cell average and the finite-volume techniques. Note that the cell-average technique predicts the zeroth and the first moments of the PSD exactly, irrespective of the number of grid points chosen for the discretization. While finite-volume schemes conserve the mass, no constraints are imposed on the prediction of number.

The numerical results reflect the same behavior for both test cases, therefore we show here the comparisons for the first case only. We consider binary breakage with monodisperse particles of size unity as an initial condition and a linear selection function. Accuracy is assessed by direct comparison of the total number and particle number density in Fig. 6.7(a) and (b), respectively. The computation has been performed by dividing the volume domain into 31 sections with the rule $x_{i+1/2} = 2x_{i-1/2}$. As expected, due to the preserving properties of the cell-average technique, the prediction of the total number by the cell-average technique is highly accurate, while the prediction by the finite-volume technique is poor. Figure 6.7(b) shows the underprediction of the number density by the finite-volume scheme at the small volumes. Even with comparable or smaller absolute errors in PSD by the

Tab. 6.6 Binary breakage with the finite-volume scheme and $S(v) = v^2$, $t = 2000$

Grid points, I	Error, L_1	EOC
61	0.8526	–
122	0.2200	1.95
244	0.0551	2.00
488	0.0138	2.00

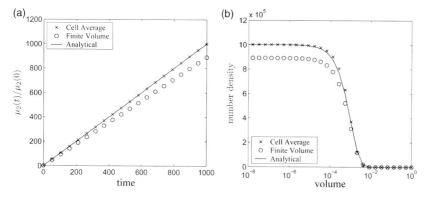

Fig. 6.7 Comparison of numerical results (a) Total number, (b) Particle size distribution, grid points = 31, $S(v) = v$, $t = 1000$.

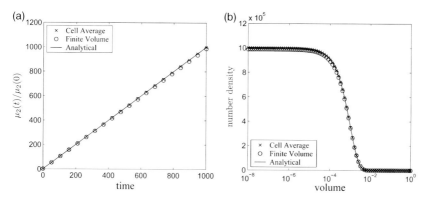

Fig. 6.8 Comparison of numerical results (a) Total number, (b) Particle size distribution, grid points = 91, $S(v) = v$, $t = 1000$.

finite-volume scheme, the prediction of the total number was still poorer. This does not change, in principle, by grid refinement to $I = 91$ (Figs. 6.8(a) and (b)).

6.3
Pure Aggregation

6.3.1
Population-balance Equation

In this section we turn to numerical techniques for solving aggregation population-balance equations. Aggregation appears in physics (colloidal particles), meteorology (drops in atmospheric clouds, aerosols), chemistry (reacting polymers, soot, pharmaceuticals, fertilizers). The temporal change of particle-number density in a spatially homogeneous physical system is described by the population balance equation developed by Hulburt and Katz (1964)

$$\frac{\partial n(t,x)}{\partial t} = \frac{1}{2}\int_0^x \beta(t, x-\epsilon, \epsilon) n(t, x-\epsilon) n(t, \epsilon) d\epsilon \\ - n(t,x) \int_0^\infty \beta(t, x, \epsilon) n(t, \epsilon) d\epsilon \qquad (6.39)$$

where $t \geq 0$. The first term represents the birth of particles of size x as a result of the coagulation of particles of sizes $(x - \epsilon)$ and ϵ. Here, we shall refer to size as the particle volume. The second term (death term) describes the merging of particles of size x with any other particles. The process is governed by the coagulation kernel β representing properties of the physical medium. It is non-negative and satisfies the symmetry condition $\beta(t, \epsilon, x) = \beta(t, x, \epsilon)$. Analytical solutions of the PBE are scarce, but very useful in order to assess the accuracy of numerical schemes.

6.3.2
Numerical Methods

Among various numerical techniques (successive approximations, Laplace transforms, moments, weighted residuals, discrete formulations using point masses, finite-volume Monte Carlo), we discuss here only the point-mass formulation of the PBE and the finite-volume scheme. The point-mass formulations are well known in process engineering because they are simple to implement and produce exact numerical results for selected properties. On the other hand, the finite-volume schemes are well suited for solving conservation laws.

6.3.2.1 The Fixed-pivot Technique
As already mentioned, the fixed-pivot technique (Kumar and Ramkrishna 1996a) not only preserves the number and mass of particles, but it can also be generalized for the preservation of any two desired properties of the population. It divides the entire size range into small cells of arbitrary size. The range contained between two sizes $x_{i-1/2}$ and $x_{i+1/2}$ is called the ith cell. The particle population in this size range is represented by a size x_i, called a grid point, such that $x_{i-1/2} < x_i < x_{i+1/2}$. A new particle of size x in the range $[x_i, x_{i+1}]$, formed either by breakup or aggregation, can be represented by assigning fractions $a(x, x_i)$ and $b(x, x_{i+1})$ to the populations at x_i and x_{i+1}, respectively. For consistency with two general properties $f_1(x)$ and $f_2(x)$, these fractions must satisfy the equations

$$a(x, x_i) f_1(x_i) + b(x, x_{i+1}) f_1(x_{i+1}) = f_1(x) \qquad (6.40)$$

$$a(x, x_i) f_2(x_i) + b(x, x_{i+1}) f_2(x_{i+1}) = f_2(x) \qquad (6.41)$$

These equations can be generalized for consistency with more than two properties by assigning the particle size x to more than two grid points. The population at representative volume x_i gets a fractional particle for every particle that is born in size range $[x_i, x_{i+1}]$ or $[x_{i-1}, x_i]$. For consistency with numbers and mass, the discrete equations for aggregation are given by

$$\frac{dN_i}{dt} = \sum_{\substack{j,k \\ x_{i-1} \leq x < x_{i+1}}}^{j \geq k} \left(1 - \frac{1}{2}\delta_{j,k}\right) \eta(x) \beta_{j,k} N_j N_k - N_i \sum_{k=1}^{I} \beta_{i,k} N_k \qquad (6.42)$$

with

$$\eta(x) = \begin{cases} \dfrac{x_{i+1} - x}{x_{i+1} - x_i}, & x_i \leq x < x_{i+1} \\[6pt] \dfrac{x - x_{i-1}}{x_i - x_{i-1}}, & x_{i-1} \leq x < x_i \end{cases} \qquad (6.43)$$

where $x = x_j + x_k$. The technique is flexible, but has the disadvantage of overprediction of the number density in the large size range when applied on coarse grids. Therefore, Kumar and Ramkrishna (1996b) also presented a moving-pivot technique. The latter is more complex and results in a system of stiff differential equations, which is difficult to solve.

6.3.2.2 The Cell-average Technique

The cell-average technique is based on the exact prediction of any two moments of the distribution and was originally developed (Kumar et al. 2006a) for conservation of total number and mass in aggregation processes. Application to breakage PBE has already been treated in the previous section. Here we focus on aggregation, starting with a formulation that exactly predicts the zeroth and the first moment.

First, we discretize the computational domain into small contiguous cells. The size of cells can be chosen arbitrarily. Then, we fix a representative size for each cell, e.g. the center of the cell as easiest choice. The next step is to compute the total birth rate in each cell and the volume average of particles. The total birth and the death in a cell are given by

$$B_{\text{agg},i} = \frac{1}{2} \int_{x_{i-1/2}}^{x_{i+1/2}} \int_0^x \beta(t, x - \epsilon, \epsilon) n(t, x - \epsilon) n(t, \epsilon) d\epsilon dx \qquad (6.44)$$

and

$$D_{\text{agg},i} = \int_{x_{i-1/2}}^{x_{i+1/2}} n(t, x) \int_0^\infty \beta(t, x, \epsilon) n(t, \epsilon) d\epsilon dx \qquad (6.45)$$

Since particles are assumed to be concentrated at representative sizes x_i, the number density can be expressed as

$$n(t, x) = \sum_{j=1}^{I} N_j(t)\delta(x - x_j) \tag{6.46}$$

Substituting $n(t, x)$ into Eq. 6.44 we obtain

$$\hat{B}_{agg,i} = \sum_{\substack{j,k \\ x_{i-1/2} \leq (x_j+x_k) < x_{i+1/2}}}^{j \geq k} \left(1 - \frac{1}{2}\delta_{j,k}\right) \beta_{j,k} N_j N_k \tag{6.47}$$

Thus, $B_{agg,i}$ is the net rate of addition of particles to cell i by coagulation of particles in lower cells. The net flux of volume M_i into cell i as a result of these coagulations is therefore given by

$$\hat{M}_{agg,i} = \sum_{\substack{j,k \\ x_{i-1/2} \leq (x_j+x_k) < x_{i+1/2}}}^{j \geq k} \left(1 - \frac{1}{2}\delta_{j,k}\right) \beta_{j,k} N_j N_k (x_j + x_k) \tag{6.48}$$

Consequently, the average volume of all newborn particles in the ith cell can be evaluated as

$$\bar{a}_i = \frac{\hat{M}_{agg,i}}{\hat{B}_{agg,i}} \tag{6.49}$$

The remaining steps are now the same as in the previous section. The modified birth term takes the form

$$\begin{aligned}\hat{B}_{agg,i}^{CA} &= \hat{B}_{agg,i-1}\lambda_i^-(\bar{a}_{i-1})H(\bar{a}_{i-1} - x_{i-1}) + \hat{B}_{agg,i}\lambda_i^-(\bar{a}_i)H(x_i - \bar{a}_i) \\ &\quad + \hat{B}_{agg,i}\lambda_i^+(\bar{a}_i)H(\bar{a}_i - x_i) + \hat{B}_{agg,i+1}\lambda_i^+(\bar{a}_{i+1})H(x_{i+1} - \bar{a}_{i+1})\end{aligned} \tag{6.50}$$

The corresponding expression for the death is simple and can be obtained by substituting number density of Eq. 6.45 into the death term from Eq. 6.46 to give

$$\hat{D}_{agg,i}^{CA} = N_i \sum_{k=1}^{I} \beta_{i,k} N_k \tag{6.51}$$

The final set of discrete equations can be written as

$$\frac{dN_i}{dt} = \hat{B}_{agg,i}^{CA} - \hat{D}_{agg,i}^{CA} \tag{6.52}$$

The Heaviside function H and the fractions $\lambda_i^{\pm}$ are the same as previously defined. The above formulation gives the exact prediction of the zeroth and the first moments. It should be noted that the choice of two properties is not trivial in the cell-average technique. Let us therefore also consider the more general case of exact prediction of the zeroth and the rth moment. In this case it is easy to modify the equations 6.48, 6.49 to

$$\hat{M}_{agg,i} = \sum_{\substack{j,k \\ x_{i-1/2} \leq (x_j+x_k) < x_{i+1/2}}}^{j \geq k} \left(1 - \frac{1}{2}\delta_{j,k}\right) \beta_{j,k} N_j N_k (x_j + x_k)^r \tag{6.53}$$

$$\bar{a}_i = \left(\frac{\hat{M}_{agg,i}}{\hat{B}_{agg,i}}\right)^{1/r} \tag{6.54}$$

Assuming that the average value $\bar{a}_i$ is bigger that x_i, the fractions must be calculated from

$$\begin{aligned} a + b &= \hat{B}_{agg,i} \\ ax_i^r + bx_{i+1}^r &= \hat{B}_{agg,i}\bar{a}_i^r \end{aligned} \tag{6.55}$$

Solving these equations, the expressions for λ will be given as

$$\lambda_i^{\pm}(x) = \frac{x^r - x_{i\pm 1}^r}{x_i^r - x_{i\pm 1}^r} \tag{6.56}$$

The final formulation (6.52) remains the same. Up to now we have considered the zeroth and one extra moment. The arbitrary choice of both moments or more moments gives serious difficulties in calculating averages of particles and assigning them to neighboring nodes. Nevertheless, the first two moments are of special interest for many applications, so that we refrain here from discussion of the exact prediction of more moments.

Numerical results: In this section we compare the cell-average with the fixed-pivot technique for analytically solvable test problems. With reference to Kumar et al. (2006a), only a few comparisons are presented. All computations were carried out in the programming software MATLAB on a Pentium4 machine with 1.5 GHz and 512 MB RAM. The set of ordinary differential equations resulting from the discretized techniques is solved using a Runge–Kutta fourth- and fifth-order method with adaptive step-size control.

Tables 6.7(a) and (b) show the EOC test of the cell-average and fixed-pivot techniques, respectively. It is computed for a test problem with exponential initial condition and the sum coagulation kernel. The degree of aggregation is chosen to be 0.95. As can be seen from the tables, the convergence of the two methods differs significantly. The FP method is only of first order, whereas the CA method is clearly of second order also for this problem.

Tab. 6.7 EOC of the cell-average and fixed-pivot techniques for exponential initial distribution and sum kernel, $I_{agg} = 0.95$

(a) The fixed-pivot technique

Grid points, I	Error, L_1	EOC
30	2.43E–2	–
60	1.16E–2	1.06
120	6.31E–3	0.87
240	2.84E–3	1.14

(b) The cell-average technique

Grid points, I	Error, L_1	EOC
30	2.54E–2	–
60	9.38E–3	1.44
120	2.93E–3	1.68
240	7.63E–4	1.94

Furthermore, the complete particle size distribution and the second moment have been compared (Fig. 6.9). To this purpose, we take the initial particle-size distribution as monodisperse with dimensionless size unity and the size-independent kernel. The prediction of the particle-size distribution as well as of the second moment is excellent by the cell-average technique. At very short times the predictions of the second moment, shown in Fig. 6.9(a), by the FP technique as well as by the cell-average technique are the same, but at later times prediction by the cell-average technique is considerably better. Figure 6.9(b) shows the improvement of the overprediction in PSD at large particle sizes by the cell-average technique.

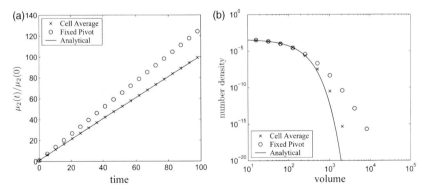

Fig. 6.9 Comparison of numerical results (a) Second moment, (b) Particle size distribution, with analytical results for monodisperse initial condition and constant kernel, $I_{agg} = 0.98$.

6.3.2.3 The Finite-volume Scheme

Filbet and Laurenot (2004) applied the finite-volume scheme for solving the aggregation PBE. This is performed (Makino et al. 1998; Tanaka et al. 1996) by changing the form of the PBE Eq. 6.39 to the mass-conservation law

$$x\frac{\partial n(t,x)}{\partial t} = -\frac{\partial J(n(t,x))}{\partial x} \tag{6.57}$$

where the flux is given as

$$J(n)(x) = \int_0^x \int_{x-u}^\infty u\beta(t,u,v) n(t,u) n(t,v) dv du \tag{6.58}$$

Now, Eq. 6.57 can be solved using the finite-volume approach as

$$g_i^{n+1} = g_i^n - \frac{\Delta t}{\Delta x_i}(F_{i+1/2}^n - F_{i-1/2}^n) \tag{6.59}$$

where g_i^n is an approximation of the mean value of mass $g(t^n,x) = xn(t^n,x)$ in cell i. The numerical flux function $F_{i+1/2}^n$ is given by

$$F_{i+1/2}^n = \sum_{k=1}^{i} \Delta x_k g_k^n \left[\sum_{j=\alpha_{i,k}}^{I} \int_{x_{j-1/2}}^{x_{j+1/2}} \frac{\beta(x',x_k)}{x'} dx' g_j^n + \int_{x_{i+1/2}-x_k}^{x_{\alpha,i}-1/2} \frac{\beta(x',x_k)}{x'} dx' g_{\alpha_{i,k}-1}^n \right] \tag{6.60}$$

where the integer $\alpha_{i,k}$ corresponds to the index of the cell such that $x_{i+1/2} - x_k$ belongs to the cell $\alpha_{i,k} - 1$. The scheme is second-order accurate in the volume coordinate.

Numerical results: In order to check the accuracy of the scheme we tested it with the same problem as in Fig. 6.9. A coarse grid, $x_{i+1/2} = 2x_{i-1/2}$, is chosen for the computation. The numerical results obtained from the finite-volume scheme are compared with the analytical results and the numerical results by the cell-average technique. Figure 6.10 presents the first three moments of the PSD. As expected, the prediction of the first two moments by the cell-average technique is exact, see Fig. 6.10(a). The prediction of the first moment by the finite-volume scheme is good, while a small underprediction is observed in the zeroth moment at the beginning of the simulation. The temporal change of the second moment is plotted in Fig. 6.10(b). Once again, the prediction of the second moment by the cell-average technique is excellent, while the finite-volume scheme gives poor results. A comparison of the complete PSD at the final time is depicted in Fig. 6.11 with advantages for the cell-average technique. For fine grids, the numerical results of both techniques come very close to each other.

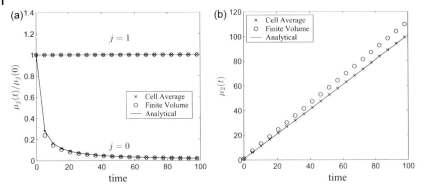

Fig. 6.10 A comparison of numerical results (a) First two moments, (b) Second moment, with analytical results for mono-disperse initial condition and constant kernel, $I_{agg} = 0.98$.

It may be concluded that the cell-average technique is a simple, accurate and general method for solving aggregation or breakage equations. The technique is independent of the grid, provides excellent predictions of the first three moments and PSD and is easy to program. We will see in the following that the coupling of aggregation and breakage processes is a trivial task for the cell-average technique.

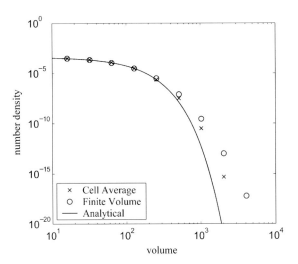

Fig. 6.11 A comparison of complete particle size distribution for monodisperse initial condition and constant kernel, $I_{agg} = 0.98$.

6.4 Pure Growth

6.4.1 Population balance Equation

The population balance equation for pure growth is

$$\frac{\partial n(t,x)}{\partial t} + \frac{\partial [G(x)n(t,x)]}{\partial x} = 0 \tag{6.61}$$

a first-order hyperbolic PDE well known for causing serious difficulties in numerical solution.

6.4.2 Numerical Methods

Finite-volume schemes are a good approach to such equations, since they automatically incorporate conservation of number in the growth process. After a brief discussion of the finite-volume schemes numerical problems with these methods will be mentioned. Our interest is in formulating a semidiscretized form of Eq. 6.61 so that the method of lines (MOL) can easily be applied. The reasons for the MOL approach are ease of coupling with further processes like aggregation, breakage, etc., and the availability of ODE integrators. We discretize the domain into equal spatial cells $\Omega_i = [x_{i-1/2}, x_{i+1/2}]$ and define as nodes $x_i = (x_{i-1/2} + x_{i+1/2})/2$ the centers of the cells. Direct integration of Eq. 6.61 over each cell gives the semidiscrete formulation

$$\frac{dN_i(t)}{dt} = G(x_{i-1/2})n(t,x_{i-1/2}) - G(x_{i+1/2})n(t,x_{i+1/2}), \quad i = 1, \ldots I \tag{6.62}$$

Various methods for the numerical solution of Eq. 6.61 can be obtained from different choices of the approximation of $n(t, x_{i-1/2})$ and $n(t, x_{i+1/2})$ in terms of $N_i(t)$. The easiest approximation $n(t, x_{i+1/2}) \approx N_i(t)/h$ gives the first-order upwind difference discretization

$$\frac{d\hat{N}_i(t)}{dt} = \frac{1}{h}\left(G(x_{i-1/2})\hat{N}_{i-1} - G(x_{i+1/2})\hat{N}_i\right) \tag{6.63}$$

Here, the numerical approximations of N_i are denoted by $\hat{N}_i$. The choice

$$n(t, x_{i+1/2}) \approx \frac{1}{2h}(N_i(t) + N_{i+1}(t))$$

gives the second-order central discretization

$$\frac{d\hat{N}_i(t)}{dt} = \frac{1}{2h}\left(G(x_{i-1/2})(\hat{N}_{i-1}(t) + \hat{N}_i(t)) - G(x_{i+1/2})(\hat{N}_i(t) + \hat{N}_{i+1}(t))\right) \tag{6.64}$$

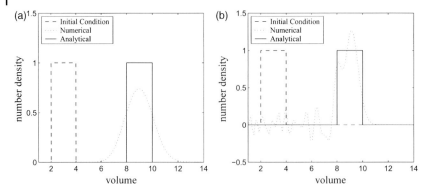

Fig. 6.12 PSDs for pure growth with constant growth rate at $t = 6$ obtained on 100 uniform fixed grid. (a) First-order upwind scheme, (b) Second-order central scheme.

The PBE Eq. (6.61) is solved using the first-order upwind discretization and the second-order central discretization for $G(v) = 1$ and

$$(0, x) = \begin{cases} 1, & 2 \leq x \leq 4 \\ 0, & \text{otherwise} \end{cases} \quad (6.65)$$

Computation is carried out on 100 uniform fixed cells at $t = 6$. For the integration of both discrete formulations, the MATLAB ode45 solver has been used. This is a one-step solver based on an explicit Runge–Kutta (4,5) formula. The results are plotted in Fig. 6.12. The first-order upwind scheme smears, the second-order scheme produces oscillations and seems to be unstable. The second-order scheme is more accurate than the first-order scheme only for very smooth solutions. Most of the higher-order schemes suffer from oscillations and produce negative values that are unrealistic. The first-order upwind scheme is a nonoscillatory alternative, but it is also very diffusive and too inaccurate for real application.

A class of more sophisticated discretization methods that preserve positivity are the so-called *flux-limiting methods*. These methods, also known as *high-resolution schemes*, are obtained by modifying the fluxes of a higher-order discretization. The idea is to apply a high-order flux in smooth regions and a low-order flux near discontinuities. This is achieved by measuring the smoothness of the data. High-resolution methods can be found in [30–33]. Here, we present one example introduced by Koren [30]. In this method the number density at the boundary is approximated as

$$n(t, x_{i+1/2}) \approx \frac{1}{h} \left(N_i(t) + \frac{1}{2} \phi(\theta_{i+1/2})(N_i(t) - N_{i-1}(t)) \right) \quad (6.66)$$

The limiter function ϕ is set to

$$\phi(\theta) = \max\left(0, \min\left(2\theta, \min\left(\frac{1}{3} + \frac{2}{3}\theta, 2\right)\right)\right) \quad (6.67)$$

where the parameter θ is defined as

$$\theta_{i+1/2} = \frac{N_{i+1} - N_i + \epsilon}{N_i - N_{i-1} + \epsilon} \qquad (6.68)$$

with a very small constant ϵ to avoid division by zero. Using this method we computed the solution of Eq. 6.61 for the test problem under the same conditions as before, see Fig. 6.13. The scheme gives a better accuracy than the first-order upwind without producing negative values. Some other high-resolution schemes for growth PBE are applied in Lim et al. (2002), Gunawan et al. (2004). It is known that finite-volume methods may be even more efficient for hyperbolic problems. However, our choice of such methods is restricted by our interest in the MOL approach. Additionally, we would like to exchange the homogeneous grids used up to now by nonhomogeneous grids, as in the previous cases of breakage and aggregation. Unfortunately, the extension of the finite-volume scheme to nonuniform grids is not a trivial task (Hundsdorfer and Verwer 2003).

The complexity of the finite-volume scheme on nonhomogeneous grids and restricted compatibility with MOL motivate us to discuss a different class of discretized methods that rely on the correct prediction of selected moments. Hounslow et al. (1988) seem to have been the first to propose a method that is consistent with the first three moments with respect to particle length. Their discrete formulation for Eq. 6.61 considering volume as the internal coordinate can be expressed (Park and Rogak 2004) as

$$\frac{d\hat{N}_i}{dt} = \frac{aG_{i-1}\hat{N}_{i-1}}{x_{i-1}} + \frac{bG_i\hat{N}_i}{x_i} + \frac{cG_{i+1}\hat{N}_{i+1}}{x_{i+1}} \qquad (6.69)$$

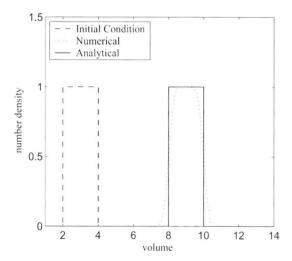

Fig. 6.13 PSDs for pure growth with constant growth rate at $t = 6$ obtained on 100 uniform fixed grid using flux-limiter method.

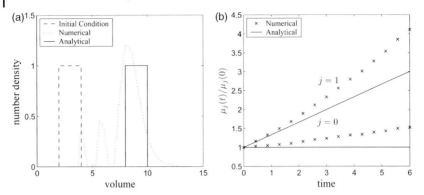

Fig. 6.14 Pure growth with constant rate at $t = 6$ obtained from Eq. 6.70 on 40 nonuniform logarithmic fixed grid. (a) Particle size distribution, (b) Moments.

where a, b and c are constants chosen so that the discretization predicts correctly the first three moments. The final equation is

$$\frac{d\hat{N}_i}{dt} = \frac{1}{x_i}\left(\frac{r}{r^2-1}G_{i-1}\hat{N}_{i-1} + G_i\hat{N}_i - \frac{r}{r^2-1}G_{i+1}\hat{N}_{i+1}\right) \tag{6.70}$$

where $r = x_{i+1}/x_i$ and $G_i = G(x_i)$. We have tested this discretization on our model problem of Eq. 6.65 under the same conditions as in Figs. 6.12 and 6.13. As suggested by Hounslow et al. (1988) negative values are set to zero, which causes an increase of the total number and mass. Unstable behavior of the model and inaccurate particle size distribution can be seen in Fig. 6.14(a). The temporal overprediction of the total number and volume due to suppressing negative values is evident in Fig. 6.14(b).

Following Hounslow et al. (1988), an effort to overcome these difficulties has been made by Park and Rogak (2004) based on the formulation

$$\frac{d\hat{N}_i}{dt} = \frac{a_{i-1}G_{i-1}\hat{N}_{i-1}}{x_{i-1}} + \frac{b_i G_i \hat{N}_i}{x_i} + \frac{c_{i+1}G_{i+1}\hat{N}_{i+1}}{x_{i+1}} \tag{6.71}$$

$G_i = G(x_i)$ and the coefficients a_i, b_i and c_i have been calculated by imposing conditions of correct prediction of the first two moments and the slope of the size distribution to

$$b_i = \begin{cases} -\dfrac{r}{r-1}\mathrm{erf}\left(\dfrac{1}{4}\dfrac{d\ln\hat{N}_i}{d\ln x_i}\right), & \dfrac{d\ln\hat{N}_i}{d\ln x_i} \le 0 \\[1em] -\dfrac{r}{r-1}\mathrm{erf}\left(\dfrac{1}{4}\dfrac{d\ln\hat{N}_i}{d\ln x_i}\right), & \dfrac{d\ln\hat{N}_i}{d\ln x_i} > 0 \end{cases} \tag{6.72}$$

$$a_i = \frac{r - b_i(r-1)}{r^2 - 1} \tag{6.73}$$

and

$$c_i = -(a_i + b_i) \tag{6.74}$$

The term $d\ln\hat{N}_i/d\ln x_i$ in b_i has been calculated using the second-order central scheme. Equation 6.71 can also be rewritten in conservative form as

$$\frac{d\hat{N}_i}{dt} = \frac{a_{i-1}G_{i-1}\hat{N}_{i-1} - c_i G_i \hat{N}_i/r}{x_{i-1}} - \frac{a_i G_i \hat{N}_i - c_{i+1}G_{i+1}\hat{N}_{i+1}/r}{x_i} \tag{6.75}$$

We have also implemented this model on our test problem. Similar to the results obtained by the technique of Hounslow et al., the prediction of PSD, Fig. 6.15(a), shows oscillations and some overshoot. Nevertheless, this model improves the prediction of moments, as shown in Fig. 6.15(b).

Due to serious problems, like inaccuracy, diffusive nature, instability, restriction to special grids, etc., it is difficult to recommend the best method for a general growth problem. There are of course still more sophisticated methods in the literature, but they are difficult to implement and to couple with other processes. Nevertheless, the flux-limiter method Eq. 6.66 seems an acceptable choice for uniform grids. On the other hand, as mentioned before, geometric grids are preferable for aggregation-breakage problems.

All the difficulties in numerical solutions stated above can be overcome by the *moving sectional methods* proposed by Gelbard (1990) as well as Kim and Seinfeld (1990). They can be classified as Lagrangian-type methods. In the moving sectional

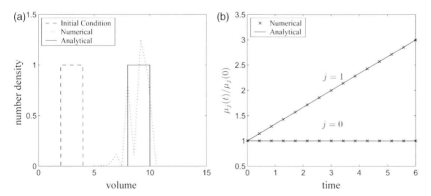

Fig. 6.15 Pure growth with constant rate at $t = 6$ obtained from Eq. 6.75 on 40 nonuniform logarithmic fixed grid. (a) Particle-size distribution, (b) Moments.

method all particles initially within a section remain there during computation, i.e.

$$\frac{dN_i(t)}{dt} = 0 \tag{6.76}$$

and the section boundaries move as particles grow. Therefore, we have the following governing equations for the boundaries

$$\frac{dx_{i\pm 1/2}(t)}{dt} = G(x_{i\pm 1/2}) \tag{6.77}$$

In order to be consistent with moments, the section representative has to move as

$$\frac{dx_i(t)}{dt} = G(x_i) \tag{6.78}$$

For simple growth functions like $G(x) = x$ or $G(x) = G_0$ the representative (cell center) can be calculated as

$$x_i(t) = \frac{x_{i-1/2} + x_{i+1/2}}{2} \tag{6.79}$$

without setting the differential Eq. 6.78.

This technique is free of diffusion and instability independently of the type of grid. Numerical results are extremely accurate and very close to analytical solutions. Our test problem with a constant growth rate is rather trivial for moving sectional methods. Hence, the numerical results are the same as the analytical solution, as shown in Fig. 6.16, where only 15 uniform grids have been used for the computation. For other applications of moving sectional methods to model and practical problems see Gelbard (1990), Kim and Seinfeld (1990), Kumar and Ramkrishna (1997), Spicer et al. (2002), Tsantilis et al. (2002), Tsantilis and Pratsinis (2004). The moving

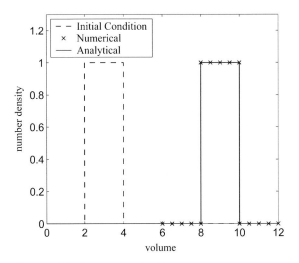

Fig. 6.16 PSDs for pure growth with constant growth rate at $t = 6$ obtained on 15 uniform moving grid points.

sectional method is the easiest approach to solve pure growth problems. However, its combination with nucleation, aggregation and breakage is not a trivial task. We will discuss this issue in the next sections.

A slightly different idea of hybrid grids was formulated by Jacobson (1997) and Jocobson and Turco (1995) who fixed the boundaries of the sections and moved their representative size. By fixing boundaries other processes like nucleation, aggregation, etc. can easily be treated. If the representative crosses the upper boundary, then all particles in the section are moved and averaged with the particles of the next section. Numerical diffusion is low and satisfactory results can be obtained. However, cells emptying at a particular time can cause fluctuations in PSD. To overcome this difficulty one may delete such cells by extending their neighbors, at the price of more complexity in the implementation.

6.5
Combined Aggregation and Breakage

In this section we will apply the cell-average technique to simultaneous aggregation and breakage. Conventionally, the easiest way to couple the processes is just to add both discretizations to

$$\frac{dN_i}{dt} = \hat{B}^{CA}_{agg,i} + \hat{B}^{CA}_{break,i} - \hat{D}^{CA}_{agg,i} - \hat{D}^{CA}_{break,i} \tag{6.80}$$

where CA, agg, and break stand for cell average, aggregation and breakage, respectively. This formulation is of course consistent with the first two moments, since this holds for breakage and aggregation separately. However, efficiency and accuracy can further be improved by using the idea of cell average for the combined problem and considering only the total birth and death. Thereby, we treat both processes together in a similar way as we have treated the individual processes. First, we collect all the newborn particles in a cell independently of the events that make them appear in a cell, namely aggregation or breakage. Then we take the volume average of all newborn particles in the cell. The remaining steps are the same as for the individual processes. Mathematically, we construct the discrete formulation for the coupled problem as

$$\frac{dN_i}{dt} = \hat{B}^{CA}_{agg+break,i} - \hat{D}^{CA}_{agg+break,i} \tag{6.81}$$

Here, the terms $B^{CA}_{agg+break,i}$ and $D^{CA}_{agg+break,i}$ represent the birth and death of particles in the cell i due to both aggregation and breakage. Thus, the birth term for the cell-average technique will be computed as

$$\hat{B}^{CA}_{agg+break,i} = \hat{B}_{agg+break,i-1} \lambda^-_i(\bar{a}_{i-1}) H(\bar{a}_{i-1} - x_{i-1}) + \hat{B}_{agg+break,i} \lambda^-_i(\bar{a}_i) H(x_i - \bar{a}_i)$$
$$+ \hat{B}_{agg+break,i} \lambda^+_i(\bar{a}_i) H(\bar{a}_i - x_i) + \hat{B}_{agg+break,i+1} \lambda^+_i(\bar{a}_{i+1}) H(x_{i+1} - \bar{a}_{i+1})$$
$$\tag{6.82}$$

where

$$\hat{B}_{agg+break,i} = \hat{B}_{agg,i} + \hat{B}_{break,i} \tag{6.83}$$

The death term is simply given by

$$\hat{D}_{agg+break,i}^{CA} = \hat{D}_{agg,i} + \hat{D}_{break,i} \tag{6.84}$$

The truly new ingredient is that the average value $\bar{a}_i$ will be computed as

$$\bar{a}_i = \frac{\hat{M}_{agg,i} + \hat{M}_{break,i}}{\hat{B}_{agg,i} + \hat{B}_{break,i}} \tag{6.85}$$

The terms $\hat{B}_{agg,i}$, $\hat{B}_{break,i}$, $\hat{D}_{agg,i}$, $\hat{D}_{break,i}$, $\hat{M}_{agg,i}$, $\hat{M}_{break,i}$ and also the λ_i are the same as in the previous sections. Next we will test the coupled formulation and see that it is not only more accurate but also computationally less expensive. This is evident since we are distributing particles only once, while in the conventional approach particles are distributed twice, once for aggregation and once for breakage.

Numerical results: We compare the fixed-pivot and the cell-average techniques with analytically solvable problems proposed by Patil and Andrews (1998) using the solutions provided by Lage (2002). These refer to a special case of simultaneous aggregation and breakage where the number of particles stays constant with uniform binary breakage $b(u,v) = 2/v$, linear selection function $S(v) = S_0 v$ and constant aggregation kernel $\beta(u,v) = \beta_0$. Two types of initial conditions are applied

$$n(x,0) = N_0 \left[2\frac{N_0}{v_0}\right]^2 x \exp\left(-2\frac{N_0}{v_0}x\right) \tag{6.86}$$

and

$$n(x,0) = N_0 \frac{N_0}{v_0} \exp\left(-\frac{N_0}{v_0}x\right) \tag{6.87}$$

Note that apart from conservation of mass these problems have the special property of balancing aggregation and breakage to also conserve the total number. The data according to Eq. 6.87 are a steady-state solution.

Figure 6.17(a) and (b) compare particle-size distributions for the initial condition Eq. 6.86 at times $t = 0.4$ and $t = 6$ and show that the cell-average technique is very accurate. In contrast, an overprediction is observed at small volumes by the fixed-pivot technique.

Let us now consider the steady-state data of Eq. 6.87 in Fig. 6.18. The prediction by the cell-average technique is both excellent and stable. On the contrary, the overprediction by the fixed-pivot technique grows with time, revealing that this technique is not stable with respect to steady states. The same numerical results for both the initial conditions at the final time are plotted on a log-log scale in Fig. 6.19.

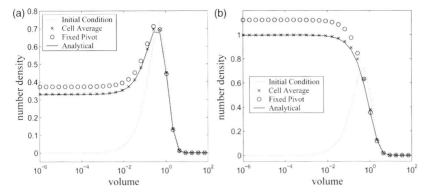

Fig. 6.17 A comparison of particle size distribution for binary breakage and aggregation at constant number of particles for initial condition Eq. 6.86. (a) $t = 0.4$, (b) $t = 6$.

We shall now confront conventional coupling (Eq. 6.80) by cell-average with the new cell average coupling (Eq. 6.82), without considering the fixed-pivot technique. Since the numerical results are indistinguishable from the analytical results for the steady-state initial condition of Eq. 6.87, we only compare for the first unsteady initial condition Eq. (6.86). The respective results are plotted in Fig. 6.20. There, the L_1 error between the numerical and analytical results of PSD is reported. Due to the simplicity of the test problem, the difference between the two types of coupling is not significant, apart from somewhat better results in the steep part of the PSD and from a lower L_1 error by cell-average-type coupling by Eq. 6.81. However, a considerable difference has been observed in computational time. The cell-average-type coupling of Eq. 6.81 takes about 38 s CPU time, while conventional coupling after Eq. 6.80 takes 52 s CPU time on a Pentium4 machine with 1.5 GHz and 512 MB RAM.

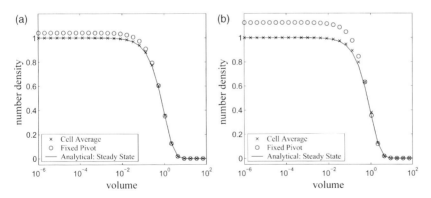

Fig. 6.18 A comparison of particle-size distribution for binary breakage and aggregation at constant number of particles for initial condition Eq. 6.87. (a) $t = 0.4$, (b) $t = 6$.

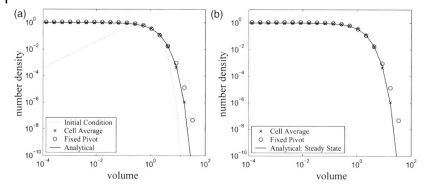

Fig. 6.19 A comparison of particle size distribution at $t = 6$ for binary breakage and aggregation at constant number of particles plotted on log-log scale. (a) Initial condition Eq. 6.86, (b) Initial condition Eq. 6.87.

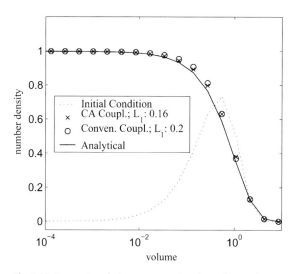

Fig. 6.20 Comparison between conventional coupling and new cell-average coupling for the test case.

To summarize, the cell-average technique provides accurate results on even coarse grids. The new coupling makes the technique not only more accurate, but also computationally less expensive for combined aggregation and breakage.

6.6
Combined Aggregation and Nucleation

In Section 6.5 we have considered particles appearing in cells independently of the processes to couple aggregation and breakage. A similar approach will be applied here for simultaneous aggregation and nucleation. The process of nucleation may be introduced into the modeling in two ways. A common approach is via a boundary condition at particle size 0. Here we want to consider the alternative of a source located near particle size 0. In a continuum theory both are valid approaches.

Conventional coupling of the processes after Kumar and Ramkrishna (1997) leads to

$$\frac{dN_i}{dt} = \text{discretization due to aggregation} + \int_{x_{i-1/2}}^{x_{i+1/2}} B^{\text{nuc}}(t,x)dx \tag{6.88}$$

In this formulation we may gain or lose mass if the nucleation does not take place exactly at the representative sizes x_i. The fixed-pivot technique is now applicable. However, to use the advantages of cell-average techniques we have to, instead, discretize as

$$\frac{dN_i}{dt} = B^{CA}_{\text{agg+nuc},i} - D^{CA}_{\text{agg},i} \tag{6.89}$$

Since nucleation is the birth of particles, there is no nucleation term to be considered in the death term. An analogous contribution to the death term would result by harvesting of certain particle sizes. Now, the birth by nucleation and aggregation can be described as

$$B^{CA}_{\text{agg+nuc},i} = \hat{B}_{\text{agg+nuc},i-1}\lambda_i^-(\bar{a}_{i-1})H(\bar{a}_{i-1}-x_{i-1}) + \hat{B}_{\text{agg+nuc},i}\lambda_i^-(\bar{a}_i)H(x_i-\bar{a}_i)$$
$$+ \hat{B}_{\text{agg+nuc},i}\lambda_i^+(\bar{a}_i)H(\bar{a}_i-x_i) + \hat{B}_{\text{agg+nuc},i+1}\lambda_i^+(\bar{a}_{i+1})H(x_{i+1}-\bar{a}_{i+1}) \tag{6.90}$$

where

$$\hat{B}_{\text{agg+nuc},i} = \hat{B}_{\text{agg},i} + \int_{x_{i-1/2}}^{x_{i+1/2}} B^{\text{nuc}}(t,x)dx \tag{6.91}$$

The average values are computed as

$$\bar{a}_i = \frac{\hat{M}_{\text{agg},i} + \int_{x_{i-1/2}}^{x_{i+1/2}} xB^{\text{nuc}}(t,x)dx}{\hat{B}_{\text{agg},i} + \int_{x_{i-1/2}}^{x_{i+1/2}} B^{\text{nuc}}(t,x)dx} \tag{6.92}$$

All other notation has already been defined in previous sections. Treating both nucleation and aggregation as particle birth in cells in a similar fashion we conserve the mass and are consistent with number. If nucleation is monodisperse and produces the smallest particles, then the smallest representative can be chosen to match exactly the size of the the monodisperse nuclei and the above formulation

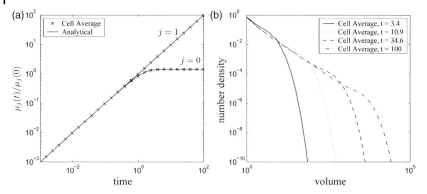

Fig. 6.21 A comparison of solutions with simultaneous nucleation and aggregation, (a) Moments, (b) Particle size distribution.

can be rewritten as

$$\frac{dN_1}{dt} = \hat{B}_{agg,1}^{CA} - \hat{D}_{agg,1}^{CA} + B^{nuc}(t, x_1) \tag{6.93}$$

and

$$\frac{dN_i}{dt} = \hat{B}_{agg,i}^{CA} - \hat{D}_{agg,i}^{CA}, \quad i = 2, 3, \ldots I$$

The formulation according to Eq. 6.89 may be used for a general polydisperse nucleation and the formulation according to Eq. 6.93 for monodisperse nucleation. **Numerical results:** We have tested the formulation for simultaneous nucleation and aggregation on a simple case of zero initial population with constant aggregation kernel and monodisperse nucleation. For this case, analytical solutions for the first two moments have been reported by Alexopoulos and Kiparissides (2005). The numerical solutions of the first two moments, plotted in Fig. 6.21(a), are in excellent agreement with the analytical results. Since an analytical solution for the complete PSD is not available, only numerical results are depicted in Fig. 6.21(b).

6.7
Combined Growth and Aggregation

For simultaneous aggregation and growth, we use the cell-average technique for aggregation and a Lagrangian approach for growth. Although the growth process can be coupled in a similar fashion like breakage and nucleation, the numerical diffusion becomes so significant that we prefer the moving-cells approach. The final formulation for growth with aggregation, breakage or simultaneous aggregation and breakage takes the form

$$\frac{dN_i}{dt} = \text{discretized form of agg/break/agg + break} \tag{6.94}$$

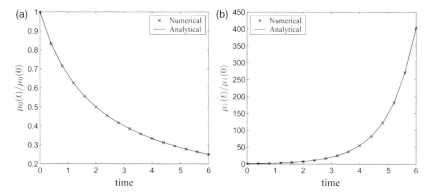

Fig. 6.22 A comparison of moments with simultaneous growth and aggregation, $\beta(u, v) = \beta_0$ and $G(v) = G_0 v$, (a) Zeroth moment, (b) First moment.

together with

$$\frac{dx_i}{dt} = G(x_i) \qquad (6.95)$$

and

$$\frac{dx_{i+1/2}}{dt} = G(x_{i+1/2}) \qquad (6.96)$$

The first equation is for the rate of change of particles due to aggregation or breakage, while the last two equations describe the motion of representatives and boundaries, respectively.

Numerical results: We have tested the scheme with an exponential initial condition

$$n(0, x) = \frac{N_0}{v_0} \exp\left(-\frac{x}{v_0}\right) \qquad (6.97)$$

combined with the constant aggregation kernel $\beta(u, v) = \beta_0$ and the linear growth rate $G(x) = G_0 x$. The analytical solution for this problem has been derived by Ramabhadran et al. (1976). Not only the numerically calculated moments (Fig. 6.22), but also the complete PSD (Fig. 6.23), are in excellent agreement with the analytical results.

6.8
Combined Growth and Nucleation

We have seen in previous sections that the coupling of growth with other processes produces excellent results, just by inclusion of extra ODEs to describe the growth of

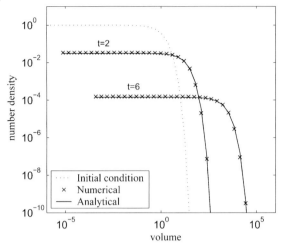

Fig. 6.23 A comparison of PSD with simultaneous growth and aggregation, $\beta\,(u,\,v) = \beta_0$ and $G(v) = G_0\,v$.

particles. This is not the case when nucleation is present, because the smallest representative sizes together with their boundaries will move and there may not be a cell to place subsequently formed nuclei. This definitely forces us to add more cells as the PSD moves to the right and therefore leads to serious difficulties in numerics. An effort in this direction was made by Kumar and Ramkrishna (1997). Their formulation was later improved by Spicer et al. (2002) by considering the effect of the mass of newly formed nuclei on the movement of cell representatives. The final set of equations for consistency of the first two moments is

$$\frac{dN_i}{dt} = B^{\text{nuc}} \delta_i \tag{6.98}$$

and

$$\frac{dx_i}{dt} = G(x_i) + \frac{1}{N_i}(x_{\text{nuc}} - x_i) B^{\text{nuc}} \delta_i \tag{6.99}$$

where

$$\delta_i = \begin{cases} 1, & x_{\text{nuc}} \in [x_{i-1/2}, x_{i+1/2}] \\ 0, & \text{elsewhere} \end{cases} \tag{6.100}$$

For further details of numerical implementation and comparison with analytical results, readers are referred to Kumar and Ramkrishna (1997), Spicer et al. (2002).

6.9
Multidimensional Population Balances

In many aggregation processes there are several important properties that may be described by particle density distribution. Therefore, a one-dimensional population balance equation where particle size is assumed to be the only significant particle property is not adequate to simulate such processes. A two-dimensional PBE for aggregation that is an extension of the one-dimensional PBE is given as (Lushnikov 1976)

$$\frac{\partial f}{\partial t} = \frac{1}{2} \int_0^x \int_0^y \beta(t, x-\epsilon, \epsilon, y-\gamma, \gamma) f(t, x-\epsilon, y-\epsilon) f(t, \epsilon, \gamma) d\gamma d\epsilon$$
$$- \int_0^\infty \int_0^\infty \beta(t, x, \epsilon, y, \gamma) f(t, x, y) f(t, \epsilon, \gamma) d\gamma d\epsilon$$

(6.101)

where x and y are the two extensive properties of the particle and β is an aggregation kernel. The two-dimensional particle number density is denoted by f. The first term corresponds to the birth of particles due to aggregation and the last term describes the death by collision and adhesion to other particles.

Numerical solution of the above PBE is difficult due to the double integral and nonlinear behavior. Several numerical techniques can be found in the literature (Immanuel and Doyle 2005; Xiong and Pratsinis 1993; Trautmann and Wanner 1999; Laurenzi et al. 2002), but either they have problems regarding the preservation of properties of the distribution or they are computationally very expensive. In order to overcome this problem, Hounslow et al. (2001) reduced the model into a system of two one-dimensional equations. The reduced model is computationally less expensive. However, it cannot capture the complete two-dimensional behavior of the population, since it assumes that particles of the same size contain the same amount of the second property. First, we summarize here the numerical methods for the reduced model. Then the extension of the cell-average technique for the complete two-dimensional case will be presented.

6.9.1
Reduced Model

Hounslow et al. (2001) consider two density functions to be computed: the number density n as a function of the first variable x and the total amount m of the second property y as a function of x by

$$n(t, x) = \int_0^x f(t, x, y) dy \tag{6.102}$$

and

$$m(t, x) = \int_0^x y f(t, x, y) dy \tag{6.103}$$

The dynamic equations for the rate of change of n(t, x) and m(t, x) can easily be obtained (Kumar and Ramkrishna 1996b; Hounslow et al. 2001) by integrating the two-dimensional equation with respect to y as

$$\frac{\partial n(t,v)}{\partial t} = \frac{1}{2}\int_0^v \beta(t,v-\epsilon,\epsilon)n(t,v-\epsilon)n(t,\epsilon)d\epsilon - n(t,v)\int_0^\infty \beta(t,v,\epsilon)n(t,\epsilon)d\epsilon \quad (6.104)$$

and

$$\frac{\partial m(t,v)}{\partial t} = \int_0^v \beta(t,v-\epsilon,\epsilon)m(t,v-\epsilon)n(t,\epsilon)d\epsilon - m(t,v)\int_0^\infty \beta(t,v,\epsilon)n(t,\epsilon)d\epsilon \quad (6.105)$$

Hounslow et al. (1988) developed a discretized method for solving Eq. 6.104 and later extended the same idea to the solution of Eq. 6.105, see Hounslow et at. (2001). However, the discretized PBEs were applicable only for special geometric grids of type $x_{i+1/2} = 2x_{i-1/2}$. Moreover, the discretization of Eq. 6.105 was not consistent with particles mass. In a recent paper Peglow et al. (2005) overcame this inconsistency in the discretized model by introducing correction factors and extended it to adjustable grids of the type $x_{i+1/2} = 2^{1/q}x_{i-1/2}$. They used the approach of Litster et al. (1995) and Wynn (1996) for the extension to adjustable grids. Introducing $N_i = \int_{x_{i-1/2}}^{x_{i+1/2}} n(t,x)dx$ and $M_i = \int_{x_{i-1/2}}^{x_{i+1/2}} m(t,x)dx$, the final extended and consistent model is given by the formula

$$\begin{aligned}\frac{dM_i}{dt} =& \sum_{j=1}^{i-S_1} \frac{2^{(j-i+1)/q}}{2^{1/q}-1}\beta_{i-1,j}(M_{i-1}N_j + N_{i-1}M_j)K_1 + \beta_{i-q,i-q}N_{i-q}M_{i-q} \\ &+ \sum_{p=1}^{q-1}\sum_{j=i+1-S_p}^{i+1-S_{p+1}} \frac{2^{1/q} - 2^{(j-i)/q} - 2^{-p/q}}{2^{1/q}-1}\beta_{i-p,j}(M_{i-p}N_j + N_{i-p}M_j)K_2 \\ &+ \sum_{p=2}^{q}\sum_{j=i-S_p-1}^{i-S_p} \frac{2^{(j-i+1)/q} - 1 + 2^{-(p-1)/q}}{2^{1/q}-1}\beta_{i-p,j}(M_{i-p}N_j + N_{i-p}M_j)K_3 \\ &+ \sum_{j=1}^{i-S_1+1}\left(1 - \frac{2^{(j-i)/q}}{2^{1/q}-1}\right)\beta_{i,j}N_iM_jK_4 \\ &- \sum_{j=1}^{i-S_1+1} \frac{2^{(j-i)/q}}{2^{1/q}-1}\beta_{i,j}M_iN_jK_5 - \sum_{j=i-S_1+2}^{I}\beta_{i,j}M_iN_j \end{aligned} \quad (6.106)$$

where

$$S_p = \text{Int}\left[1 - \frac{q\ln(1 - 2^{-p/q})}{\ln 2}\right] \quad (6.107)$$

and

$$K_1 = \frac{2^{(i-j)/q}}{1 + 2^{(i-j-1)/q}} \quad (6.108)$$

$$K_2 = K_3 = \frac{2^{(i-j)/q}}{1 + 2^{(i-j-p)/q}} \tag{6.109}$$

$$K_4 = \frac{1}{1 + 2^{(j-i)/q}}, \tag{6.110}$$

$$K_5 = \frac{2^{1/q} - 1}{2^{(j-i)/q}} - \frac{-2^{i/q} + 2^{(2i-j)/q}(2^{1/q} - 1)}{2^{i/q} + 2^{j/q}} \tag{6.111}$$

Int[x] is the integer part of x. Let us call this formulation the modified discretized tracer PBE (DTPBE).

Equation 6.106 can be used to calculate M_i and N_i simultaneously. Remember, though, that the above formulation is limited to grids of type $x_{i+1/2} = 2^{1/q} x_{i-1/2}$. Furthermore, we will see later that the numerical results are overpredicting. Next, we present a new discretization that is more general and accurate than the above formulation. It is based on the application of the cell-average technique to the equations under consideration.

The cell-average discretization of the Eq. 6.104 has already been given in Section 6.3. Kumar et al. (2006b) extended the cell-average technique for the computation of Eq. 6.105. The final formulation is

$$\begin{aligned}\frac{dM_i}{dt} &= \hat{B}_{T,i-1}\lambda_i^-(\bar{a}_{i-1})\eta_{T,i}(\bar{a}_{i-1})H(\bar{a}_{i-1} - x_{i-1}) + \hat{B}_{T,i}\lambda_i^-(\bar{a}_i)\eta_{T,i}(\bar{a}_i)H(x_i - \bar{a}_i) \\ &+ \hat{B}_{T,i}\lambda_i^+(\bar{a}_i)\eta_{T,i}(\bar{a}_i)H(\bar{a}_i - x_i) + \hat{B}_{T,i+1}\lambda_i^+(\bar{a}_{i+1})\eta_{T,i}(\bar{a}_{i+1})H(x_{i+1} - \bar{a}_{i+1}) \\ &- M_i \sum_{j=1}^{I} \beta(x_i, x_j) N_j\end{aligned} \tag{6.112}$$

where

$$\eta_{T,i}(a) = x_i/a \tag{6.113}$$

and

$$\hat{B}_{T,i} = \sum_{\substack{j,k \\ x_{i-1/2} \leq (x_j + x_k) < x_{i+1/2}}}^{j \geq k} \left(1 - \frac{1}{2}\delta_{j,k}\right) \beta_{j,k}(M_j N_k + N_j M_k) \tag{6.114}$$

All the remaining parameters have already been defined in Section 6.3. Now, the discretized Eq. 6.112 is coupled with the discretization Eq. 6.114 of the Eq. 6.104. These new discretized equations will be compared with the discrete formulation 6.106 by application to one analytically solvable problem in the next section.

Numerical results: Hounslow et al. (2001) considered the following problem to test their discretization

$$\frac{\partial n(t,v)}{\partial t} = \frac{1}{2}\int_0^v \beta(t, v-\epsilon, \epsilon) n(t, v-\epsilon) n(t, \epsilon) d\epsilon$$
$$-n(t,v)\int_0^\infty \beta(t,v,\epsilon) n(t,\epsilon) d\epsilon + B^0 \delta(v) - \frac{n(t,v)}{\tau}; \qquad (6.115)$$
$$n(t, 0^-) = 0, \quad \frac{\partial n(t,v)}{\partial t}\bigg|_{t=0} = 0$$

$$\frac{\partial m(t,v)}{\partial t} = \int_0^v \beta(t, v-\epsilon, \epsilon) m(t, v-\epsilon) n(t, \epsilon) d\epsilon$$
$$-m(t,v)\int_0^\infty \beta(t,v,\epsilon) n(t,\epsilon) d\epsilon - \frac{m(t,v)}{\tau}; \qquad (6.116)$$
$$m(t, 0^-) = 0, \quad m(0, v) = \delta(v - v_0)$$

Ilievski and Hounslow (1995) provided the solutions of decay of total tracer mass and tracer-weighted mean particle volume

$$\bar{v}_T(t) = \frac{\int_0^\infty v m(t,v) dv}{\int_0^\infty m(t,v) dv} \qquad (6.117)$$

for the size-independent kernel $\beta(u,v) = \beta_0$, the sum kernel $\beta(u,v) = \beta_0 \times (u+v)$, and the product kernel $\beta(u,v) = \beta_0 \times u \times v$.

We take only the sum kernel in our comparison here. In this case, $\bar{v}_T$ gives meaningful values up to one-third degree of aggregation (1978). The decay of total mass of tracer $M_T = \int_0^\infty m(t,x) dx$, takes the form

$$M_T = M_0 \exp(-t/\tau) \qquad (6.118)$$

The present numerical computations have been made with $B_0 = \tau = v_0 = \beta_0 = 1$ with a very coarse grid $x_{i+1/2} = 2x_{i-1/2}$. The maximal degree of aggregation is taken to be 1/4. Figure 6.24(a) shows that both formulations predict exactly the same total tracer mass. In Fig. 6.24(b), the numerical and analytical temporal changes of $\bar{v}_T$ are compared. The numerical solution by the cell-average technique is in very close agreement with analytical results, while the previous discrete formulation gives an overprediction at large times.

6.9.2
Complete Model

We now briefly present the extension of the cell-average technique to solve the complete two-dimensional PBE Eq. 6.101 as provided by Kumar et al. (2006). The entire two-dimensional property domain is divided into small cells, see Fig. 6.25. The particles within a cell $C_{i,j}$ are assumed to be concentrated at a representative

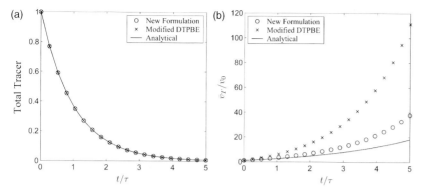

Fig. 6.24 A comparison of numerical results with analytical results, (a) Total tracer mass, (b) Tracer-weighted mean particle volume, for a sum aggregation kernel in Ilievski and Hounslow's problem, $I_{agg} = 1/4$.

node $P_{i,j}$. The idea is to take the property average of all newborn particles within the cell and then assign them to the neighboring nodes such that prechosen properties are exactly preserved. Let us define the discrete number density N_{ij}, i.e. the total number of particles in a cell, by integrating the number density over both properties as

$$N_{ij} = \int_{x_{i-}}^{x_{i+}} \int_{y_{j-}}^{y_{j+}} f(t, x, y) \mathrm{d}x \mathrm{d}y \qquad (6.119)$$

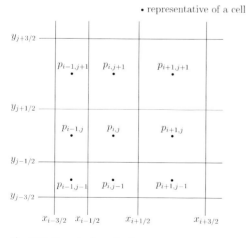

Fig. 6.25 Domain discretization.

For notational convenience $x_{i\pm 1/2}$ and $y_{j\pm 1/2}$ are replaced by $x_{i\pm}$ and $y_{i\pm}$, respectively. The total birth and the death in a cell $C_{i,j}$ can be computed as

$$B_{ij} = \sum_{\substack{k,l \\ x_{i-} \leq (x_k+x_l) < x_{i+}}}^{k \geq l} \sum_{\substack{m,n \\ y_{j-} \leq (y_m+y_n) < y_{j+}}}^{m \geq n} \left(1 - \frac{1}{2}\delta_{k,l}\delta_{m,n}\right) \beta_{km,ln} N_{km} N_{ln} \qquad (6.120)$$

and

$$D_{ij} = \sum_{p=1}^{l_x} \sum_{q=1}^{l_y} \beta_{ij,pq} N_{ij} N_{pq} \qquad (6.121)$$

Similar to the one-dimensional case, the birth B_{ij} has to be modified in order to make the formulation consistent with selected moments. The net flux of the property x into the cell i as a result of aggregations is therefore given by

$$M_{x,i} = \sum_{\substack{k,l \\ x_{i-} \leq (x_k+x_l) < x_{i+}}}^{k \geq l} \sum_{\substack{m,n \\ y_{j-} \leq (y_m+y_n) < y_{j+}}}^{m \geq n} \left(1 - \frac{1}{2}\delta_{k,l}\delta_{m,n}\right) \beta_{km,ln} N_{km} N_{ln} (x_k + x_l) \qquad (6.122)$$

Similarly, the net flux of the second property can be computed as

$$M_{y,i} = \sum_{\substack{k,l \\ x_{i-} \leq (x_k+x_l) < x_{i+}}}^{k \geq l} \sum_{\substack{m,n \\ y_{j-} \leq (y_m+y_n) < y_{j+}}}^{m \geq n} \left(1 - \frac{1}{2}\delta_{k,l}\delta_{m,n}\right) \beta_{km,ln} N_{km} N_{ln} (y_m + y_n) \qquad (6.123)$$

Consequently, the average property values of all newborn particles in the ith cell can be calculated as

$$\bar{x}_i = \frac{M_{x,i}}{B_{ij}} \qquad (6.124)$$

and

$$\bar{y}_j = \frac{M_{y,j}}{B_{ij}} \qquad (6.125)$$

If these average values $\bar{x}_i$ and $\bar{y}_j$ are different from the representatives x_i and y_j, particles have to be assigned to the neighboring nodes. Note that in the case of linear (equidistant) grids in both directions, $\bar{x}_i$ and $\bar{y}_j$ are always equal to x_i and y_i. Therefore, in that case all particles will remain in the cell they belong to without any assignment to neighboring nodes. The choice of neighboring nodes depends on the position of $(\bar{x}_i, \bar{y}_j)$ in the cell $C_{i,j}$. Let us suppose that particle fractions a_1, a_2, a_3 and a_4 have been assigned to the neighboring nodes $p_{i-1,j-1}, p_{i,j-1}, p_{i,j}$ and $p_{i-1,j}$, respectively. For consistency of zeroth μ_{00} and first moments μ_{10} or μ_{01}, these fractions must satisfy the following relations

$$a_1 + a_2 + a_3 + a_4 = B_{ij}$$
$$a_1 x_{i-1} + a_2 x_i = (a_1 + a_2)\bar{x}_i$$
$$a_3 x_i + a_4 x_{i-1} = (a_3 + a_4)\bar{x}_i \qquad (6.126)$$
$$a_1 y_{j-1} + a_4 y_j = (a_1 + a_4)\bar{y}_j$$
$$a_2 y_{j-1} + a_3 y_j = (a_2 + a_3)\bar{y}_j$$

The above system of equations has the unique solution

$$a_1 = \frac{(x_i - \bar{x}_i)(y_j - \bar{y}_j)}{(x_i - x_{i-1})(y_j - y_{j-1})} B_{ij} \qquad (6.127)$$

$$a_2 = \frac{(\bar{x}_i - x_{i-1})(y_j - \bar{y}_j)}{(x_i - x_{i-1})(y_j - y_{j-1})} B_{ij} \qquad (6.128)$$

$$a_3 = \frac{(\bar{x}_i - x_{i-1})(\bar{y}_j - y_{j-1})}{(x_i - x_{i-1})(y_j - y_{j-1})} B_{ij} \qquad (6.129)$$

$$a_4 = \frac{(x_i - \bar{x}_i)(\bar{y}_j - y_{j-1})}{(x_i - x_{i-1})(y_j - y_{j-1})} B_{ij} \qquad (6.130)$$

Furthermore, we can see that the assignment process is also consistent with the first moment μ_{11}, i.e.

$$a_1 x_{i-1} y_{j-1} + a_2 x_i y_{j-1} + a_3 x_i y_j + a_4 x_{i-1} y_j = B_{ij} \bar{x}_i \bar{y}_j \qquad (6.131)$$

However, we have lost this consistency during the averaging process, i.e.

$$\sum_{\substack{k,l \\ x_{i-} \leq (x_k + x_l) < x_{i+}}}^{k \geq l} \sum_{\substack{m,n \\ y_{j-} \leq (y_m + y_n) < y_{j+}}}^{m \geq n} \left(1 - \frac{1}{2}\delta_{k,l}\delta_{m,n}\right) \beta_{km,ln} N_{km} N_{ln}(x_k + x_l)(y_m + y_n) \neq B_{ij} \bar{x}_i \bar{y}_j$$

In the total, the formulation is consistent with μ_{00}, μ_{10} and μ_{01}, but not with μ_{11}. Similarly to the one-dimensional case, it is convenient to define a function λ as

$$\lambda_{i,j}^{\pm,\pm}(x,y) = \frac{(x - x_{i\pm 1})(y - y_{j\pm 1})}{(x_i - x_{i\pm 1})(y_j - y_{j\pm 1})} \qquad (6.132)$$

Now a_1, a_2, a_3 and a_4 can be rewritten using $\lambda_{i,j}^{\pm,\pm}$ as

$$a_1 = \lambda_{i-1,j-1}^{+,+}(\bar{x}_i, \bar{y}_j) B_{ij} \tag{6.133}$$

$$a_2 = \lambda_{i,j-1}^{-,+}(\bar{x}_i, \bar{y}_j) B_{ij} \tag{6.134}$$

$$a_3 = \lambda_{i,j}^{-,-}(\bar{x}_i, \bar{y}_j) B_{ij} \tag{6.135}$$

$$a_4 = \lambda_{i-1,j}^{+,-}(\bar{x}_i, \bar{y}_j) B_{ij} \tag{6.136}$$

The next step is to collect all contributions in cell $C_{i,j}$ from the neighboring cells. There are 8 neighboring cells that may give rise to birth in cell $C_{i,j}$. The final formulation takes the form

$$\begin{aligned}\frac{dN_{ij}}{dt} &= \sum_{p=0}^{1}\sum_{q=0}^{1} B_{i-p,j-q}\lambda_{i,j}^{-,-}(x_{i-p}, y_{j-q}) H((-1)^p(x_{i-p} - \bar{x}_{i-p})) H((-1)^q(y_{j-q} - \bar{y}_{j-q})) \\ &+ \sum_{p=0}^{1}\sum_{q=0}^{1} B_{i-p,j-q}\lambda_{i,j}^{-,+}(x_{i-p}, y_{j-q}) H((-1)^p(x_{i-p} - \bar{x}_{i-p})) H((-1)^{q+1}(y_{j-q} - \bar{y}_{j-q})) \\ &+ \sum_{p=0}^{1}\sum_{q=0}^{1} B_{i-p,j-q}\lambda_{i,j}^{+,+}(x_{i-p}, y_{j-q}) H((-1)^{p+1}(x_{i-p} - \bar{x}_{i-p})) H((-1)^{q+1}(y_{j-q} - \bar{y}_{j-q})) \\ &+ \sum_{p=0}^{1}\sum_{q=0}^{1} B_{i-p,j-q}\lambda_{i,j}^{+,-}(x_{i-p}, y_{j-q}) H((-1)^{p+1}(x_{i-p} - \bar{x}_{i-p})) H((-1)^q(y_{j-q} - \bar{y}_{j-q})) \\ &- \sum_{p=1}^{I_x}\sum_{q=1}^{I_y} \beta_{i,j,pq} N_{ij} N_{pq}\end{aligned} \tag{6.137}$$

Clearly, the two-dimensional formulation is a direct extension of the one-dimensional case. In this way we can also treat more than two-dimensional population balances.

Numerical results: In this section we test the proposed numerical formulation for problems where analytical results are available (Lushnikov 1976; Laurenzi et al. 2002; Gelbard and Seinfeld 1978). Two types of aggregation kernels, the constant kernel $\beta(t, x, \epsilon, y, \gamma) = \beta_0$ and the sum kernel $\beta(t, x, \epsilon, y, \gamma) = \beta_0(x + \epsilon + y + \gamma)$ have been considered. Since the numerical results show a similar behavior, we present them here only for the constant kernel.

Two different types of monodisperse particles with the same initial concentration have been considered as the initial condition. The computation has been made to a high degree of aggregation (0.98), with 20 grid points in each property direction. A low-order MATLAB ode23 solver is used for integration. The numerical result for

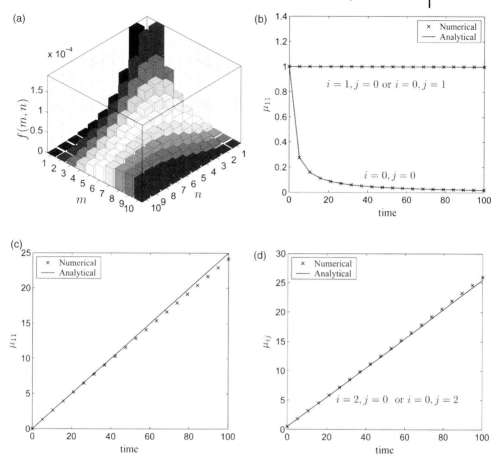

Fig. 6.26 A comparison of numerical and analytical results, (a) Evolution of particle size distribution, (b) Temporal change of total number and mass, (c) Temporal change of μ_{11} moment (d) Temporal change of higher moments.

the complete number size distribution is shown in Fig. 6.26(a). It is symmetrical, since we have taken the same initial concentration of both types of particles. Different moments are plotted in Figs. 6.26(b), (c) and (d). As we know that the formulation is consistent with the two moments μ_{00} and μ_{01} or μ_{10}, we expect the correct prediction of those moments in the numerical results. The prediction of μ_{00} and μ_{01} or μ_{10}, Fig. 6.26(b), is, indeed, excellent, while a slight underprediction has been observed in μ_{11}, see Fig. 6.26(c). This is due to the fact that we lose the consistency with respect to μ_{11} during averaging. Figure 6.26(d), the temporal change of second moment, clearly reflects the ability of the scheme to predict higher moments with good accuracy.

Additional Notation used in Chapter 6

a_1, a_2, a_3, a_4	fraction of particles	–
b	breakage function	m^{-3}
B	birth rate	$m^{-6}s^{-1}$
D	death rate	$m^{-6}s^{-1}$
f	multidimensional number density function	$m^{-3}(\prod_j [i_j])^{-1}$
F	mass flux	s^{-1}
g	volume density function	–
G_j	growth rate along internal coordinate j	$[i_j]/s$
h	enthalpy	J
H	Heaviside function	–
I	total number of cells	–
I_{agg}	degree of aggregation	–
l	mass of liquid	kg
n	number density function	m^{-6}
N	number	m^{-3}
q	discretization parameter	–
r	ratio x_{i+1}/x_i	–
S	selection function	s^{-1}
t	time	s
u, v, x	volume of particles	m^3
V	rate of change of volume	s^{-1}
$\mathbf{v}$	velocity	$([\mathbf{i}]/s, [\mathbf{e}]/s)$
$\mathbf{x}$	phase space	$([\mathbf{i}], [\mathbf{e}])$
$\mathbf{x}_e$	vector of external coordinates	$[\mathbf{e}]$
$\mathbf{x}_e^j$	component j of external coordinates	$[e_j]$
$\mathbf{x}_i$	vector of internal coordinates	$[\mathbf{i}]$
$\mathbf{x}_i^j$	component j of internal coordinates	$[i_j]$

Greek letters

α	integer	–
β	aggregation kernel	$m^3 s^{-1}$
$\delta(x)$	Dirac-delta distribution	–
δ_i	a function defined by Eq. 6.100	–
δ_{ij}	Kronecker delta	–
η	particle distribution function	–
θ	a function defined by Eq. 6.67	–
μ_r	rth moment	$m^{3r}m^{-3}$
τ	age of particles	s
ϕ	limiter function	–

Subscripts

agg	aggregation
break	breakage
i, j	index
in	inlet
nuc	nucleation
out	outlet
T	tracer

Superscripts

–	mean or average values
ˆ	numerical approximations
ana	analytical
num	numerical

Acronyms

CA	cell average
DTPBE	discretized tracer population balance equation
EOC	experimental order of convergence
FP	fixed pivot
MOL	method of line
ODE	ordinary differential equation
PBE	population balance equation
PSD	particle-size distribution

References

Adetayo, A. A., Litster, J. D., Pratsinis, S. E., Ennis, B. J., 1995. Population balance modelling of drum granulation of materials with wide size distribution. *Powder Technology* **82**: 37–50.

Alexopoulos, A. H., Kiparissides, C. A., 2005. Part II: Dynamic evolution of the particle size distribution in particulate processes undergoing simultaneous particle nucleation, growth and aggregation. *Chemical Engineering Science* **60**: 4157–4169.

Burgschweiger, J., Tsotsas, E., 2002. Experimental investigation and modelling of continuous fluidized bed drying under steady-state and dynamic conditions. *Chemical Engineering Science* **57**: 5021–5038.

Dubovskii, P. B., Galkin, V. A., Stewart, I. W., 1992. Exact solutions for the coagulation-fragmentation equation. *Journal of Physics A: Mathematical and General* **25**: 4737–4744.

Everson, R. C., Eyre, D., Campbell, Q. P., 1997. Spline method for solving continuous batch grinding and similarity equations. *Computers and Chemical Engineering* **21**: 1433–1440.

Eyre, D., Everson, R. C., Campbell, Q. P., 1988. New parameterization for a discrete batch grinding equation. *Powder Technology* **98**: 256–272.

Filbet, F., Laurenot, P., 2004. Numerical simulation of the Smoluchowski coagulation equation. *SIAM Journal of Scientific Computing* **25**: 2004–2028.

Gelbard, F., 1990. Modelling multicomponent aerosol particle growth by vapor condensation. *Aerosol Science and Technology* **12**: 399–412.

Gelbard, F., Seinfeld, J. H., 1978. Coagulation and growth of a multidimensional aerosol. *Journal of Colloid and Interface Science* **63**: 472–479.

Gunawan, R., Fusman, I., Braatz, R. D., 2004. High resolution algorithms for multidimensional population balance equations. *AIChE Journal* **50**: 2738–2749.

Hill, P. J., Ng, K. M., 1995. New discretization procedure for the breakage equation. *AIChE Journal* **41**: 1204–1216.

Hounslow, M. J., Pearson, J. M. K., Instone, T., 2001. Tracer studies of high-shear granulation: II. Population balance modelling. *AIChE Journal* **47**: 1984–1999.

Hounslow, M. J., Ryall, R. L., Marshall, V. R., 1988. A discretized population balance for nucleation, growth and aggregation. *AIChE Journal* **38**: 1821–1832.

Hulburt, H. M., Katz, S., 1964. Some problems in particle technology. A statistical mechanical formulation. *Chemical Engineering Science* **19**: 555–578.

Hundsdorfer, W., Verwer, J., 2003. *Numerical Solution of Time-Dependent Advection-Diffusion-Reaction Equations*, Springer Verlag.

Ilievski, D., Hounslow, M. J., 1995. Agglomeration during precipitations: II. Mechanism deduction from tracer data. *AIChE Journal* **41**: 525–535.

Immanuel, C. D., Doyle, F. J., 2005. Solution technique for a multi-dimensional population balance model describing granulation processes. *Powder Technology* **156**: 213–225.

Jacobson, M. Z., 1997. Development and application of a new air pollution modeling system - ii. Aerosol structure and design. *Atmospheric Environment* **31**: 131–144.

Jacobson, M. Z., Turco, R. P., 1995. Simulating condensational growth, evaporation and coagulation of aerosols using a combined moving and stationary size grid. *Aerosol Science and Technology* **22**: 73–92.

Kettner, C., Peglow, M., Metzger, T., Tsotsas, E., 2006. Distributed product quality in fluidized bed drying. *Proceedings of 15th International Drying Symposium.*

Kim, Y. P., Seinfeld, J. H., 1990. Simulation of multicomponent aerosol condensation by the moving sectional method. *Journal of Colloid and Interface Science* **135**: 185–199.

Koren, B., 1993. A robust upwind discretization for advection, diffusion and source terms. *Numerical Methods for Advection–Diffusion Problems* **45**: 117–138.

Kostoglou, M., Karabelas, A. J., 2002. An assessment of low-order methods for solving the breakage equation. *Powder Technology* **127**: 116–127.

Kostoglou, M., Karabelas, A. J., 2004. Optimal low order methods of moments for solving the fragmentation. *Powder Technology* **143–144**: 280–290.

Kumar, J., Peglow, M., Warnecke, G., Heinrich, S., Mörl, L., 2006a. Improved accuracy and convergence of discretized population balances: The cell average technique. *Chemical Engineering Science* **61**: 3327–3342.

Kumar, J., Peglow, M., Warnecke, G., Heinrich, S., Mörl, L., 2006b. A discretized model for tracer population balance equation: Improved accuracy and convergence. *Computers and Chemical Engineering Science* **30**: 1278–1292.

Kumar, J., Peglow, M., Warnecke, G., Heinrich, S., Tsotsas, E., Mörl, L. Numerical solutions of a two dimensional population balanace equation for aggregation. In *The Fifth World Congress on Particle Technology*, Orlando, FL, USA, April 2006, AIChE.

Kumar, S., Ramkrishna, D., 1996a. On the solution of population balance equations by discretization – i. a fixed pivot technique. *Chemical Engineering Science* **51**: 1333–1342.

Kumar, S., Ramkrishna, D., 1996b. On the solution of population balance equations by discretization – ii. a moving pivot technique. *Chemical Engineering Science* **51**: 1333–1342.

Kumar, S., Ramkrishna, D., 1997. On the solution of population balance equations by discretizationiii. nucleation, growth and aggregation of particles. *Chemical Engineering Science* **52**: 4659–4679.

Kurganov, A., Tadmor, E., 2000. New high-resolution central schemes for nonlinear conservation laws and convectiondiffusion equations. *Journal of Computational Physics* **160**: 141–182.

Lage, P. L. C., 2002. Comments on 'an analytical solution to the population balance equation with coalescence and breakage the special case with constant number of particles' by D. P. Patil and J. R. G. Andrews [Chemical Engineering Science 53(3) 599–601]. *Chemical Engineering Science* **19**: 4253–4254.

Laurenzi, I. J., Bartels, J. D., Diamond, S. L., 2002. A general algorithm for exact simulation of multicomponent aggregation processes. *Journal of Computational Physics* **177**: 418–449.

Lee, K., Matsoukas, T., 2000. Simultaneous coagulation and break-up using constant-N monte carlo. *Powder Technology* **110**: 82–89.

Leveque, R. J., 1990. *Numerical Methods for Conservation Laws*, Birkhuser-Verlag.

Lim, Y., Lann, J., Meyer, X., Joulia, X., Lee, G., Yoon, E., 2002. On the solution of population balance equations (pbe) with accurate front tracking methods in practical crystallization processes. *Chemical Engineering Science* **57**: 3715–3732.

Litster, J. D., Smit, D. J., Hounslow, M. J., 1995. Adjustable discretized population balance for growth and aggregation. *AIChE Journal* **41**: 591–603.

Lushnikov, A. A., 1976. Evolution of coagulating systems iii. Coagulating mixtures. *Journal of Colloid and Interface Science* **54**: 94–101.

Makino, J., Fukushige, T., Funato, Y., Kokubo, E., 1998. On the mass distribution of planetesimals in the early runaway stage. *New Astronomy* **3**: 411–417.

Mishra, B. K., 2000. Monte Carlo simulation of particle breakage process during grinding. *Powder Technology* **110**: 246–252.

Park, S. H., Rogak, S. N., 2004. A novel fixed-sectional model for the formation and growth of aerosol agglomerates. *Journal of Aerosol Science* **35**: 185–199.

Patil, D. P., Andrews, J. R. G., 1998. An analytical solution to continuous population balance model describing floc coalescence and breakage – a special case. *Chemical Engineering Science* **53**: 599–601.

Peglow, M., Kumar, J., Warnecke, G., Heinrich, S., Mörl, L., Hounslow, M. J., 2005. Improved discretized tracer mass distribution of Hounslow et al. *AIChE Journal* **52**: 1326–1332.

Ramabhadran, T. E., Peterson, T. W., Seinfeld, J. H., 1976. Dynamic of aerosol coagulation and condensation. *AIChE Journal* **22**: 840–851.

Ramkrishna, D., 2000. *Population Balances*, Academic Press, New York.

Randolph, A. D., Larson, M. A., 1978. *The Theory of Particulate Processes*, Academic Press, New York.

Schaafsma, S. H., Vonk, P., Segers, P., Kossen, N. W. F., 1998. Description of agglomerate growth. *Powder Technology* **97**: 183–190.

Spicer, P. T., Chaoul, O., Tsantilis, S., Pratsinis, S. E., 2002. Titania formation by TiCl4 gas phase oxidation, surface growth and coagulation. *Journal of Aerosol Science* **33**: 17–34.

Tan, H. S., Salman, A. D., Hounslow, M. J., 2004. Kinetics of fluidised bed melt granulation IV. Selecting the breakage model. *Powder Technology* **143–144**: 65–83.

Tanaka, H., Inaba, S., Nakaza, K., 1996. Steady-state size distribution for the self-similar collision cascade. *Icarus* **123**: 450–455.

Trautmann, T., Wanner, C., 1999. A fast and efficient modified sectional method for simulating multicomponent collisional kinetics. *Atmospheric Environment* **33**: 1631–1640.

Tsantilis, S., Kammler, H. K., Pratsinis, S. E., 2002. Population balance modeling of flame synthesis of titania nanoparticles. *Chemical Engineering Science* **57**: 2139–2156.

Tsantilis, S., Pratsinis, S. E., 2004. Narrowing the size distribution of aerosol-made titania by surface growth and coagulation. *Journal of Aerosol Science* **35**: 405–420.

Vanni, M., 1999. Discretization procedure for the breakage equation. *AIChE Journal* **45**: 916–919.

Vanni, M., 2000. Approximate population balance equations for aggregationbreakage processes. *Journal of Colloid and Interface Science* **221**: 143–160.

Watano, S., Fukushima, T., Miyanami, K., 1996. Heat transfer and rate of granule growth in fluidized bed granulation. *Chemical and Pharmaceutical Bulletin* **44**: 572–576.

Wynn, E. J. W., 1996. Improved accuracy and convergence of discretized population balance of Lister et al. *AIChE Journal* **22**: 2048–2086.

Xiong, Y., Pratsinis, S. E., 1993. Formation of agglomerate particles by coagulation and sintering – part i. a two-dimensional solution of the population balance equation. *Journal of Aerosol Science* **24**: 282–300.

Ziff, R. M., McGrady, E. D., 1985. The kinetics of cluster fragmentation and depolymerisation. *Journal of Physics A: Mathematical and General* **18**: 3027–3037.

Ziff, R. M., 1991. New Solutions to the fragmentation equation. *Journal of Physics A: Mathematical and General* **24**: 2821–2828.

7
Process-systems Simulation Tools
Ian C. Kemp

7.1
Introduction

7.1.1
Summary of Contents

Drying technology, like all areas of life, has seen a software boom in the last 25 years. Programs for dryers have been developed by many different groups, including universities, equipment vendors, end users and software companies.

This chapter reviews the different types of software that can be usefully used in design and performance calculations, selection and troubleshooting of dryers. In particular, modern process-systems simulation tools capable of treating not only liquid, but also solids processes including drying are outlined. The role of drying in such simulators and the different possible levels of model sophistication (heat and mass balance, short-cut kinetic approaches, detailed dryer design modules) are discussed. Supporting functions like expert systems for dryer selection and problem solving are reviewed. Furthermore, possibilities for interconnection of flow sheets with data banks for physical properties, library equation solvers, or computational fluid dynamics tools are addressed.

Drying software falls into five broad types:
1. calculation programs, including numerical models of dryers (Sections 7.2–7.6), and ancillary calculations, e.g. psychrometrics and processing experimental data (Section 7.7),
2. process simulators (Section 7.8),
3. expert systems and other decision-making tools (Section 7.9),
4. information delivery (online libraries or knowledge bases) (Section 7.10),
5. software for dryer control systems and instrumentation, which is related to process plant design and manufacture and is outside the scope of this review.

Modern Drying Technology. Edited by Evangelos Tsotsas and Arun S. Mujumdar
Copyright © 2007 WILEY-VCH Verlag GmbH & Co. KGaA. All rights reserved.
ISBN: 978-3-527-31556-7

Each of these types is complementary in building up a clear overall understanding of a drying process. For solids processing in industry, reliable qualitative information is as important as accurate numerical models. In addition to these, it is often important to link to broader generic programs, such as physical properties databanks.

The chapter concludes by investigating why relatively little commercial software is available for dryers, in contrast to many chemical engineering unit operations, such as distillation and heat exchange, and suggests possible ways forward.

7.1.2
The Solids Processing Challenge

In solids drying, there has often been a great gulf between academic theory and industrial design practice. Traditionally, practical dryer design has tended to be based on simple correlations and scale-up from pilot-plant tests, rather than on rigorous theoretical models. Moreover, subjects such as selection and troubleshooting were largely neglected. This dichotomy has been reflected in software programs. On the one hand, dryer manufacturers and users have developed in-house programs based on simple correlations, sometimes including other aspects such as mechanical design. On the other hand, numerous computer-based models have been developed in university research projects, but have very rarely been tested on industrial data or used for practical improvement of industrial dryers. Moreover, little commercial software has been developed to "bridge the gap".

The main reason for this is that drying, like all processes involving solids, is much more difficult to model than fluid-phase (liquid and gas) processes. Physical properties of fluids can be obtained easily from databanks, and are uniquely defined for given temperatures and pressures, while the system is controlled by equilibrium thermodynamics. In contrast, physical properties of solids are highly dependent on solid structure, they can vary for the same material made by different processes, or even between different batches from the same process. For example, drying kinetics can differ by orders of magnitude for the same chemical substance depending on particle size, porosity, polymorph, etc. Drying is also inherently a nonequilibrium, rate-controlled process; equilibrium-based analysis can predict that the moisture within a particle will flash off instantly when exposed to hot dry air, which certainly does not happen!

The simultaneous heat, mass and momentum transfer processes give a highly nonlinear set of governing equations. Numerous parameters affect drying processes, and even basic ones may be difficult to measure (e.g. Sauter mean diameter of a particle-size distribution). Some are hard to quantify at all (e.g. stickiness and handling characteristics, represented by a number of secondary parameters of varying relevance). As a result, it is difficult to evaluate the parameters required for dryer models and software with sufficient accuracy for practical use.

The consequences for industrial design were shown in a famous study by Merrow and coworkers (1981, 1985); while 95% of processes involving fluids alone (liquids and gases) reached nameplate capacity within a year, barely two-thirds of processes involving solids (including dryers) did so. Experience shows that many rigorous theoretical models are extremely difficult to apply in practice because the many necessary parameters cannot be obtained with sufficient accuracy or reliability. In particular, prediction of drying kinetics from first principles and published data, without experimental measurements, is highly unreliable, especially for formulated products. One consequence of this is that scale-up using experimental data from small-scale tests or pilot plants is often more reliable than a design-mode calculation using physical properties from databanks or theoretical estimates.

7.1.3
Types of Software for Dryers

Drying software may be, in essence, quantitative (calculation programs and process simulators) or qualitative (expert systems and knowledge bases).

Calculation programs include programs for dryer design, performance rating and scale-up, plus methods of processing drying kinetics data and drawing psychrometric charts. Users may either write their own software or use commercial programs such as spreadsheets, math solvers or computational fluid dynamics (CFD) software. Process simulators, such as Aspen Plus, HYSYS, Prosim and Batch Plus, set the dryer in its context in the overall process flow sheet, and show how it links with the general heat and mass balance. However, optimizing a dryer is pointless if the wrong type of dryer has been selected. Expert systems and wizards can help with selection, problem solving and other aspects of process design that are not amenable to numerical calculation. Likewise, effective design, troubleshooting and debottlenecking of a process requires qualitative information and in-depth understanding in addition to numerical simulation. These are provided by knowledge resources: textbooks, papers and journals such as *Drying Technology*, which are nowadays available online, and by the web-based *Process Manual*. Combining all these types of software allows engineers to understand their dryers thoroughly and take a holistic approach to optimization.

Software has been produced over the last 20 years by a wide range of groups, both academic and commercial. Menshutina and Kudra (2001) have previously given an extensive review of available packages. Software produced by Aspen Technology (previously SPS, AEA and Hyprotech) will be highlighted throughout this chapter. Another noteworthy program was the multipurpose dryer simulator, DRYPAK, produced in the 1990s by Professor Pakowski of Lodz Technical University, Poland.

7.2
Numerical Calculation Procedures

Conventional standalone calculation programs are the largest group of drying software.

7.2.1
Categorization of Dryer Models

Dryer models can be categorized in several different ways:
- The mode of calculation; design, performance or scale-up.
- The complexity of the model; it can be considered to be at one of four levels (the layering approach, Fig. 7.1).
- The physical processes considered; heat and mass balance, heat and mass transfer, drying kinetics and equilibria.
- The level of information generated; basic heat and mass balance, approximate sizing, overall dryer dimensions, localized conditions in the dryer.

Kemp and Oakley (2002) concentrated on the first two of these distinctions, describing them as follows:

(i) The *purpose* or *mode* of the calculation.
   a. *Design* of a new dryer to perform a given duty.
   b. For an existing dryer, calculation of *performance* under a different set of operating conditions.
   c. *Scale-up* from laboratory-scale or pilot-plant experiments to a full-scale dryer.

   These affect the type of inputs and outputs needed by the calculation.
   As pointed out above, predicting drying kinetics from first principles is extremely difficult. Experimental work is almost always necessary in order to design a dryer accurately, and scale-up calculations (c) are more reliable than design based only on thermodynamic data (a). The experiment is used to verify the theoretical model and find the difficult-to-measure parameters; the full-scale dryer can then be modeled accurately.

(ii) The *level of complexity* of the calculation.
   Level 1. Simple *heat and mass balances* (Section 7.3) give some useful information, but say nothing about the required equipment size or the performance that a given dryer is capable of.
   Level 2. *Scoping* or approximate calculations (Section 7.4) give rough sizes and throughputs (mass flow rates)

Layer 1: heat and mass balance

Layer 2: scoping (approximate) design

Layer 3: scaling (overall) methods

Layer 4: detailed (localized) design

Fig. 7.1 The layering model.

for dryers, using simple data and making some assumptions.

Level 3. *Scaling* calculations (Section 7.5) give overall dimensions and performance figures for dryers by scaling up drying curves from small-scale or pilot-plant experiments.

Level 4. *Detailed* methods (Section 7.6) track the local conditions of the solids (and gas) as drying progresses. Naturally, these methods require a lot more input data and much more complex modeling techniques. CFD (Chapter 5) and the other advanced methods in Chapters 1–6 fall into this category.

7.2.2
Equipment and Material Model

An alternative categorization was developed by SPS (the Separation Processes Service consortium) at Harwell Laboratory, UK. They also divided dryer models into two parts; an *equipment* model and a *material* model (Reay 1989, Fig. 7.2). The equipment model includes factors that depend on the type of dryer used; particle transport through the dryer, external heat transfer from the hot gas or hot surfaces to the solids, and vapor-phase mass transfer. These are all *external to the particle*, and can be modeled theoretically for the various different types of dryer. The material model covers factors dependent on the *nature of the solid* being dried; drying kinetics and heat transfer within the particle, equilibrium moisture-content relationships, product quality (e.g. trace components) and materials handling. Some of these properties can be found from databanks, but the majority are highly dependent on the nature and structure of the solid and on the upstream particle-formation process, and must be measured (materials characterization).

Relating this back to the layering concept, we find that a Level 1 heat and mass balance uses neither equipment nor material models, a Level 2 approach using unhindered drying considers only the equipment model, whereas Level 3 and 4 approaches consider both the equipment and material models. Only the last two approaches consider the falling-rate period thoroughly, but to do this, experimental

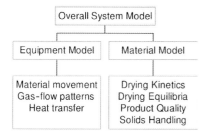

Fig. 7.2 Representation of the overall drying model as equipment and material models.

Tab. 7.1 Comparison of features of available models for dryers

Type of calculation	Physical basis of calculation	Dryer details calculated	Main uncertainties	Typical use
Heat and mass balance	Heat and mass balance only	None	Mass flowrate, moisture content, gas humidity	Performance
Scoping design; continuous convective dryers	(a) Heat and mass balance only, (b) heat transfer	Overall dimensions: (a) Cross-section/diameter, (b) length	(b) Heat transfer coefficients, falling rate drying kinetics	Design (scale-up?)
Scoping design for continuous contact dryers	Heat and mass balance, heat transfer	Overall dimensions; diameter, length, residence time	Heat transfer coefficients, temperature difference, falling rate kinetics	Design (scale-up?)
Scoping design for batch dryers	Mass and volume of solids	Overall dimensions; diameter, length	Drying time and all factors influencing it	Design (scale-up?)
Scaling (integral model)	Heat and mass balance, heat transfer, drying curve	Overall dimensions; diameter, length, residence time	Local variations inside dryer	Scale-up, performance
Detailed design; incremental model (1-D)	Heat and mass balance, heat transfer, full drying kinetics	Length (use scoping method for D); local conditions	3-dimensional flows, parameter measurement	Design, scale-up
Detailed design; CFD	Heat and mass balance, heat transfer, kinetics, 3-D flow patterns	Local conditions throughout dryer	Measuring required parameters, computing time	Design, scale-up, performance

measurements are needed. The falling-rate period is by far the most difficult period to model accurately, as shown in Section 7.2.3; unfortunately, since drying rates are lowest in this period, it tends to have the dominant effect on the required drying time.

Table 7.1 summarizes the characteristics of the available types of dryer models in terms of the physical processes modeled. At the first three levels, gas and solids flows through the dryer are considered only in terms of mean inlet and outlet conditions (total mass or volume flow rate, mean gas velocity, average inlet and outlet humidity). At the detailed design level, however, localized conditions within the dryer must be understood.

7.2.3
Parametric Models

Another important distinction is between distributed-parameter and lumped-parameter models. This is particularly significant in falling-rate drying kinetics.

Levels 2 and 3 use simple lumped-parameter models; the drying coefficient (first-order kinetics) and characteristic drying curve respectively. At Level 4, distributed-parameter models are often used.

If the falling-rate period is to be modeled effectively, it is vital to choose an appropriate drying kinetics model. Distributed parameter models for drying kinetics, such as those described in Chapter 1 and of Luikov (1935, 1966) and Whitaker (1977, 1980), use fundamental physical equations and quantities, and are clearly the most theoretically rigorous. However, a large number of nonlinear simultaneous equations are required because of the parallel competing transfer mechanisms (gas- and liquid phase diffusion, convection, capillary action and desorption), and parameters such as pore-size distribution, pore tortuosity and diffusion coefficients in solids (Chapter 2) are difficult to measure or predict accurately. Hence, the uncertainty in the result is high, and the theory often fails to yield usable numerical results.

Moreover, some distributed-parameter models tend to be divergent and are very sensitive to errors in the basic parameters. For example, Kemp and Oakley (1997), for their incremental model for pneumatic conveying (flash) dryers, found that a 10% error in Sauter mean particle size $d_{p(SM)}$ had a substantial effect on the predicted drying rates. In a real drying situation with a wide range of particle sizes and shapes in the feed, the error in measuring $d_{p(SM)}$ can easily exceed 10%. Hence, even for an apparently fundamental quantity such as particle size, the results may need to be backfitted against experiment.

Lumped parameter models are simpler and require a smaller number of equations, as the "lumped" parameters typically combine aspects of several different physical phenomena. For example, the various competing moisture transport processes in a solid may be modeled using a pseudodiffusion coefficient, treating the system as if it were controlled only by gas- or liquid-phase diffusion. Obviously, lumped parameters cannot be theoretically predicted and must be measured, and it is dangerous to extrapolate them to widely different operating conditions. However, if their physical significance and limits of applicability are clearly understood, interpolation and limited extrapolation to new conditions are often possible.

The widely used characteristic drying curve (CDC) concept, introduced by van Meel (1958), is a classic example of a lumped-parameter model. Indeed, one can regard a drying curve as an extreme example of a lumped parameter, expressing the relationship between moisture content and drying time that depends on many different factors.

Obviously the major question is; are lumped parameters, with their limited theoretical basis, adequate to describe real drying processes? Fyhr and Kemp (1998) simulated a drying system accurately by an advanced model and approximately by the characteristic drying curve. The responses of the two models to changes in operating conditions were remarkably similar. It was shown that the CDC introduced three main errors due to the simplification of the theory, but in nearly all circumstances, these errors acted in opposite directions and tended to cancel each other out. Hence, in appropriate circumstances, a simple lumped-parameter model requires less input data and may actually give a better result than a more complex, supposedly rigorous distributed-parameter model.

Another way of meeting this challenge is to use fuzzy logic, as suggested by Baker and Lababidi (2006). This could allow for the uncertainties involved in the equations, and has been successfully used in dryer-selection algorithms, as described in Section 7.9. Neural networks offer another possibility for dealing with data from complex systems with a wide range of uncertainties.

7.3
Heat and Mass Balances

Simple heat and mass balances at Level 1 (Fig. 7.3) give some useful information, especially for continuous dryers. However, dryer sizing and performance capabilities are not predicted.

A mass balance on the solvent relates the mass flow rates, moisture content and humidity directly. A more detailed overall mass balance can take into account flows of air and dry solids, and computer programs can be very helpful in automating the tedious calculations that allow for leaks and solids losses (e.g. through elutriation) and the corresponding effects on humidity and moisture content.

The heat balance has more terms, but simplified versions can give useful results quickly (for example, using the constant-enthalpy lines on a psychrometric chart gives a rapid first estimate of exhaust conditions for a convective dryer). Again, software (either standalone or simulators) can assist the calculations.

Despite their simplicity, heat and mass balances can yield important benefits, for example in troubleshooting. Kemp and Gardiner (2001) quote an example of a convective dryer that unexpectedly showed a very low thermal efficiency (below 20%) when a heat balance was performed, and the mass balance could not be reconciled as more water vapor was emerging in the exhaust gas than could possibly be present in the wet solid and inlet gas. The only possible explanation was a severe leak in the steam-heated indirect heater for the inlet gas. Pressure tests confirmed this. The repair was simple and substantial cost savings were made in the short and long term. The problem had not previously been identified because the instrumentation (particularly on humidity and solids moisture content) was insufficient to close

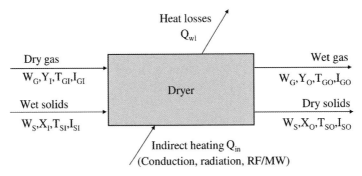

Fig. 7.3 Simple generic heat and mass balance for a continuous dryer.

the balance; this shows the value of retrofitting additional instruments to existing dryers now that sensors are becoming more reliable.

Heat and mass balances are generally less useful for batch dryers because the conditions can vary substantially through the drying cycle, so only an instantaneous balance or a very uninformative average over the batch period can be obtained.

7.4
Scoping Design Methods

Level-2 models can be described as scoping or approximate calculations, and give rough sizes and throughputs (mass flow rates) for dryers, using simple data and making some assumptions. Results are based on a heat and mass balance or on basic heat-transfer considerations, without measuring a drying curve or other experimental drying data. Constant rate (unhindered) drying is generally assumed, but an initial estimate of falling-rate drying may also be obtained by assuming first-order kinetics (which can be solved analytically). For the design mode:
- Continuous convective dryers – calculate gas mass flow rate from mass balance and find cross-sectional area for a typical gas velocity.
- Continuous-contact dryers – calculate evaporation rate from mass balance and find surface area needed to achieve required drying rate, assuming typical heat-transfer coefficients. Backcalculate dryer dimensions.
- Batch dryers with fixed batch size – calculate the dimensions of the dryer required to contain the batch and estimate drying time by reversing the methods above.

Scoping methods give surprisingly accurate estimates of the cross-sectional area of convective dryers and the volume of batch dryers, but are less effective for other calculations and can give highly overoptimistic results.

7.4.1
Continuous Convective Dryers

When designing a continuous dryer, the required solids throughput (mass flow rate) W_S and the inlet and outlet moisture content X_I and X_O are known, as is the ambient humidity Y_I. For convective dryers, if the inlet gas temperature T_{GI} is chosen, the outlet gas conditions (temperature T_{GO} and humidity Y_O) can be found, either by calculation or (more simply and quickly) by using the constant-enthalpy lines on a psychrometric chart. It may be necessary to allow for heat losses, sensible heating of solids and the heat of wetting, especially if tightly bound moisture is being removed. The gas mass flow rate W_G can now be calculated as it is the only unknown in the mass balance on the solvent:

$$W_G(Y_O - Y_I) = W_{ev} = W_S(X_I - X_O) \tag{7.1}$$

A suitable gas velocity U_G along the dryer is now chosen, typically 20 ms^{-1} for a flash dryer, 0.5 ms^{-1} for a fluidized bed, 3 ms^{-1} for a cocurrent rotary dryer. For through-circulation and dispersion dryers, the cross-sectional area A is given by $W_G = \rho_{Gl} U_G A$, and the diameter can then be calculated. The result is usually accurate within 10%, and can be further improved by better estimates of velocity and heat losses.

The method gives no information about solids residence time or dryer length. A minimum drying time can be calculated by evaluating the maximum (unhindered) drying rate N_{cr}, assuming gas-phase heat-transfer control and free moisture. Alternatively, it may be assumed that first-order kinetics apply throughout the drying process. These crude methods are not recommended, and can give serious underestimates of the required drying time, especially if unhindered drying is assumed. It is much better to measure the drying time experimentally and apply scaling methods.

7.4.2
Continuous-contact Dryers

For continuous-contact dryers, the vital factor is the area of the heat-transfer surface, A_S. This can be found from the equation:

$$A_S = \frac{W_v \Delta H_v}{\alpha_{WS} \Delta T_{WS}} \tag{7.2}$$

The latent heat of evaporation ΔH_v should allow for bound moisture and heating of solids and vapor to the final temperature. A typical wall-to-solids heat-transfer coefficient α_{WS} for the given dryer type should be used. The calculation is less accurate than the one for convective dryers. It implicitly assumes that the heat-transfer rate is the overall limiting factor. If the drying process is strongly limited by falling-rate drying kinetics, the calculated size of dryer corresponding to the given heating surface A_S may not give sufficient solids residence time to reach the desired final moisture content. Again, experimental measurement of a drying curve is strongly recommended.

7.4.3
Batch Dryers

For batch dryers where the batch size is stipulated, the problem becomes very different, as the requirement is simply that the dryer can physically contain the volume of the solids, and the dryer volume and dimensions can thus be calculated directly. The difficulty is now to calculate solids residence time. Equations 7.1 and 7.2 can be reversed, but again the calculation will be inaccurate for falling-rate drying and it is preferable to measure a drying curve and use a scaling calculation, as outlined in the next section. However, it is possible to compare the surface-area-to-volume ratio

7.4 Scoping Design Methods

Tab. 7.2 Comparative dimensions and drying times for various batch-contact dryers

Dryer type	Typical h.t.c. kWm^{-2} K^{-1}	Typical L/D	D m	L m	Total area m^2	Active area m^2	$\alpha_{ws}A_s$ kJs^{-1} K^{-1}	t_s h
Tumbler/double-cone	0.1	1.5	3.71	5.56	38.91	19.45	1.95	2.78
Vertical pan	0.1	0.5	3.71	1.85	32.37	21.58	2.16	2.51
Spherical	0.1	1	3.37	3.37	35.63	17.82	1.78	3.04
Filter-dryer	0.1	0.5	3.71	1.85	21.58	10.79	1.08	5.01
Conical agitated	0.1	1.5	3.71	5.56	34.12	21.50	2.15	2.52
Horizontal agitated	0.08	5	1.72	8.60	46.50	23.25	1.86	2.91
HA, agitator heated	0.08	5	1.72	8.60	278.99	139.49	11.16	0.48

of various types of dryer and deduce how their drying times will compare with each other.

Table 7.2 gives calculated dimensions for various batch-contact dryers of volume 20 m^3 operating at 50% volumetric fill (i.e. a batch of 10 m^3). The active area is the area of heat-transfer surface actually in contact with the solids. Assuming unhindered drying with a typical heat-transfer coefficient for each dryer type, the drying time for a given batch has been calculated.

The table shows that drying times are comparable for most types of double-cone (rotating vacuum batch) and vertical agitated dryer, and for horizontal agitated dryers with only the jacket heated. The drying time for a simple filter dryer is longer, as the bottom filter plate cannot be heated. It can also be shown that drying times for nearly all contact dryers increase approximately proportionately with diameter and with the cube root of batch size. Much larger heating surfaces in a given volume, and hence shorter drying times, are obtained by heating the internal agitator of a horizontal agitated (paddle) dryer, and this will be the preferred contact dryer type for large batches. However, if the rate of heat supply is not the limiting factor for drying and a minimum residence time is required to allow removal of tightly bound moisture, there will be little or no gain from providing very large heat-transfer areas.

7.4.4
Simple Allowance for Falling-rate Drying

The calculation method for the cross-sectional area of a continuous convective dryer in Section 7.4.1 makes no distinction between constant-rate (unhindered) and falling-rate (hindered) drying kinetics, although in practice the latter will tend to give exhaust gases with higher temperature, lower absolute humidity and lower relative humidity, as the effective number of transfer units (NTUs) is less.

The methods given in Sections 7.4.2 and 7.4.3 for continuous- and batch-contact dryers are based on heat-transfer rate and assumed constant-rate drying. However, a very approximate time for falling-rate drying can be calculated using the "drying coefficient" theory. This in effect assumes first-order falling-rate drying kinetics throughout, with drying rate proportional to the free moisture $(X - X_E)$. This

Full Scale Dryer Specification Summary

Dryer Type	Heat Transfer Coeff. [kW/m².K]	Agitation Heating Ratio [-]	Capacity Specification		Dimensions					Heat Transfer Area		
			Target Fill% [%]	Nominal Capacity [L]	Internal Diameter [m]	Height (or Length) [m]	Batch Height [m]	Aspect Ratio [-]	Actual Fill% [%]	At Capacity [m²]	At Batch Size [m²]	Effective [m² equiv]
Double Cone Dryer	0.1	0	50%	4000	2.31	2.31	0.96	1.0	50%	9.73	6.27	6.27
Vertical Pan Dryer	0.1	0	50%	4000	2.77	1.11	0.33	0.4	50%	11.80	8.91	8.91
Spherical Dryer	0.1	0	50%	4000	2.12	2.12	0.86	1.0	50%	10.09	5.72	5.72
Nutsche Filter-Dryer	0.1	0	50%	4000	2.26	2.26	0.50	1.0	44%	8.00	3.54	3.54
Conical Agitated Dryer	0.1	0	50%	4000	2.27	4.24	2.99	1.9	50%	12.33	7.77	7.77
Horizontal Agitated Dryer	0.08	0	50%	4000	1.08	5.42	0.46	5.0	50%	12.25	8.30	8.30

Pilot and Full Scale Drying Times

Dryer Type	Theory			Pilot Actual [hours]	Full Scale 'Two Stage' Scale-up [hours]
	100% Unhindered [hours]	Two Stage [hours]	100% Hindered [hours]		
Pilot Dryer	0.75	2.13	2.13	3.00	-
Double Cone Dryer	2.71	7.19	7.19	-	10.12
Vertical Pan Dryer	1.91	5.06	5.06	-	7.12
Spherical Dryer	2.97	7.89	7.89	-	11.09
Nutsche Filter-Dryer	4.80	12.72	12.72	-	17.89
Conical Agitated Dryer	2.19	5.81	5.81	-	8.17
Horizontal Agitated Dryer	2.56	6.80	6.80	-	9.56

Fig. 7.4 Spreadsheet for scoping design and scale-up of batch-contact dryers (extracts).

expression can be solved analytically. The falling-rate drying time has a logarithmic relationship to free moisture, and the ratio between falling-rate and constant-rate drying time is:

$$\frac{t_{fr}}{t_{cr}} = \frac{X_I - X_E}{X_I - X_O} \ln\left(\frac{X_I - X_E}{X_O - X_E}\right) \tag{7.3}$$

Figure 7.4 shows summary results from a spreadsheet (part of the Aspen Process Tools) for scoping design of batch-contact dryers, and scale-up based on the ratio of heat-transfer surface area to solids volume and mass. Using a typical geometry, the required heat-transfer area and dimensions for a given batch size have been calculated and the drying time for a given drying duty has then been found, initially by assuming constant-rate drying and then with first-order falling-rate kinetics (a constant rate followed by falling rate could also be calculated if the critical moisture content were given). This is a Level-2 model. Then, the actual drying time measured in a pilot-plant double-cone dryer has been supplied. Assuming the scaling factor from "predicted" to "real" is the same for all dryer types, this gives new drying times based on experimental data (final column) – a very simple Level-3 model. Although several assumptions are made, this is nevertheless useful for rapid initial sizing and ballpark prediction of drying times. The predictions will be much more reliable for dryers of the same type as the pilot-plant than for dryers of other types, as in the latter case, errors in the estimated heat-transfer coefficients for both types of dryer will be additive.

7.5
Scaling Methods

Heat and mass balances and scoping design calculations at Levels 1 and 2 can be performed using only tabulated data from the literature, but falling-rate drying can only be analysed using the major assumptions of first-order kinetics and an estimated critical moisture content. In the next two sections, we cover Level-3 and -4 models that incorporate the material model as well, requiring experimental data from drying kinetics tests.

Integral or scaling models (Level 3) treat the dryer as a complete unit, and are generally based on scaling up a batch-drying curve recorded by small-scale or pilot-plant experiments. The inclusion of real experimental drying data for the vital falling-rate period gives more reliable results than scoping calculations. These methods are used for layer dryers (tray, oven, horizontal-flow band and vertical-flow plate types) and for an overall estimate of fluidized-bed dryer performance. As in scoping design, the air velocity is again simply averaged over the dryer, or over a given section or region of it.

7.5.1
Basic Scale-up Principles

The key to all scaling models is a moisture–time drying curve expressed as a function $X(t)$. The drying curve obtained for the small scale test is $X_{test}(t)$ and we need to find how to transform this to the function for the full-scale dryer $X_{full}(t)$.

In some cases, the two functions are equal. Moyers (1994) stated the specific drying rate concept, which was that if key external conditions such as temperature and air velocity are kept the same for two geometrically similar dryers, the evaporation rate expressed as a mass flux (vapor mass flow per unit area) will also be identical. As a result, the moisture content vs. time graphs should also be identical; the required residence time in both dryers will be the same.

More common is the situation where geometrical similarity cannot be fully maintained, or operating conditions are not equivalent. In this case, the drying rates at any point on the drying curve will be different between the small-scale test and full-scale dryer. The best way to model this is often to alter the drying time Δt for a given moisture content ΔX; this in effect stretches the time axis of the drying curve. The ratio between Δt_{full} and Δt_{test} is called the normalization factor, Z. This is often easily calculated, especially for straightforward layer dryers (tray, band, vacuum agitated, rotating double-cone, etc.) and fluidized beds. For example, Oakley (2001) has outlined a number of basic scale-up criteria that are simple ratios of heat-transfer coefficients, solids mass, cross-sectional area and dryer diameter. For fluid-bed dryers, as will be seen in Section 7.5.3, the vital factors are gas velocity, bed depth and temperature driving force. These rules work well – as long as the set of criteria needed to satisfy the rules are met. It is therefore vital to understand the assumptions involved for each type of scale-up equation.

The principle of scale-up calculations is in essence to measure a batch-drying curve (or, at least, inlet and outlet moisture content) for a given residence time, then scale to new conditions or equipment. The equations are similar to those for a scoping performance calculation, but allow for the falling-rate period correctly. Hence, scaling calculations can be used to improve the result from a scoping calculation, as seen in Section 7.4.4 and Fig. 7.4, where an actual drying time was used to replace the provisional one calculated using first-order kinetics.

Most dryer manufacturers have traditionally relied on performing pilot-plant tests and scaling the results to a new set of conditions on a dryer with greater airflow or surface area. In effect, they have been using a Level-3 scaling model. Moreover, the empirical rules employed have generally been based on the external driving forces (temperature, vapor pressure or humidity driving forces). By implication, therefore, a characteristic drying curve concept is being used, scaling the external heat and mass transfer and assuming that the internal mass transfer changes in proportion.

7.5.2
Integral Model

The integral model was first suggested for fluidized beds by Vanecek et al. (1964, 1966). The mean outlet moisture content is given by summing the product of the particle moisture content and the probability that it emerges at time t:

$$\overline{X} = \int_0^\infty E(t)X(t)\mathrm{d}t \qquad (7.4)$$

Implicit assumptions are therefore;
- the drying curve function $X(t)$ must be experimentally measured,
- it must be possible to scale the curve to the new operating conditions,
- the results must be capable of scale-up from small-scale test to large-scale dryers,
- the residence time function $E(t)$ must be known.

The first three assumptions are the same as the simple scaling model. For batch and pure plug-flow dryers, $E(t)$ is a delta (spike) function; all particles have the same residence time, so only one point on the drying curve need be considered. However, for well-mixed dryers or nonideal plug flow with backmixing, the integral model can be used to obtain a moisture-content distribution and a corrected mean moisture content. For fluid beds, Reay and Allen (1982a, b) showed that these values differed substantially from those on the initial drying curve; indeed, for well-mixed units, it was virtually impossible to reach low moisture contents because of the fresh wet material constantly being mixed in.

7.5.3
Application to Fluidized-bed Dryers

All four levels of design model can be used in the design and simulation of fluid-bed dryers. A heat and mass balance (Level 1) provides the basic information, and scoping design (Level 2) gives a first estimate of bed area. The integral model (Level 3) gives an improved estimate of solids residence time and bed area based on actual drying kinetics measured in a small-scale fluidized bed, replacing the Level-2 estimate based on assumed outlet exhaust conditions. Finally, a Level-4 incremental model can be used to determine local temperatures during a batch, at different distances along a plug-flow continuous dryer and at different heights in the bed.

We will concentrate on the main Level-3 model for dryer sizing, for which the basic method was given by Reay and Allen (1982a, b) and extended by Reay (1989), McKenzie and Bahu (1991) and Bahu (1994). The drying kinetics are measured in a pilot-scale fluidized bed and the time axis of the drying curve is scaled for a given change in moisture content ΔX using the equations:

$$\text{Type A(fast-drying material)}: \quad Z = \frac{\Delta \tau_2}{\Delta \tau_1} = \frac{(m_B/A)_2 G_1 (T_{GI} - T_{wb})_1}{(m_B/A)_1 G_2 (T_{GI} - T_{wb})_2} \quad (7.5)$$

$$\text{Type B(slow-drying material)}: \quad Z = \frac{\Delta \tau_2}{\Delta \tau_1} = \frac{(T_{GI} - T_{wb})_1}{(T_{GI} - T_{wb})_2} \quad (7.6)$$

For Type-A materials, drying rate is accelerated by higher gas velocities and higher inlet gas temperatures, but reduced by increasing bed depth. For Type-B materials, drying rate is independent of gas velocity and bed depth but is still affected by gas temperature. Most materials obey Type-A rules, even in falling-rate drying; for example, iron ore, silica gel, ion exchange resin. The only material found to exhibit Type-B drying throughout was wheat, where drying is limited by the very slow diffusion through the cell wall.

The normalization factor Z allows the drying-curve function to be scaled from the initial test conditions to the chosen gas velocity, bed depth and temperature of the full-scale dryer. Gas mass velocity (flux) G is used instead of gas velocity U_G as it is independent of temperature. Likewise, bed weight per unit area (m_B/A) is preferred to bed depth z, as the former does not vary as the bed bulk density changes on fluidization, expansion and repacking.

The SPS design method has been successfully used since the late 1970s to design and debottleneck many industrial dryers. It was implemented initially as a mainframe software program in the 1980s, then as a PC DOS program, FLUBED, in the 1990s, and finally modified to a Windows program in 2001, in which form it is sold by Aspen Technology as one of the Process Tools. However, the equations were derived from experiment and the theoretical basis was not clear originally. Subsequent work reported by Kemp and Oakley (2002) has shown that Eqs. 7.5 and 7.6 are in fact limiting cases of a more general equation:

$$Z = \frac{\Delta \tau_2}{\Delta \tau_1} = \frac{(m_B/A)_2 G_1 (T_{GI} - T_{wb})_1 (1 - e^{-f.NTU.z})_1}{(m_B/A)_1 G_2 (T_{GI} - T_{wb})_2 (1 - e^{-f.NTU.z})_2} \quad (7.7)$$

For high NTUs (number of transfer units) the exponential terms tend to zero and the formula reduces to Eq. 7.5. For low NTUs the exponential term becomes significant and can be expanded, giving Eq. 7.6 after various terms cancel out. Equation 7.7 also covers the situation where a material changes from Type-A to Type-B normalization as the moisture content and drying rate fall.

As discussed by Kemp and Oakley (2002) and Kemp (2006), the driving force term in Eq. 7.5 to Eq. 7.7 has been variously expressed as $(T_{GI} - T_{wb})$, $(p_{Ys} - p_{YI})$, $(Y_{wb} - Y_I)$, $(T_{GI} - T_{as})$ and $(T_{GI} - T_{GO})$ and, depending on the situation, any of these may be the most convenient form to use. The most rigorous form is $(h_{GI} - h_{GO})$, which is also applicable where additional heat is supplied via internal heating coils or surfaces; the normalization factor is proportional to the total heat input into the bed, and it does not matter how the heat is distributed between the hot air and the coils (McKenzie and Bahu 1991).

This Level-3 model is adequate for assessing overall dryer performance in virtually all circumstances; factors such as bubble growth and local gas and particle flow patterns only need to be considered if there is a specific requirement to know local conditions. The integral model including a well-mixed residence-time distribution gives successful results for well-mixed continuous fluid beds (Reay and Allen 1982a, b). Variations in solids residence time and outlet moisture content in a nonideal plug-flow dryer can be predicted by backmixing calculations (Reay 1989; Fyhr et al. 1999). Finally, an incremental model along the dryer axis of a plug-flow dryer, or during the cycle period for a batch dryer, gives local values of gas and solids temperature and exhaust humidity (Kemp and Oakley 2002; Kemp 2006), and an incremental model including vertical steps can be used to predict local conditions at various heights within the fluidized bed (Fyhr and Kemp 1999).

7.6
Detailed Design Models

Detailed models (Level 4) give local conditions within a dryer, rather than average conditions over the complete dryer. They are distributed-parameter models giving a large amount of output information but requiring a correspondingly large amount of input data. Many in-house programs produced by universities, often as part of a PhD project and very rarely available externally, are Level 4 models. There are two key types:

- Incremental models. These track the local conditions of the gas and particles through the dryer, generally in one dimension. They are especially suitable for cocurrent and countercurrent dryers, e.g. flash (pneumatic conveying) and rotary dryers. The air conditions are usually treated as uniform across the cross section, and dependent only on axial position. This method can also be used to determine local conditions (e.g. temperature) in a dryer, such as a band or plug-flow fluid-bed unit, where a simpler model has been used to find overall drying rate.
- Complex three-dimensional models, e.g. CFD (computational fluid dynamics). These are the only effective models for gas conditions and particle motion in spray dryers because of the complex swirling flow pattern; they can also be used to find localized conditions in other dryers.

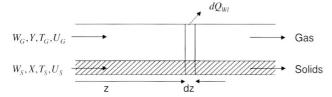

Fig. 7.5 Principle of the incremental model (one-dimensional axial flow).

7.6.1
Incremental Model

The one-dimensional incremental model (Fig. 7.5) is a key analysis tool for several types of dryer. A set of simultaneous equations is solved at a given location and the simulation moves along the dryer axis in a series of steps or increments – hence the name. The procedure may be attempted by hand if a few large steps (say 5–10) are used, but for an accurate simulation, a computer program is needed and thousands of increments may be used.

Increments may be stated in terms of time (dt), length (dz) or moisture content (dX). A set of six simultaneous equations is then solved, and ancillary calculations are also required, e.g. to give local values of gas and solids properties. The generic set of equations (for a time increment Δt) is as follows:

Heat transfer to particle: $$Q_P = \alpha_{PG} A_P (T_G - T_S) \tag{7.8}$$

Mass transfer from particle: $$\frac{-dX}{dt} = \text{function}(X, Y, T_P, T_G, \alpha_{PG}, A_P) \tag{7.9}$$

Mass balance on moisture: $$W_G \Delta Y = (-W_S \Delta X) = W_S \frac{-dX}{dt} \Delta t \tag{7.10}$$

Heat balance on particle: $$\Delta T_S = \frac{Q_P \Delta t - \Delta H_v m_P \Delta X}{m_P (C_{PS} + C_{PL} X)} \tag{7.11}$$

Heat balance for increment:
$$-\Delta T_G = \frac{W_S(C_{PS} + C_{PL}X)\Delta T_S + W_G(\Delta H_{v0} + C_{PY} T_G)\Delta Y + \Delta Q_{Wl}}{W_G(C_{PG} + C_{PY} Y)} \tag{7.12}$$

Particle transport: $$\Delta z = U_S \Delta t \tag{7.13}$$

The mass- and heat-balance equations are the same for any type of dryer, but the particle transport equation is different, and the heat- and mass-transfer correlations are also somewhat different as they depend on the environment of the particle in the gas (i.e. single isolated particles, agglomerates, clusters, layers, fluidized beds or packed beds). The mass-transfer rate from the particle is regulated by the drying kinetics, and is thus obviously material dependent (at least in falling-rate drying).

The model is effective and appropriate for dryers where both solids and gas are approximately in axial plug flow, such as pneumatic conveying and cascading rotary dryers. However, it runs into difficulties where there is recirculation or radial flow.

The incremental model is also useful for calculating variations in local conditions such as temperature, solids moisture content and humidity along the axis of a dryer (e.g. plug-flow fluidized bed), through a vertical layer (e.g. tray, band or fluid-bed dryers) or during a batch-drying cycle (using time increments, not length). It can be applied in these situations even though the integral model has been used to determine the overall kinetics and drying time. In a band dryer, both horizontal and vertical increments can be used to give a two- or three-dimensional grid.

7.6.2
Application to Pneumatic Conveying, Rotary and Band Dryers

7.6.2.1 Pneumatic Conveying Dryers
The cross-sectional area of the duct is found from the scoping design calculation, using an assumed gas velocity (well above terminal velocity, typically about 20 m/s). To find the duct length, the one-dimensional incremental model is used, tracking up or along the duct. For this type of dryer, particle motion, heat transfer and drying kinetics require specific treatment. The model is summarized by Kemp and Oakley (1997). Drying kinetics may be measured by a thin-layer method or by suspending single particles in a drying tunnel.

Figure 7.6 shows a one-dimensional incremental model (Level 4) for a flash dryer, as implemented by SPS (Harwell, UK) in the Drycon software program, written in DOS in the 1990s. A stepwise calculation gives local conditions along the dryer (in this case for gas and solids temperature). This type of model could now also be implemented using an open equation solver, such as ACM (Aspen Custom Modeler), and a similar two-dimensional incremental model could be used for band dryers or similar, allowing for variations both horizontally along and vertically through the bed.

At this level of detail, the problem of cumulative errors in data and parameters becomes more significant. It is strongly recommended that any modeling results are checked against experimental data by a scale-up or "fitting-mode" calculation. There will often be significant discrepancies, and this allows some of the more uncertain parameters in the model to be corrected to values that allow a better fit.

Kemp and Oakley (1997) showed that this flash dryer model was sensitive to small errors in certain parameters such as Sauter mean diameter (volume–surface mean) of the particles, especially in the initial acceleration zone where particle velocity and temperature are changing rapidly. It is not surprising that it is difficult to obtain equivalence between the drying kinetics measured by experiment and those in an industrial pneumatic conveying dryer, as the residence time of the particles in the latter is typically of the order of 1 s, and the temperature and flow field to which the particle surface is exposed varies widely during this period. The only moisture likely

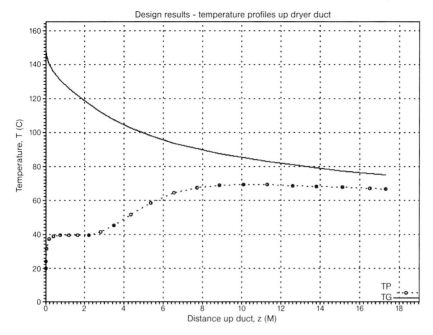

Fig. 7.6 Incremental model (Level 4) simulation of temperatures in pneumatic conveying (flash) dryer.

to be removed will be that at the surface or within a few micrometers of it. Kemp and Fernandez (2002) had some success in measuring thin-layer drying kinetics at residence times down to a few seconds, but this required the use of a specialized drying kinetics rig.

Nevertheless, by fitting the model predictions to pilot-plant experimental data, reliable results can be obtained. Figure 7.7 shows three successive fits to data for silica gel particles of 0.35 mm mean diameter. A drying curve was measured for approximately 100 g of solid at constant airflow and temperature, and corrected to the local values calculated by the program, using the Ranz–Marshall heat-transfer equation and characteristic drying curve. Fit 1 assumes that the falling-rate drying kinetics is based on initial maximum drying rate; it grossly overpredicts drying rates, thus underpredicting moisture content. Fit 2 bases kinetics on local maximum drying rate, giving an improvement but now consistently underpredicting drying. Fit 3 is based on fit 2, with the Sauter mean particle diameter reduced from 329 to 300 μm (well within experimental error for this size distribution), wall friction factor (highly uncertain and obtained by backcalculation) increased from 0.2 to 0.4, and a small increase in the external heat-transfer coefficient for heat losses. These minor changes are sufficient to give a good fit.

Kemp and Oakley (1997) also described a study on an industrial dryer, where only the inlet and outlet conditions could be recorded (gas and solids temperature, gas

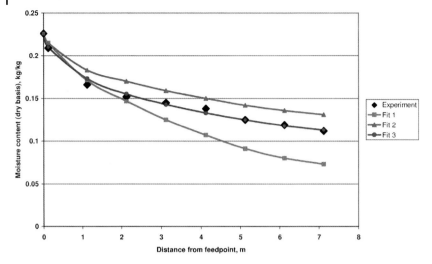

Fig. 7.7 Comparison of initial and fitted models with experimental data for pilot-plant flash dryer.

humidity and solids moisture content), but this was sufficient to allow the model to be fitted successfully. The effect of varying operating conditions was then studied, and this showed that there was a clear optimum gas velocity to obtain minimum outlet moisture content or maximum throughput, as shown in Fig. 7.8. At low gas velocities the dryer is heat- and mass-balance limited; the mass flow of air is so low that the heat

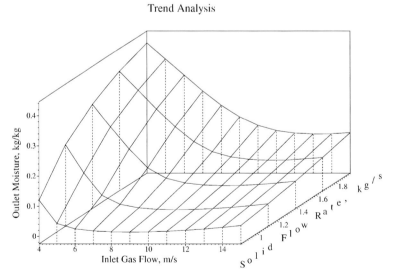

Fig. 7.8 Outlet moisture-content variation with gas and solids flow rate for pneumatic conveying dryer.

content of the air is insufficient to supply the latent heat of evaporation required to remove the moisture. At high gas velocities the kinetics become the limiting factor; the solids residence time in the duct becomes insufficient to allow the material to dry adequately. The model predicted an increase of 50% in throughput by altering operating conditions for the existing dryer, and a further 50% by increasing gas flows beyond the capacity of the existing equipment by installing new fans, ductwork and cyclones. The results from the actual changes that were made on the plant bore out these predictions. This illustrates the benefits that an appropriate Level-4 model can give.

The model can be used to improve scale-up from pilot-plant to full-size dryers, by fitting it to the pilot-plant data (thus normalizing the uncertain parameters) and then simulating the full-scale unit. This is a more rigorous approach than the empirical relationships previously used by dryer manufacturers for scale-up, such as for the change in duct length required as duct diameter increases, which has a complex nonlinear dependence on wall friction.

7.6.2.2 Cascading Rotary Dryers

Again, the one-dimensional incremental model is used, but this time tracking horizontally along the drum axis. Axial particle motion is divided into an airborne and a dense phase, as described by Matchett and Baker (1987, 1988). Drying kinetics are measured by the thin-layer method. Overall models are described and reviewed by Cao and Langrish (1999) and Kemp (2004b). A Level-4 software program, ROTARY, was developed in prototype form by SPS and the underlying model is summarized by Kemp and Oakley (1997, 2002).

Again, to obtain reliable results, a fitting mode or scale-up calculation was needed, based on measuring residence time and dryer inlet and outlet conditions for a test run. In this case, the parameters that could be backfitted were the airborne drag coefficient (for which only approximate correlations are available because of the complex nature of the spreading curtains of particles), the dense phase velocity number (which characterizes particle axial velocity due to rolling, kilning and bouncing in the bed at the bottom of the dryer) and a factor relating the airborne and dense phases residence times (affected by many parameters and typically having an error of up to 20%).

7.6.3
Advanced Methods – Computational Fluid Dynamics (CFD)

CFD provides a very detailed and accurate model of the gas phase, including three-dimensional effects and swirl. Where localized flow patterns have a major effect on the overall performance of a dryer and the particle history, CFD can give substantial improvements in modeling and in understanding of physical phenomena. Conversely, where the system is well mixed or drying is dominated by falling-rate kinetics and local gas-phase conditions are unimportant, CFD modeling will give little or no advantage over conventional methods, but will incur a vastly greater cost in computing time. The calculations are highly complex, involving solving the

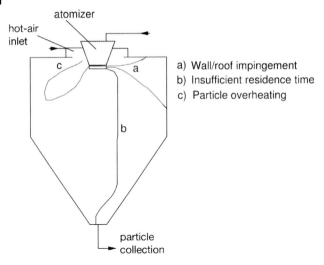

Fig. 7.9 Typical spray-dryer problems arising from poor airflow patterns.

Navier–Stokes equations over billions of grid cells, and a commercial program such as Fluent®, Ansys-CFX® or StarCD® should be used.

Advances in the application of CFD to drying are described in Chapter 5. The most noteworthy achievements have been for spray dryers, where it has helped to reveal the complex gas-flow patterns within the spray chamber, and how these influence droplet trajectories. This has allowed several practical problems in spray dryers to be identified and solved, as shown schematically in Fig. 7.9. Because small calculation errors can build cumulatively, it is essential to validate CFD and other complex theoretical models against real physical measurements, even more so than with incremental models. This has been done successfully on both pilot-plant and the industrial scale since the mid-1980s, e.g. by Livesley et al. (1992). Langrish and Fletcher (2001) gave a detailed and up-to-date review of the theory of CFD and successful applications in troubleshooting, research and evaluation of novel spray-dryer designs.

CFD has also been useful for other local 3-dimensional flows that involve rotation and swirl, e.g. around the feedpoint of pneumatic conveying dryers (see Fig. 7.10), or affect drying significantly, e.g. local overdrying and warping in timber stacks (Langrish 1996). AFSIA's 2001 meeting (AFSIA 2001) covered successful applications of CFD in modeling airflows in a wide range of dryers, including fluidized beds, superheated steam and infrared dryers, for materials such as pharmaceuticals, clays and ceramics, Emmental cheeses and smoked sausages.

In the pneumatic conveying dryer shown in Fig. 7.10, the CFX program predicted a set of observed but unexplained phenomena around the feeder, revealing a large dead zone of low gas velocity and low temperature that affected solids and gas-flow pattern, solids residence times and local drying rates (Kemp et al. 1991).

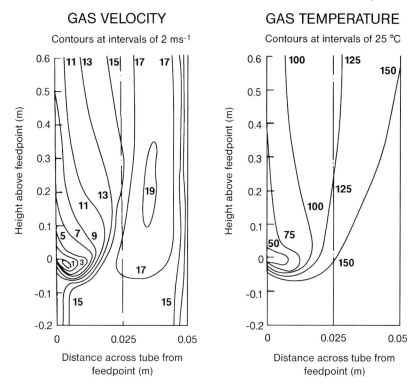

Fig. 7.10 Gas flow and temperature patterns at feedpoint of flash dryer revealed by CFD simulation.

7.7
Ancillary Calculations

7.7.1
Processing Experimental Data

An obvious candidate for computer assistance is the processing of data from drying experiments. In particular, a spreadsheet can be effectively used to handle large numbers of data points, and to plot and interconvert between the different types of drying curve (moisture–time, rate–time and rate–moisture). For example, if humidity–time data is available from a continuous humidity monitor, it is easily converted to drying rate data by mass balance, and then to moisture-time data by integration. It is essential to crosscheck the moisture range obtained by integration against initial and final moisture content measured by oven test, and to correct the former to match the latter.

The research groups working on the EU-supported QUID project (1996–2001) produced a joint paper suggesting best practice for experimental data processing (Kemp et al. 2001).

7.7.2
Humidity and Psychrometry

This is one area where a wide range of software is available, including humidity calculators available free on the Internet and programs to plot psychrometric charts, usually of the Grosvenor (temperature–humidity) type. This is because these calculations are used not only in drying, but in many other fields such as meteorology and air conditioning, giving a much greater user base.

7.7.2.1 British Standard BS1339 for Humidity Calculations

An important development in recent years has been the revision and updating of British Standard BS 1339 (BS 1339, 2004) on Humidity and Dewpoint, with the aim of producing both standards and calculation software based on internationally agreed best practice. The standard is in three parts, with Part 1 covering terms, definitions and formulae, and Part 3 a descriptive Guide to Humidity, including practical measurement methods and some of the instrumentation used. However, the biggest change in approach over previous standards comes in Part 2, the most recently published part, on humidity-calculation methods, which uses spreadsheet-based software instead of the previous printed tables.

The former BS 1339 and BS 4833 contained extensive lookup tables where the user could search by hand for a derived quantity based on two input variables. Even with many pages of tables, this covered only a small fraction of the possible permutations and combinations.

The new standard takes an entirely different approach, based on the capabilities of modern software. Firstly, the calculations are provided as a Microsoft Excel® spreadsheet, thus providing the material in a form familiar to users and compatible with other spreadsheets and programs (e.g. MS Word®).

However, the most significant advance over existing standalone programs is that the calculations are provided in the form of functions, over 20 in all. These allow an output to be generated from an appropriate group of inputs, e.g. to generate relative humidity from dry-bulb temperature, total applied pressure and wet-bulb temperature (or absolute humidity, or dewpoint). Where a single function is not available, almost any desired transformation can be achieved by combining two functions in series, usually with absolute humidity (mixing ratio) as the intermediate parameter.

Furthermore, the user can take his own spreadsheet and incorporate any of the functions into it to do the required calculations in situ. Hence, he can use current best practice formulae to replace previous, less accurate versions.

If lookup tables are still required, these can also be generated easily from the functions, with the advantage that the user can choose the precise range of conditions in which he is interested. An example of a table generated in this way is shown in Tab. 7.3.

The software does not construct psychrometric charts, as other programs are already widely available for doing this. The spreadsheet is also only applicable to the air–water system, but formulae for other vapor–gas systems are given in Parts 1 and 3.

Tab. 7.3 Generated values of absolute humidity (mixing ratio) at given temperatures and relative humidities

Relative humidity, %	Dry-bulb temperature, °C										
	0	10	20	30	40	50	60	70	80	90	100
0	0	0	0	0	0	0	0	0	0	0	0
10	0.00038	0.00076	0.00146	0.00265	0.00463	0.00778	0.01266	0.02003	0.03096	0.04695	0.07020
20	0.00076	0.00153	0.00292	0.00533	0.00932	0.01576	0.02585	0.04140	0.06516	0.10156	0.15826
30	0.00114	0.00230	0.00440	0.00803	0.01409	0.02394	0.03959	0.06423	0.10315	0.16588	0.27200
40	0.00152	0.00307	0.00587	0.01075	0.01893	0.03233	0.05393	0.08870	0.14558	0.24276	0.42455
50	0.00191	0.00384	0.00736	0.01349	0.02385	0.04094	0.06891	0.11497	0.19328	0.33625	0.63988
60	0.00229	0.00462	0.00885	0.01626	0.02884	0.04979	0.08456	0.14326	0.24731	0.45242	0.96678
70	0.00267	0.00539	0.01035	0.01906	0.03391	0.05887	0.10094	0.17381	0.30900	0.60064	1.52228
80	0.00306	0.00617	0.01186	0.02188	0.03906	0.06821	0.11810	0.20690	0.38012	0.79631	2.67506
90	0.00344	0.00695	0.01338	0.02472	0.04428	0.07780	0.13610	0.24287	0.46301	1.06653	6.50846
100	0.00382	0.00773	0.01490	0.02759	0.04960	0.08766	0.15499	0.28209	0.56084	1.46395	∞

7.7.2.2 Plotting Psychrometric Charts

As noted earlier, several commercial programs are available (e.g. from ASHRAE) to plot Grosvenor psychrometric charts for the air–water system around normal atmospheric pressures. However, it is possible to construct a psychrometric chart for any gas–vapor system at any pressure if the necessary physical properties (specific and latent heats and vapor pressure data) are given. Aspen's PSYCHIC program (part of the Process Tools) does this for Grosvenor (temperature–humidity), Mollier (humidity–enthalpy) and Bowen (enthalpy–humidity) charts. The Mollier chart presents a significant programming challenge as there are no horizontal lines; the constant enthalpy lines are sloping and the constant temperature loci are very shallow curves. Figure 7.11 shows a simplified Mollier chart from PSYCHIC (the full version relies on color to distinguish between the different lines) for the nitrogen–acetone system under reduced pressure, representing a typical pharmaceutical system.

7.7.3
Physical-properties Databanks

It is frequently suggested that a database should be compiled for drying kinetics and equilibrium moisture content for various materials, to allow scoping calculations to be "improved". This approach would potentially be very unreliable for drying kinetics of formulated products such as chemicals and pharmaceuticals. Firstly, there is often

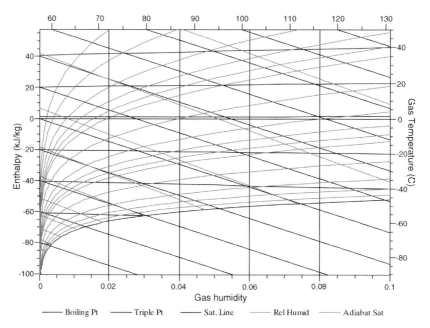

Fig. 7.11 Mollier psychrometric chart from PSYCHIC for nitrogen–acetone system at 10 kPa (100 mbar) pressure.

no information in the literature about the particular material to be dried. Secondly, the kinetics depend strongly on the pore structure, which is highly dependent on the upstream formulation process. Drying rates can easily vary by two or more orders of magnitude for the same material depending on the particle structure – number, size distribution and tortuosity of the pores (see Chapter 2). Even for the same particles made in the same process, different batches may show significantly different drying kinetics if, for example, the upstream crystallization process conditions vary. Equilibrium moisture content is more a function of the material composition and less of the structure, but even here, smaller pore sizes give higher surface tension effects, higher moisture binding and increased equilibrium moisture content.

For processes where the raw material is already supplied as a solid, the extent of variability in internal structure is considerably less, and the literature values may give a reasonable first estimate. For agricultural products (e.g. grain, hay), drying rates are reasonably comparable for a given material. However, being a natural product, there are still significant variations with seasonal weather, harvesting time and storage conditions. Building materials such as brick and concrete have been analysed by Moropoulou et al. (2005); for their sample of materials, there was an approximately inverse relationship between drying time and mean pore size (or, conversely, drying coefficient was roughly proportional to pore diameter), and they provided some data that could be used for a first estimate for these materials. However, the pore diameter will depend on the pressing or forming process. On the other hand, for formulated products such as pharmaceuticals, polymers and solid-phase chemicals, literature values could not be used safely for even an approximate estimate of drying kinetics.

7.8
Process Simulators

These allow the dryer to be modeled as part of the overall process flow sheet or recipe. However, current major simulators for continuous processes, such as Aspen Plus, HYSYS and Prosim, have very limited capabilities for solids, and use simplified models at levels 1 or 2. This is adequate to study the main interactions between dryers and the rest of the process, which need to be considered in the earliest stages of process design, but an interface is needed with more rigorous Level-3 or 4-models for detailed design or analysis of the dryer.

7.8.1
Current Simulators and their Limitations

Process simulators, including Aspen Plus, HYSYS and Batch Plus, allow the dryer to be modeled as part of the overall process flow sheet or recipe. For example, the effect of a change in production rate or in upstream conditions on the dryer and the overall flow sheet can be found automatically. However, the major simulators for continuous processes, such as Aspen Plus, HYSYS and Prosim, have had extremely limited capabilities for solids, as they were originally designed for fluid processes, and

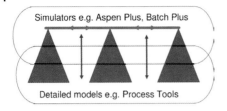

Fig. 7.12 Layering approach to simulation of dryers and overall solids process.

improvement has been relatively slow. Batch Plus, which was designed from the outset to handle pharmaceuticals and similar processes, handles solids more effectively.

All the above simulators use simplified models at levels 1 or 2. As shown by Kemp (2004a), and illustrated in Fig. 7.12, this level of detail is in fact adequate to study the main interactions between dryers and the rest of the process, but not for detailed design or analysis of the dryer.

7.8.2
Potential Developments

The ideal simulator would have a full range of rigorous unit operation models at Levels 3 and 4, not only for drying but for all major unit operations. However, writing such a simulator would be an extremely complex and time-consuming task, and commercial limitations have so far prevented software houses achieving this, for reasons discussed in Section 7.11.

A more flexible approach is to enable simulators to interface with standalone Level 3 or 4 models, e.g. for fluid-bed dryers. The development of the CAPE-OPEN standards, which allow different programs to interface with each other even if written by different groups, could be a significant step forward in enabling this. This was the approach taken by the Solidsim consortium based in Germany, who developed a process simulator that included all key solids properties and linked it with software written by various groups for a range of unit operations (Werther 2005). Initial results were promising, but the program has not yet been fully commercialized; as pointed out in Section 7.11, the step from research to commercial program is a difficult one, particularly for multiauthor programs.

Even interfacing current standalone models to existing simulators is difficult. Apart from the need to write the interface, the simulator itself may not have the necessary internal parameters to capture the results produced by the unit operation model. Many simulators did not hold a particle-size distribution in their internal dataset, and if such information was supplied from one unit operation (e.g. a dryer or grinding mill), it would not be passed on to the next one (e.g. a cyclone). Special programming was necessary to capture this information and enable it to "bypass" other unit operations without being lost. This has been done, for example, in some solids unit operations in HYSYS, which are classified as "extensions".

Moreover, an ideal solids simulator should be able to have properties that vary with size fraction; e.g. fines often have a higher moisture content than coarse particles, and if they are removed by elutriation, the fines fraction will have a different moisture content to the remaining solids. It is difficult and expensive to retrofit this into an existing simulator. In contrast, Solidsim (Werther 2005) had this capability built in from the outset.

7.9 Expert Systems and Decision-making Tools

Where a problem is not amenable to conventional numerical modeling, an expert system may be appropriate. This is particularly the case where the requirement is for decision making rather than equipment sizing or performance simulation. These decisions often rely on qualitative rather than quantitative data. Potential exists in two key areas in drying; selection and troubleshooting.

7.9.1 Dryer Selection

Dryer selection is of great practical importance in industry. Proposed selection procedures have included tree-search methods (Van't Land 1984, 1991) and complex rule-based algorithms (Kemp 1999; Kemp and Bahu 1995) which have been implemented as expert systems (Kemp et al. 1997, 2004; Baker and Lababidi 2000, 2001), combined with provision of qualitative information to help the engineer. Significant experience has now been gained with the procedures and software, and this suggests refinements and alternative approaches that are reported below.

7.9.1.1 Tree-search Algorithms

A tree search is a question and answer session where, depending on the user's answer, the consultation follows a particular branch of the tree. A good example is Van't Land (1984, 1991), who proposed a decision tree with 6 questions for batch dryers and 7 for continuous. The main route through the tree is by "No" answers; "Yes" answers indicate a particular recommended dryer type. This approach can yield some useful suggestions, but it is too prescriptive; for example, where a fluidized-bed dryer is possible, it is recommended, but may not in fact be the best choice. The decisions are not as clear-cut as a simple "Yes/No" choice suggests. Nevertheless, this decision tree gives useful pointers to how a more thorough tree-type algorithm could be developed.

7.9.1.2 Matrix-type Rule-based Algorithms

The algorithm developed by SPS (Kemp 1997; Kemp and Bahu 1995) uses this approach. Each dryer is awarded numerical merit scores ranging from 0 (completely infeasible) to 1 (no drawbacks) on criteria such as solids throughput and particle size. There are approximately 40 rules, and over 50 dryer types and subtypes are included. Typically, a dryer scores 1 on a rule if it is completely suitable, or a progressively lower score if it is only partly suitable for the given material or process operating parameters (e.g. solids throughput); this is implicitly a form of fuzzy logic. At the beginning, the choice between batch and continuous dryers, and the required form of feed and product, is considered, mainly to eliminate any obviously inappropriate choices and save time (conversely, both batch and continuous options can be kept open if desired). At the end, an overall merit score can then be calculated for each dryer. The structure of the algorithm is illustrated schematically in Fig. 7.13.

Hand calculation is tedious and hence the method has been implemented in a rule-based expert system, DRYSEL (Kemp 1999; Kemp et al. 1997), which ranks over 60 types and subtypes of dryers using approximately 50 rules. Assigning precise scores for each rule is difficult and a fuzzy-logic approach is appropriate, as also noted by Baker and Lababidi (2000, 2001).

In the first version of the expert system, input was by a conventional entry form. It was found that a question and answer consultation was also helpful, as it enabled the user to see how the choices evolved as each question was answered. Both input formats are now available, and the form input helps with "what-if" calculations, e.g. how does the ranking change if we double the throughput? Output is in several forms; an equipment tree, a numerically ranked list of overall scores, detailed scores on individual rules and a consultation history. Figure 7.14 shows a sample screen, in Q&A mode with equipment tree view.

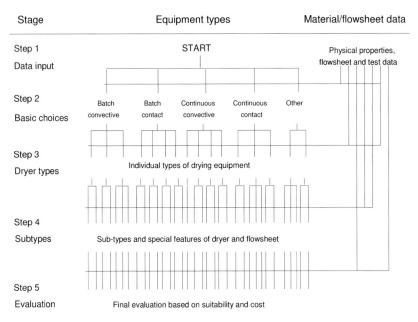

Fig. 7.13 Matrix-type selection algorithm.

7.9 Expert Systems and Decision-making Tools

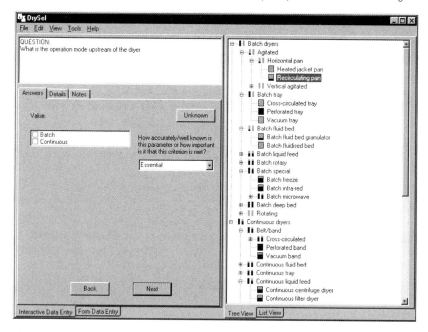

Fig. 7.14 Typical output screen from DRYSEL.

The overall numerical merit scores are a simple and clear indication of the "best" dryers, but experience shows that they tend to mislead the inexperienced user into relying too heavily on the ranking. The error bars on the final score are large because of the cumulative "fuzziness" on each rule – it is impossible to assign exact numerical values that apply in all situations. Hence, a clear-cut distinction can only be drawn if the overall merit scores are substantially different – say 0.8 and 0.2, not 0.8 and 0.7. This is very difficult to represent; logarithmic plots have been tried, but with only moderate success. Users may believe that a dryer with a marginally better score has a clear advantage, whereas the only firm conclusion should be that there is a group of potentially feasible dryers.

Hence, the aim of the expert system should be to present the user with a list of options and their comparative advantages and disadvantages, rather than a direct recommendation for one dryer. There are two reasons for this. Firstly, the choice is rarely clear-cut; for particles of around 500 μm, there may be ten or more good choices of dryer, whereas for 5 μm particles, almost nothing will score well and it is a case of selecting the "least worst". Secondly, the main factor in the choice between several similar dryers will usually be user preference; he may have a specialized requirement outside the scope of the general rules, or may know from experience that a particular dryer functions well or badly, or may already have other dryers of a particular type (in which case he will replicate these rather than selecting a completely new type, allowing operator and maintenance familiarity and common spare parts, unless the advantages of an alternative type are overriding). Hence a selection program should be an "aid to decision making" rather than a "black box" making oversimplified choices.

7.9.1.3 Qualitative Information

Heuristics – simple advice on which dryer to choose and why – have of course been available for many years. This information is useful, but often incomplete or conflicting. As an increasing number of experienced engineers retire or move between companies, the need for complete and reliable information becomes more crucial.

The text and explanation of the individual rules, and the score on each one for a particular dryer, is a byproduct of the SPS matrix approach. It has been found in practice that this is more valuable than the overall merit scores. The user is told what are the potential drawbacks of a dryer (the rules where it has scored less than 1). He can then study this information and consider how serious the problem really is for his particular plant and product. The process engineer is thus left in control, but is able to avoid potential pitfalls and ensure that no unusual dryer options are overlooked. Also, many rules are complex, e.g. a convective dryer handling flammable solvent will be safer with nitrogen as the carrier gas than with air, but the closed system will cost more. This information is difficult to express numerically; it needs to be given to the designer as qualitative advice.

The biggest difficulty with this approach is finding a way of representing it clearly and simply. One recent prototype form of the DRYSEL advisory software has a Web front end (Fig. 7.15) and emphasizes the text of the rules and the individual scores, rather than the overall score. Compare the level of explanation provided about the question with that in Fig. 7.14. Links are provided to the Aspen Process Manual, which gives the exact text of a rule and the reasoning behind it.

7.9.1.4 Alternative Tree-search Approach

Instead of the complex matrix system, an improved tree-search approach can be envisaged. Instead of trying to select a single dryer, a group of similar dryer types would generally be indicated. For example, for small batch-drying tasks with a solid feed, where agitation is desirable to give reasonable drying rates, the choice tends to lie between vertical pan, conical (Nauta), tumbler (double-cone) and spherical dryers. The user is guided through the question-and-answer consultation and thus sees the principles involved in choosing the dryers. Some key questions would be:

1. Required form of feed and product.
2. Is particle size increase or reduction required?
3. Batch or continuous (based mainly on mass flow rate).
4. Particle size range.
5. Is the solvent flammable, or are the solids toxic?
6. Are the particles fragile, and can attrition be allowed?
7. What is the temperature limit of the particles?

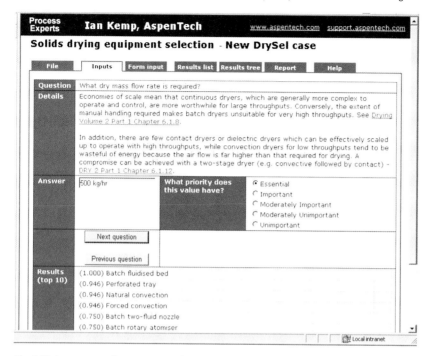

Fig. 7.15 Input screen for prototype web expert system for dryer selection.

These questions also appear in the SPS matrix algorithm (Kemp and Bahu 1995) and (most) in van't Land's (1991) procedure, but the order is different. For example, for a typical pharmaceutical, question 1 distinguishes between liquid feeds (requiring filter dryers, thin-film or spray dryers) and solid feeds, and film and sheet dryers are eliminated. In question 2, a size increase would point to agglomerators and granulators and a size reduction to mills and some high attrition dryers. For Q3, the mean throughput would typically be below 50 kg/h and the rest of the process would be batch, indicating that continuous dryers can be eliminated. In question 4, particles of several millimeters would best be dried by through-circulation or crosscirculation tray dryers, whereas for 100 micrometer particles, convective dispersion (including fluidized beds) or contact dryers are a good choice. Q5 would show whether a nitrogen-inerted system or stringent gas cleaning is required, both factors that favor contact dryers rather than convective. In Q6, fragile particles would need tray dryers, with no agitation but correspondingly poor mixing and low heat-transfer and drying rates; otherwise, mechanical agitation or rotation is preferred. Q7 indicates whether vacuum or atmospheric drying is preferable in contact dryers. Thus, for a 50-kg batch of 100-micrometer particles, not unusually fragile, with acetone as the solvent and a wet cake feed, the group of four agitated batch dryer types listed above will come out as a good choice. The selection of a specific dryer from the group can then be made, based on solids properties, customer preference or capital cost.

The logic is tricky to express in a tree diagram as the same questions need to be asked on several branches of the tree; for example, Q5 and Q6 will apply to virtually all particulate materials, whatever the answers to Q3 and Q4. This confirms the value of the "matrix" approach in the SPS method, and the benefits of describing dryers in terms of key characteristics (batch/continuous, agitated/static, contact/convective) and selecting on this basis.

7.9.2
Troubleshooting and Problem Solving in Dryers

Solving practical problems on dryers is another area that is of great industrial importance but has been relatively little studied in research. As pointed out by Gardiner (1996) and Kemp and Gardiner (2001), problems are often case specific and generic methods are hard to find, so that dryer troubleshooting presents an even greater challenge than selection. However, they proposed a systematic approach to problem solving that helps to categorize the problem and generate a shortlist of possible causes and solutions.

The procedure could in principle be implemented as a simple expert system, as for selection, based on a tree-search algorithm. The techniques could also be used for performance improvement and debottlenecking as well as problem solving. However, systematic classification of dryer problems is more difficult than that of drying equipment, so it has been hard to develop an algorithm that goes beyond a basic level. Kemp 2005 has suggested a possible structure. In the problem definition stage, it is extremely useful to categorize the problem, as the different broad groups require different types of solution. Five main categories of dryer problems can be identified, of which the first three are the most common in practice:

1. drying performance (outlet moisture content too high, throughput too low),
2. materials handling (stickiness, blockages, deposits, lump formation, etc.),
3. product quality (wide range of possible problems depending on specification),
4. mechanical breakdown (catastrophic sudden failure),
5. safety, health, and environmental (SHE) issues.

Performance problems can be further categorized as:
- heat- and mass-balance deficiencies (not enough heat input to do the evaporation),
- drying kinetics (drying too slowly, or solids residence time in dryer too short),
- equilibrium moisture limitations (reaching a limiting value, or regaining moisture in storage).

The pneumatic conveying dryer shown in Fig. 7.3 is a classic example of a system that is heat-balance limited at low gas flows and kinetics limited at high velocity, with an intermediate optimum.

No commercial program has so far been produced for dryer troubleshooting, although the Aspen expert system shell could in principle be used. Indeed, the Aspen Process Tools already include a similar program, CRYSES, for solving problems with particle-size distributions from crystallizers.

7.10 Knowledge Bases and Qualitative Information

7.10.1 Internet Websites

The obvious software-based information source is the Internet, particularly for drying-equipment manufacturers. As a good starting point, there is now an interlinked network of key worldwide drying sites, including those of the International Drying Symposia (IDS), the EFCE Drying Working Party and Professors Mujumdar and Pakowski, the last-named being the official website of the IDS Secretariat and providing references to past papers. Website addresses tend to change frequently, but a web search (e.g. with Google) will usually reveal the new location. Some present addresses are:

- EFCE Drying Working Party site: currently hosted at http://www.uni-magdeburg.de/ivt/efce. May change with chairmen; a permanent link via the main EFCE site is at http://www.efce.info/wp.html.
- Drying Technology journal: http://www.dekker.com/servlet/product/productid/DRT.
- International Drying Symposium site: http://wipos.p.lodz.pl/idsap/.
- Web-based dryer catalogue (University of Lodz): http://wipos.p.lodz.pl/idsap/webdryer.html.
- Arun Mujumdar's personal site: http://www.geocities.com/drying_guru/.
- British Standards Institution (for BS1339): http://www.bsi-global.com/index.xalter.

7.10.2 The Process Manual Knowledge Base

A different approach is given by the Aspen Process Manual, which is a comprehensive online knowledge base covering 10 unit operations, mainly in the field of solids and separation processes, and including drying. This is a commercial program sold under license. It can be divided conceptually into two parts:
- the platform – the front end that delivers the information,
- the content – the information itself.

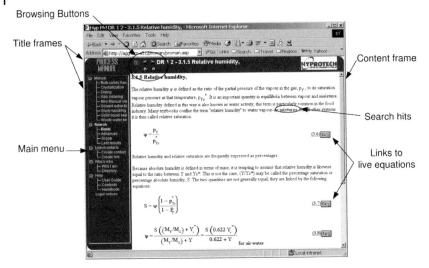

Fig. 7.16 A typical Process Manual page (intranet version).

The front end is web based and available in two formats; Intranet, where the program is mounted on a company's central server, and Internet, delivered via the web from Aspen servers. The interfaces are largely identical, and the main components are the same as for standard web browsers such as Internet Explorer. As would be expected, a powerful search engine is included.

The content is derived from the work of SPS (separation processes service) and other R&D "clubs" over more than 25 years. Each Technical Area contains detailed information ranging from fundamental technology and data correlation techniques, through equipment selection and design, to plant operation and troubleshooting. Selected equations have been converted to "live equations" that allow the user to perform the calculations directly. Figure 7.16 shows a sample page. A more detailed description of the Process Manual is given by Kemp et al. 2004 and information may also be obtained from the Aspen website www.aspentech.com and the Process Manual site www.processmanual.com. The Internet version front end can also be viewed on www.internet.processmanual.com.

Experience with users suggests that knowledge bases, when available, are used more frequently than specialized calculation programs and expert systems, because they can give qualitative answers to such a wide range of questions.

7.11
Commercialization of Drying Software

One might expect that, given the huge developments in computing power over recent years, many new drying software programs would have appeared, giving increasingly accurate and comprehensive simulations of dryers. In fact, apparent progress has been disappointingly slow, especially in terms of commercially available software packages. This section tries to explore this paradox, and to look at the consequences.

7.11.1
Barriers to Drying-software Development

For heat exchangers, more than one comprehensive commercial package has been continuously available for some years (e.g. HTFS/B-Jac, HTRI), and these cover both thermal and mechanical design of all major types. In contrast, for dryers there has been little software available commercially, and it has covered only a few types and a relatively small range of calculations. Why is there a difference? Four reasons can be suggested, of which two may be considered "science based" and two "market based";

1. complexity of the calculations;
2. difficulties in modeling solids;
3. limited market and lack of replicability;
4. changes in operating system software.

The science-based aspects will be apparent from the previous sections and will be discussed only briefly. The market-based aspects, however, are of major significance, but have not been treated above, and are therefore considered here in more depth.

7.11.1.1 Complexity of the Calculations

Drying analysis, with its multiple nonlinear equations, presents a nontrivial programming problem. Moreover, the many different factors involved (heat and mass transfer, particle motion, etc.) vary between different dryer types.

For example, consider a detailed simulation of a fluid-bed dryer. Ideally this should include: a convective heat and mass balance, scoping design, scale-up integral model based on experimental data, processing of the experimental drying curves, an incremental model for local conditions, and appropriate gas-phase physical property calculations (due to variation with temperature), plus possibly fluidization calculations, terminal velocity, entrainment and freeboard height, and heat transfer to/from immersed coils or heating surfaces. Other major types will be able to use some, but not all, of these equations. For example, a flash dryer will need a different particle motion equation based on entrainment in the gas, and a cascading (direct) rotary dryer will need a third form considering both airborne and dense phases.

7.11.1.2 Difficulties in Modeling Solids

As pointed out in Section 7.1.2, the numerous parameters involved, their variability and the difficulty in measurement all make modeling of drying and other solids processes much harder than their gas- and liquid-phase equivalents.

The usefulness of most commercial process simulators is limited by their inability to handle solids rigorously. Generic software can be adapted to include the necessary equations, e.g. spreadsheets for Level-1 and -2 models and math solvers to solve the incremental model at Level 4, but the user still has to do a substantial amount of customization to add all the necessary equations to model his particular system. Checking for errors and debugging then becomes a challenging and time-consuming task.

7.11.1.3 Limited Market and Lack of Replicability

Heat exchangers are widely used throughout many industries, and the same basic types (shell-and-tube, plate, recuperative) are used across-the-board with few major differences between applications. A heat-exchanger design and simulation program can sell to hundreds of customers, and large customers such as contractors or multinational process companies may regularly use the program for many different units.

In contrast, dryers are relatively specialized equipment. Even large processing companies may only have a few, and will not design or purchase a new dryer for many years. Often, a process may contain many heat exchangers, but only a single dryer. The different types need to be modeled in very different ways. Hence, any company that has made the major effort to model a dryer rigorously (Levels 3 and 4) and develop a software program will find only a handful of sales. More generic programs (Level 1 and 2) may have wider applicability, but cannot perform the dryer-specific calculations needed for any accurate attempt at sizing or performance modeling.

At the very least, separate models are needed for all the following groups:
– convective tray and band dryers,
– contact tray and band dryers,
– agitated and rotating contact dryers,
– continuous rotary dryers,
– fluidized beds,
– pneumatic conveying (flash) dryers,
– spray dryers,
– drum and thin-film dryers,
– radio frequency and microwave dryers.

Each of these also needs to be split into subgroups, for example the different types of agitated/rotary contact dryer covered by the spreadsheet described earlier.

The one area where the replicability problem does not apply is humidity. The user base is much wider because humidity calculations are needed in many other applications, e.g. air conditioning, cooling towers, horticulture, museums, etc. Moreover, solids are not involved, so the calculations have much less uncertainty. As a result, several programs are available, both free and under licence, as described in Section 7.7.2.

7.11.1.4 Changes in Operating-system Software

This is a problem faced by all programs intended for long-term use – as DOS changed to Windows, and as Windows evolved through 95, 98, ME, 2000 and XP, the programs have had to be rewritten, debugged and retested with the new operating system.

Unfortunately, this reinforces the market and replication problem. For software with high sales, a large enough development team can be supported to allow both adapting the program to the latest platform and improving the underlying models. In contrast, sales of a specialized program or suite may only support one or two developers, and their time may be totally consumed by keeping up with the operating-system changes. There is an interesting contrast here to textbooks, which can continue to sell at low levels for many years as long as the knowledge does not go out of date, with occasional reprinting if extra copies are needed. With software, even if the underlying model does not change at all, the program can become unusable in a few years (backward compatibility is limited, and it is usually the user interface that goes out of date first).

As a direct result, we can see that in some respects, drying software has gone backwards since the 1990s. For example, DRYPAK gave an excellent level of sophistication for a generic program, but was written in DOS, and the major reprogramming effort required to convert it to Windows has not been available. Likewise, by the 1990s SPS had developed a substantial suite of simulation programs for drying, including Level-4 models for pneumatic conveying and rotary dryers; these again were in DOS and the manpower has never been available to convert them to Windows. The only ex-SPS detailed design program for dryers converted to Windows was FLUBED (Level 3), as described in Section 7.5.3.

The best forward plan was to write a completely new program (DRYSIM) using an incremental model, with additional modules for calculations unique to specific dryer types (e.g. particle motion for flash or rotary dryers). This would have used the ACM (Aspen Custom Modeler) platform that includes a suitable simultaneous equation solver. However, even with ACM already available, the resources could not be found to develop the program.

The expert system shell used for DRYSEL had to be completely rewritten for the different methods of interactive data handling used by Windows. In this case, the actual algorithm had to be changed, and this brought both gains and losses in the calculations of results and the data input/output. All the expert system rules had to be rewritten and rechecked (and in some cases, completely reformulated). Further changes in input/output design were required when the program was converted into a web-based format to be compatible with the Process Manual.

Finally, for commercial programs, the increasing demands for user-friendliness tend to increase the complexity of the user interface and the amount of planning, testing and debugging required. Early mainframe programs had batch input; the user supplied a list of numerical values for the required parameters, in a strictly specified order (originally on punched cards or paper tape, later as a numeric data file). The program was then run, the user received the results, modified the input data and resubmitted it for another run. If the data was wrongly entered, the program simply failed to compute (or gave ridiculous answers). Interactive data input increased programming complexity by an order of magnitude; allowance had to be made for users entering data in varying order, performing "what-if" calculations and over-writing previous data, and entering infeasible or incompatible numbers that had to be error trapped. This has outweighed the time savings made by, for example, the availability of standard packages for solving simultaneous equations rather than having to manually copy or rewrite a standard subroutine (e.g. Runge–Kutta). Ironically, some of the old batch-type programs will still run on current PCs if a suitable data file is constructed, whereas more recent programs may not function at all if some key part of the user interface is incompatible with current operating systems.

The interesting conceptual point is that batch programs required relatively low programming time but were more inconvenient and time consuming to run. Interactive programs give advantages to the user in speed and user friendliness but have a huge penalty in programming time. Therefore, for programs that will only be used occasionally by a small number of experts, it is more cost effective to provide only a very basic user interface with no error trapping. This is, of course, the approach that will be taken for a company's in-house program or a student writing software to perform the calculations for his doctoral project. For programs that will be used widely and regularly, such as heat-exchanger design software, the time, effort and money required to develop a good user interface can be justified. However, almost no drying programs will ever fall into this category. Therefore, users may have to accept a lower standard of user interface, reliability and debugging on drying software, or alternatively pay a very high cost premium; otherwise it is hard to see how any software house, large or small, could afford to produce a drying program to modern standards.

7.11.2
The Future: Possible Ways Forward

Can we see ways of implementing drying software successfully and making it more widely available commercially, given the limitations above?

One way would be a "wrapper" that would give a modern Windows interface to old programs (e.g. those produced by SPS, or companies' in-house software). Even here, some manual work would be necessary in the conversion, but the effort would be much less than writing a modern user interface from scratch. A number of groups have experimented with this approach; one that has reached full commercialization is EASA.

The inherent problem of low replicability is difficult to solve. For all the power of modern computers and programming languages, none have really significantly reduced the amount of work to program a complex model and, more significantly, to debug it and make it work fully reliably. This must be taken into account to formulate realistic plans for any drying software intended to be made available commercially and effectively maintained over the years to come.

An obvious ideal would be for various people to develop specialized models for different dryer types that could be "plugged in" to standard process-simulation packages. The limitations of current commercial simulators in handling solids continue to be a major barrier, and progress is painfully slow. The Solidsim initiative (Werther 2005) gave a promising way forward; the idea was to develop a rigorous simulator capable of handling solids, with a Cape-Open interface to allow any other group (university, equipment manufacturer, end user) to plug in their specialized models. However, although results during the two-year funding period (2003–5) were promising, the experience seems to have re-emphasized the difficulties in developing software for solids, as commercialization of the next stage of the project does not seem to have progressed rapidly. In addition, where software modules are being developed by different groups, it is very difficult to achieve interfacing and debugging at the levels of reliability normally required for commercial packages. Users may have to accept a lower reliability level, but will not then wish to pay commercial prices, and will be reluctant to use the software for critical applications (for both safety and commercial reasons).

7.12
Conclusions

7.12.1
Range of Application of Software in Drying

It can be seen that there is potential to apply software in all aspects of drying operations, from initial process development and equipment selection and design through to commissioning, optimization, debottlenecking, and troubleshooting of existing plants. It can also be used by all departments within a company, from research and development through to operations. Table 7.4 summarizes appropriate software, both of calculation and expert-system formats, for use in various situations. Knowledge bases are useful in all cases.

Although many drying programs have been written to perform simulations and calculations in design mode, this may not match well to industrial needs. Design and selection tools are only required when a new dryer is being installed, which for end-user companies will only be at intervals of many years. The main users of design and scale-up programs are equipment manufacturers, who usually have their own in-house methods. Simulation and performance-rating tools are useful on existing

Tab. 7.4 Software for design of new dryers (D) and optimizing performance of existing plants (P)

Operation	Appropriate software
Initial simple block flow sheet (D)	Development-type simulator (Batch Plus)
Equipment Selection (D)	Selection expert system (Drysel)
Heat and mass balance (P)	Simulator (Aspen Plus, etc.), Level-1 calculation
Experiments and measurements (D, P)	Data processing spreadsheet
Scoping design/performance (approximate sizing) (D,P)	Level-2 calculation program (Dryscope)
Detailed design and simulation of specific dryers (D,P)	Level-3/4 calculations (Flubed, Drycon, CFD)
Troubleshooting and debottlenecking (P)	Troubleshooting expert system

plants, and a dryer is only designed once but will then operate for maybe 20–50 years. Even so, when a model has been developed for a particular dryer, it does not need to be rerun often. In contrast, troubleshooting needs may arise without warning at any time, so if an effective program were available, it might be quite widely used! Again, there will be many different circumstances where information is required quickly on some particular subject, and a knowledge base will be used more than all the other types of software put together.

7.12.2
Overall Conclusion

There is potential to apply software in all aspects of drying operations, from initial process development and equipment selection and design through to commissioning, optimization, debottlenecking, and troubleshooting of existing plants. It can also be used by all departments within a company, from research and development through to operations.

However, although implementation as software seems an obvious extension of dryer modeling, progress has been relatively slow. Various in-house and research software has been written, but only a handful of commercial programs have been successfully launched and updated. The barriers to successful development of drying software are as much commercial as scientific. In particular, the market size is inadequate to pay for the development of complex programs, and interactive user interfaces need frequent and costly rewriting as operating systems change.

Because of the complexity of drying and solids processing operations, and the difficulty in determining key parameters, detailed and rigorous mathematical models are not necessarily the best way to analyze dryers. Simplified methods, consistent with the level of data available and the objective of the calculation, may be more appropriate. Spreadsheets often provide an effective way to implement these simplified models.

Many dryer models work better in scale-up mode, fitted to even limited experimental data, than for design from scratch. Software should reflect this and be usable for scale-up calculations, including input of existing process conditions.

Decision-making algorithms and tools for selection and problem solving can be useful, and in-depth qualitative information is essential to develop a clear understanding of the key factors in the process.

The dryer must also be considered in the context of the entire process flow sheet, not just as a unit operation. This needs to be done in the early stages of development, and simulators are valuable for this, preferably with data links to programs for individual unit operations that can be used for later optimization.

Future developments seem likely to be in low-cost models such as spreadsheet implementations and in detail improvements to generic simulators to allow them to handle solids better. Market considerations make it unlikely that detailed design models will become available as commercial programs, unless either industry is prepared to make a very substantial investment, or a reduction from current standards in user-friendliness and debugging is accepted.

Additional Notation used in Chapter 7

C_P	specific heat capacity	J kg^{-1} K^{-1}
G	mass velocity of gas (mass flux, or mass flow per unit area)	kg m^{-2} s^{-1}
H	enthalpy	J kg^{-1}
N	drying rate (rate of change of moisture content)	kg s kg^{-1}
N_{cr}	unhindered drying rate at critical moisture content	kg s kg^{-1}
U	velocity	m s^{-1}
W	mass flow rate (dry basis)	kg s^{-1}
Z	normalization factor	–

Greek letters

τ	solids residence time or required drying time	s
ϕ	factor allowing for humidity not being a true driving force	–

Subscripts and superscripts

0	at datum conditions
B	for bed of material
full	for full-scale dryer
G	for gas
I	at inlet
L	for liquid/moisture
O	at outlet
P	for particles
S	for solids
test	for small scale test
v	for evaporation
W	at wall or heated surface
Wl	wall heat losses
Y	for vapor

References

AFSIA, 2001. Cahier No. 19, Sechage et aeraulique. *Proceedings of meeting on 22–23 March 2001*, Poitiers, France. AFSIA-ESCPE, 69616 Villeurbanne cedex.

Bahu, R. E., 1994. Fluidised bed dryer scale-up. *Drying Technol*. 12(1&2): 329–339.

Baker, C. G. J., Lababidi, H. M. S., 2000. Development of a fuzzy expert system for selection of batch dryers for foodstuffs. *Drying Technol*. 18(1&2): 117–141.

Baker, C. G. J., Lababidi, H. M. S., 2001. Developments in computer-aided dryer selection. *Drying Technol*. 19(8): 1851–1874.

Baker, C. G. J., Lababidi, H. M. S., 2006. Fuzzy modeling – an alternative approach to the simulation of dryers. *Proceedings of 15th International Drying Symposium (IDS2006)*, Budapest, Volume A, 258–264.

BS 1339 (Revised). Humidity and dewpoint. Part 1 (2002): Terms, definitions and formulae. Part 2 (2006): Psychrometric calculations and software. Part 3 (2004): Guide to the measurement of humidity. British Standards Institution, Gunnersbury, UK.

Cao, W. F., Langrish, T. A. G., 1999. Comparison of residence time models for cascading rotary dryers, *Drying Technol.* **17**(4&5): 825.

Fyhr, C. and Kemp, I. C., (1998), Comparison of different kinetics models for single particle drying kinetics, *Drying Technology* **16**(7) 1339–1369.

Fyhr, C., Kemp, I. C., 1999. Mathematical modeling of batch and continuous well-mixed fluidised bed dryers, *Chemical Engineering and Processing* **38**: 11–18.

Fyhr, C., Kemp, I. C., Wimmerstedt, R., 1999. Mathematical modeling of fluidised bed dryers with horizontal dispersion. *Chemical Engineering and Processing* **38**: 89–94.

Gardiner, S. P., 1996. Dryer Troubleshooting. SPS Drying Manual Volume XI (Dryer Operations), Part 3. Available only to licencees of the SPS Process Manual.

Kemp, I. C., 1999. Progress in dryer selection techniques. *Drying Technol.* **17**(7&8): 1667–1680.

Kemp, I. C., 2004a. Drying in the context of the overall process. *Drying Technol.* **22**(1&2): 377–394. Paper originally presented at IDS 2002, Beijing, August 2002.

Kemp, I. C., 2004b. Comparison of particle motion correlations for cascading rotary dryers. *Proceedings of 14th International Drying Symposium (IDS)*, Sao Paulo, Volume B, pp. 790–797.

Kemp, I. C., 2005. Troubleshooting. Process Manual, DRY 1 Part 4, Section 15.4 (commercial and academic licences available from Aspen Technology).

Kemp, I. C., 2006. Developments in design, scale-up and optimisation procedures for fluidised bed dryers. *Proceedings of 15th International Drying Symposium (IDS2006)*, Budapest, Volume A, 508–515.

Kemp, I. C., Bahu, R. E., 1995. A new algorithm for dryer selection. *Drying Technol.* **13**(5–7): 1563–1578 (also in Drying'94, *Proceedings of 9th IDS, Gold Coast, Australia*, pp. 439–446).

Kemp, I. C., Bahu, R. E., Oakley, D. E., 1991. Modeling vertical pneumatic conveying dryers. *Drying '91 (7th Int. Drying Symp. Prague, Czechoslovakia, Aug. 1990)*, edited by A. S. Mujumdar et al., Elsevier, pp. 217–227.

Kemp, I. C., Snowball, I. A. G., Bahu, R. E., 1997. An expert system for dryer selection. *Proc. ECCE-1, Florence*, Vol. 2, pp. 1067–1070 (summary). *AIDIC Conf. Series*, Vol. 2, pp. 175–182 (full version).

Kemp, I. C., Fernandez, M., 2002. A new method to determine drying kinetics and volatile retention in pneumatic conveying dryers. Paper presented at *13th IDS, Beijing, August 2002.*

Kemp, I. C., Fyhr, C., Laurent, S., Roques, M., Groenewold, C., Tsotsas, E., Sereno, A., Bonazzi, C., Bimbenet, J., Kind, M., 2001. Methods for processing experimental drying kinetics data. *Drying Technol.* **19**(1): 15–34.

Kemp, I. C., Gardiner, S. P., 2001. An outline method for troubleshooting and problem-solving in dryers. *Drying Technol.* **19**(8): 1875–1890 (also in *Proceedings 12th IDS, 2000, Noordwijkerhout,* [on CD-ROM], paper 271).

Kemp, I. C., Hallas, N. J., Oakley, D. E., 2004. Developments in Aspen Technology drying software. *Proceedings of 14th Intl Drying Symposium (IDS)*, Sao Paulo, August 2004, Volume B, pp. 767–774.

Kemp, I. C., Oakley, D. E., 1997. Simulation and scale-up of pneumatic conveying and cascading rotary dryers. *Drying Technol.* **15**(6–8): 1699–1710 (also in *Proceedings of 10th IDS, 1996*, A, pp. 250–258).

Kemp, I. C., Oakley, D. E., 2002. Modeling of particulate drying in theory and practice, *Drying Technol.* **20**(9): 1699–1750.

Kemp, I. C., Oakley, D. E., Bahu, R. E., 1991. Computational fluid dynamics modeling of vertical pneumatic conveying dryers. *Powder Technol.* **65**(1–3): 477–484.

Langrish, T. A. G., 1996. Flow sheet simulations and the use of CFD simulations in drying technology, *Drying '96, Proc. 10th. Int. Drying Symp. (IDS '96), Krakow, Poland*, Lodz Technical University (publ.), edited by C. Strumillo, Z. Pakowski, Vol. A, pp. 40–51.

Langrish, T. A. G., Fletcher, D. F., 2001. Spray drying of food flavors and applications of CFD in spray dryers. *Chemical Engineering and Processing* **40**: 345–354.

Livesley, D. M., Oakley, D. E., Gillespie, R. F., Elhaus, B., Ranpuria, C. K., Taylor, T., Wood, W., Yeoman, M. L., 1992. Development and validation of a computational model for spray-gas mixing in spray dryers. *Drying '92, Proc. 8th Intl. Drying Symp (IDS '92)*, Montreal, Canada, edited by A. S. Mujumdar, Elsevier, pp. 407–416.

Luikov, A. V., 1935. The drying of peat. *Ind. Eng. Chem.* **27**: 40–69.

Luikov, A. V., 1966. Heat and mass transfer in capillary-porous bodies. Pergamon Press, Oxford, UK.

McKenzie, K. A., Bahu, R. E., 1991. Material model for fluidised bed drying. *Drying '91 (Selected papers from 7th Int. Drying Symp. Prague, Czechoslovakia, Aug. 1990)*, edited by A. S. Mujumdar et al., Elsevier, pp. 130–141.

Matchett, A. J., Baker, C. G. J., 1987. Particle residence times in cascading rotary dryers. Part 1 – Derivation of the two-stream model. *J. Separ. Proc. Technol.* **8**: 11–17.

Matchett, A. J., Baker, C. G. J., 1988. Particle residence times in cascading rotary dryers. Part 2 – Application of the two-stream model to experimental and industrial data. *J. Separ. Proc. Technol.* **9**: 5.

Menshutina, N. V., Kudra, T., 2001. Computer aided drying technologies. *Drying Technol.* **19**(8): 1825–1850 (also in Proceedings 12th IDS, 2000).

Merrow E. W., 1985. Linking R&D to Problems Experienced in Solids Processing. *Chem. Eng. Processing*, May 1985: 14–22.

Merrow E. W., K. E. Phillips, C. W. Myers, 1981. Understanding Cost Growth and Performance Shortfalls in Pioneering Process Plants. In: Rand Corporation Report, Section V.

Moropoulou, A., Karoglou, M., Krokida, M. K., Maroulis, Z. B., Saravacos, G., 2005. Prediction of hygrometric properties of building materials, based on the average pore radius. *Proceedings of 7th World Congress of Chemical Engineering*, Glasgow, UK (on CD), Paper O70–003.

Moyers, C. G., 1994. Scale-up of layer dryers; a unified approach. *Drying Technol.* **12**(1&2): 393–416.

Oakley, D. E., 2001. Design of Layer and Contact Dryers.SPS Drying Manual Volume VII,Part 4. Available only to licencees of the SPS/Process Manual.

Reay, D., 1989. A scientific approach to the design of continuous flow dryers for particulate solids. *Multiphase Science and Technology*, edited by G. F.Hewitt, J. M.Delhaye, N.Zuber, Hemisphere Pub. Corp., Vol. 4, pp. 1–102.

Reay, D., Allen, R. W. K., 1982a. Predicting the performance of a continuous well-mixed fluid-bed dryer from batch tests. *Proc. 3rd Int. Drying Symp.*, Birmingham, edited by J. C. Ashworth, Drying Research Ltd., Vol. II, pp. 130–140.

Reay, D., Allen, R. W. K., 1982b. The effect of bed temperature on fluid bed batch-drying curves. *J. Sep. Process Technol.* **3**(4): 11–13.

vanMeel, D. A., 1958. Adiabatic convection batch drying with recirculation of air. *Chem. Eng. Sci.* **9**: 36–44.

Vanecek, V., Picka, J., Najmr, S., 1964. Some basic information on the drying of granulated NPK fertilisers. *Int. Chem. Eng.* **4**(1): 93–99.

Vanecek, V., Picka, J., Najmr, S., 1966. *Fluidised Bed Drying*. Leonard Hill, London.

van'tLand, C. M., 1984. Selection of industrial dryers. *Chem. Eng.* **91**(5): 53–61.

van'tLand, C. M., 1991. *Industrial Drying Equipment: Selection and application*. Marcel Dekker, New York.

Werther, J., 2005. SolidSim – a new system for the flow sheet simulation of solids processes. Paper presented at *World Congress of Chemical Engineering*, Glasgow, July 2005.

Whitaker, S., 1977. Simultaneous heat, mass and momentum transfer in porous media. *Advances in Heat Transfer* 13: 119–203. Academic Press, New York, USA.

Whitaker, S., 1980. Heat and mass transfer in porous media. *Advances in Drying*, edited by A. S.Mujumdar, Vol. 1, pp. 23–61, Hemisphere, Washington D.C., USA.

Index

a

adaptive Lagrangian time step   170
advection terms, population
  balances   213
aerodynamic interaction, stochastic
  collisions   186
age of particles   211
agglomeration   187–189, 191
  –particle tracking   170
  –population balances   212
  –spray-dryer simulation   161,
  199
aggregation   225–233
  –and growth   244–245
  –and nucleation   243–244
aggregation-breakage coupling
  239–243
aggregation, population balances   231
air conservation   6
air flow
  –spray-dryer simulation   160
  –stack model   46
airtight sensors   26
algebraic turbulence models   157
algorithm description, droplet
  drying   179
alternative tree-search approach   292
alumina, continuous
  thermomechanical drying   107
ancillary calculations   283–287
Archimedes buoyancy   168
Arnoldi process   17
Arnoldi vectors   18
Aspen process manual   295
assumed elastic behavior   115–120
atomization, spray-dryer
  simulation   162, 193
averaged internal equations
  –balance   113

  –continuous thermomechanical
    drying   107
  –Darcy's law   110
  –energy conservation   112–113
averaging theorem   3, 5

b

balance equations   213–215
  –averaged internal   113
  –global   126–130
  –stack model   46
basic scale-up principles   273–274
Basset-Bousinesq-Oseen (BBO) particle
  tracking   169
Basset history force   168
batch-contact dryers   271
batch dryers   270–271
batch lumber kiln   45, 47
BBO see Basset–Bousinesq–Oseen
beech, volume averaging   42
bimodal networks   97
  –heat transfer   91
  –isothermal drying models   86
bimodal pore-size distributions   95–100
binary breakage, population balance
  equation   220–221
body deformations   143–144
Bond number   75
Borel's convolution formula   142
bound-water diffusion   6
bound-water flux   7
boundary conditions
  –convective drying   113–114, 145
  –generalized Darcy's law   7
  –moving sectional methods   237
  –spray-dryer simulation   160
boundary layers
  –drying curves   85
  –isothermal drying models   60–62

Modern Drying Technology. Edited by Evangelos Tsotsas and Arun S. Mujumdar
Copyright © 2007 WILEY-VCH Verlag GmbH & Co. KGaA. All rights reserved.
ISBN: 978-3-527-31556-7

–thickness 69
breakage-aggregation coupling 239–243
breakage function 214
breakage process, finite-volume scheme 222
British standards (BS), humidity calculations 284–286
burning period, microwave heating 31

c

calculation programs, software types 263
calculations complexity, development barriers 297
capillary number
– length scales 76
– wettability effects 84
capillary pressure 143
– heat transfer 88
– isothermal drying models 64
– length scales 75
capillary pumping
– isothermal drying models 64–65
– pore structure influence 98
carpet plots
– radio-frequency heating 34
– softwood drying behavior 29, 32
cascading rotary dryers 281
CDC see characteristic drying curve
CDRP see constant drying rate period
cell-average mechanism 219
cell-average technique 227–230, 239
– population balance equation 216–222
CFD see computational fluid dynamics
characteristic drying curve (CDC) concept 267
characteristic lengths 76
chemical potential
– balance equations 129
– convective drying 145
– rate equations 134
– thermohydromechanical drying 139
circumferential stresses
– convective drying 148–159
– microwave drying 151
Clausius–Clapeyron equation 175
clusters 63
– isothermal drying models 72
– labeling 65

coagulation, pure aggregation 228
coalescence 187–189
collector efficiency, stochastic collisions 186
collision number 198
collisions
– dry particles 161, 189–192
– particle 181–192
– probability 185
– stochastic modelling 182–187
commercialization, drying software 296–301
complex three-dimensional models 276
comprehensive drying 1–52
computational fluid dynamics (CFD) 281–283
– collisions of particles 181–192
– dissipation rate 156
– droplet-drying models 173–181
– Euler–Lagrange approach 162–173
– examples 192–200
– prediction of product properties 200–203
– spray-dryer simulation 155–208
concrete, high-temperature convective drying 19–25
condensation, heat transfer 89
connectivity of pores 59, 94
conservation
– air 6
– energy 7
– numerical resolution technique 114
– water 6
conservation equations 15–17, 107–112, 121–122, 156–158, 192
– Reynolds-averaged 156
conservation law
– balance equations 128
– finite-volume scheme 222
constant drying flux period 8
constant drying rate phase 8–9, 140
constituent velocity of moisture 126
constitutive equations 130–132
continuity equation
– gas phase water vapor 4
– liquid phase 3
– volume averaged 5
continuous contact dryers 270–271
continuous convective dryers 269–270
continuous fluidised bed dryer 210–211
continuous models 100–101
continuous-phase source terms 163
continuous thermohydromechanical models 125–154
continuous thermomechanical models 103–124
– simulation 114–120

control volume 45
  –deformable 127
control-volume finite-element (CVFE) discretization procedure 12–14
  –discretization process 17
convective drying 18–29, 107
  –boundary conditions 113–114
  –heat transfer 58
  –high-temperature 19–25
  –IDC 33, 35
  –isothermal drying models 60
  –kaolin cylinder 144–149
  –low-temperature 7–10
  –radio-frequency heating 34
convective flux 24
cooling effect, surface evaporation 91
coordinate space, total 210
coordination number, pore structure influence 94–95
corner films, film flow 80
corner flows 93
coupling
  –breakage aggregation 239
  –Euler–Lagrange approach 159
  –stack model 46
creep 142
creep strain, thermohydromechanical drying 134
cumulative size distributions, stochastic collisions 183
Curie-Prigogine symmetry principle 133
CV face 14–15
CVFE see control-volume finite-element

d

damping coefficient, microwave heat source 137
Darcy's law 7
  –averaged internal equations 110
data structures, isothermal drying models 59–60
databanks, physical properties of dryers 287
decision-making tools 289–295
decomposition, volume-averaging models 5
decreasing drying rate phase 9
deformable materials, drying behaviour 34–37
deformation gradient tensor 36

deformations, thermohydromechanical drying 125
density, volume-averaging models 4
depth scalar, generalized Darcy's law 7
design models, detailed 276–277
desorption, averaged internal equations 113
destruction, microwave drying 151–152
deviation variables, volume-averaging models 5
dielectric properties, microwave heat source 137
  –thermohydromechanical drying 126
differential equations, thermohydromechanical drying 134–141
diffusion
  –fluid phase modeling 163
  –identity drying card 33
  –isothermal drying models 58
  –milk drying 178
  –moving sectional methods 237
  –volume-averaging models 6
diffusion coefficient
  –milk drying 176
  –particle morphology 202
diffusion tensors 14
diffusivity
  –identity drying card 33
  –vapor phase 5
  –volume-averaging models 6
Dirac-delta distribution 215
direct numerical simulations (DNS) 155
discontinuous Heaviside function 218
discretization
  –finite-volume scheme 223
  –procedure 13–14
  –pure breakage 215
  –pure growth 234
dispersed phase results 195–200
dispersed phase source terms 163
dispersed two-phase flows 159
dispersive transport, volume-averaging models 5
displacement, balance equations 127
displacement fields, periodic 41
dissipation rate 156
DNS see direct numerical simulations
domain discretization 251
double coordinates system, homogenization 38
drag coefficient 170
drag forces
  –Lagrangian particle tracking 167
  –VD collisions 191
driving forces 146

droplet density   178
droplet-drying models   173–181
   – numerical implementation   178
   – review   175–176
droplet evaporation   175
droplet interaction   190
droplet size   160
droplets dominated by viscous forces (VD droplets)   188
dry particles
   – collisions   161, 189–192
   – solid   184
dry patches distribution   71
dry solids, enthalpy   181
dryer designs   157
   – software   301
dryer geometry   192–193
dryer length   270
dryer models
   – categorization   264
   – comparison   266
dryer selection   289–294
dryers
   – continuous contact   270–271
   – continuous convective   269–270
   – continuous fluidised bed   210
   – scale up   157
   – troubleshooting and problem solving   294–295
drying
   – at high temperature   10
   – convective   7–10
   – definition   103
   – pore level   57–102
   – quality   50–51
drying air, isothermal drying models   61
drying algorithm   66–69
drying behavior
   – highly deformable materials   34–37
   – potato   36
   – softwood   25–29
drying body   141–144
drying conditions   45
drying conservation equation   15
drying curves
   – boundary layer influence   85
   – convective drying   147
   – length scales   77–79
   – normalization   275
   – pore structure influence   95–96
drying front   21, 71
drying kinetics   57–102
drying models   1–52
drying periods
   – constant rate   140
   – dynamic equilibrium   9
   – falling rate   140
   – microwave heating   30
   – transition for light concrete   20
drying process duration
   – severe conditions   24
   – soft conditions   22
drying software   261
   – commercialization   296–301
   – development barriers   297–298
drying technology   155–208
drying theory and practice, gap   262
dual-scale models   47
Duhamel-Neuman equation   132
dynamic equilibrium, drying phases   9

e

ECM   *see* equilibrium moisture content
EDECAD   *see* efficient control and design of agglomeration during spray drying
effective diffusivity tensor, volume-averaging models   6
effective stress   130–131
efficient control and design of agglomeration during spray drying (EDECAD)   176
elastic behavior   115–120
elastic case   104
elastic deformations   130
electric conductivity, microwave heat source   135
electric field intensity, microwave heat source   136
electromagnetic heating, less common drying configurations   30
EMC   *see* equilibrium moisture content
energy absorption, microwave heat source   137
energy balance   129
energy conservation
   – averaged internal equations   112–113
   – volume-averaging models   7
enthalpic period, microwave heating   30
enthalpy
   – continuous fluidised bed dryer   211
   – desorption   113
   – dry solids   181
   – per unit mass   113
   – pore network models   87

–stack model   46
–vapor   113
entropy
   –balance equations   128
   –time derivative   134
EOC *see* order of convergence
equations
   –averaged internal   107–113
   –balance   46, 50, 113, 126–129, 213–215, 277
   –Clausius–Clapeyron   175
   –conservation   15–17, 107–112, 121–122, 156–158, 192
   –constitutive   130–132
   –continuity   3–5
   –differential   134–141
   –drying conservation   15
   –Duhamel-Neuman   132
   –gas-phase water vapor continuity   4–5
   –Gibbs   130–131, 134
   –Gibbs–Duhem   132, 140
   –heat-balance   277
   –Heaviside function   229
   –Hoshen-Kopelman algorithm   65
   –liquid phase continuity   3
   –macroscopic set   6–7
   –mass-balance   277
   –mass-conservation   109–112
   –Maxwell   30, 135
   –momentum   6, 156
   –momentum conservation   103–105, 109–114, 120–122
   –particle transport   277
   –population balance   210, 214–225, 233
   –pressure linked   165
   –rate   132–134
   –Reynolds   156
   –Reynolds-averaged Navier-Stokes   155, 163
   –state   108, 131
   –thermomechanical   141–144
   –total momentum conservation   110
   –transport   1–5, 82, 163–166
   –two-equation models   157
   –water-vapor transport   1
   –wave   136
equilibrium contact angle, wettability effects   83
equilibrium moisture content (ECM)   26
equilibrium solvent concentration, droplet drying   180
equilibrium vapor pressure reductioin, isothermal drying models   58
equipment models, drying software   265–266
error progression, fluid phase modeling   165
Euler–Euler approach   158
Euler–Lagrange approach   162–173
   –flow chart   174
   –multiphase flow modeling   159
   –particle collisions   181
Euler's description, balance equations   126
Eulerian correlation tensor   172
Eulerian time scale   170
evaporation , 140
   –film flow   80
   –solvent   201
   –thermohydromechanical drying   135
evaporation rates   63, 85
   –heat transfer   88
experimental data processing   283
experimental order of convergence (EOC)   220
expert systems   289–295
external coordinates, population balances   210
external pressure decrease   10

f

falling-rate drying   271–272
falling-rate period   140
FDI *see* flow direction indicator
fictitious particles   183–184
field forces, Lagrangian particle tracking   167
film flow, pore networks   79–83
finite-volume scheme   222–225, 231–232
first drying period   85–87
   –drying phases   8
   –heat transfer   90
fixed-pivot technique
   –aggregation nucleation coupling   244
   –pure aggregation   226–227
   –pure breakage   215
flow chart
   –Euler–Lagrange approach   174
   –isothermal convective drying   67
   –menisci identification   74
flow direction indicator (FDI)   15
flow-field pattern, spray-dryer simulation   193
fluctuation velocity   171
fluid-dynamic forces   166
fluid-phase modeling   163–166
fluid-phase results, spray-dryer simulation   193–195

fluidized-bed dryers 274–276
flux limiting, tensor evaluation 14
flux-limiting methods, pure growth 234
fluxes, continuous thermomechanical models 113
Fourier's law 112
free liquids, isothermal drying models 58
french fry, moisture content field 37
friction
 – characteristic lengths 76
 – drag force 167
 – length scales 75

g

gap, drying theory and practice 262
gas-flow patterns 158, 283
gas permeability tensor 6
gas-phase properties, droplet drying 179
gas-phase water vapor continuity equation 4–5
gas-phase water-vapor transport equations 1
gas pores, mass balances 63
gas pressure 58, 64
gas velocity contour 195–196
gaseous pressure, total 10
Gauss divergence theorem 13
Gauss-Ostrogradsky theorem 128
Gaussian random number 184
gels, continuous thermomechanical drying 106
general transport theorem 3
generalized Darcy's law 7
geometry, spray dryer 192
Gibbs' equation 130–131, 134
Gibbs–Duhem equation 140
Gibbs–Duhem relationship 132
glass-transition temperature, particle collisions 189
global balance equations 126–130
globally convergent Newton's method 16
GMRES 17
gravitational effects, length scales 75–76
gravitational force, Lagrangian particle tracking 169
gravity modeling 71–74
gravity potential
 – isothermal drying models 72
 – rate equations 134
growth and aggregation 244–245
growth and nucleation 245–247

h

Haines jump 74
heartwood, drying behavior 25–29
heat and mass balances 268–269
heat and mass transfer
 – differential equations 134–141
 – rate equations 132–134
heat and vapor transfer 8
heat-balance equations 277
heat capacity 87
heat conservation, numerical resolution technique 114
heat damage, product properties prediction 201
heat-damage index number 201
heat flow, milk drying 177
heat-flow rate 88
heat transfer
 – bimodal network 91
 – capillary pressure 88
 – condensation 89
 – continuous contact dryers 270
 – convective drying 58
 – evaporation rates 88
 – first drying period 90
 – isothermal drying models 58
 – milk drying 176
 – pore network models 87–92
 – surface evaporation 91
 – thermohydromechanical drying 132–134
 – time discretization 89
 – vapor condensation 89
heating period, microwave 30
heating power, microwave 139
Heaviside function
 – discontinuous 218
 – pure aggregation 229
high-resolution schemes, pure growth 234
high-temperature drying 10–13
 – convective 19–29
highly deformable materials, drying behaviour 34–37
homogenization 37–42
homogenized properties 40
Hoshen-Kopelman algorithm 65
humidity calculations 284–287
 – British standards 284
hydraulic conductivity, film flow 80
hydrophilic porous media 84
hydrophobic porous media 84

i

identity drying card (IDC) 32–33
  –typical diagrams 35
impact efficiency, stochastic collisions 186
incremental models 276–278
induction, microwave heat source 136
industrial design, drying software 262
inertia parameter, stochastic collisions 186
information loss 165
inner iterations 17–18
instantaneous strains 142
integral models, scale up principles 274–276
interfacial tension forces 187
internal coordinates, population balances 209
internal energy, balance equations 129
internal moisture transfer 26
internal overpressure 6
internal pressure, drying phases 10
internet websites, knowledge bases 295–296
intrinsic averages 3
invasion percolation, pore structure influence 98
irreversible deformations
  –constitutive equations 130
  –rate equations 132
isothermal convective drying, flow chart 67
isothermal drying models
  –boundary layers 60–62, 69
  –capillary pressure 64
  –data structures 59–60
  –pore network models 58–87
iterations
  –inner 17–18
  –outer 16–17

j

Jacobian matrix 17

k

kaolin cylinder, convective drying 144–152
κ–ε turbulence models 157, 163
Kelvin effect
  –isothermal drying models 58
  –length scales 76
kinematic viscosity 156
kinetic energy transport 166

knowledge bases, process-systems simulation 295–296
Knudsen effect
  –isothermal drying models 58
  –length scales 76
Kolmogorov length scale 156
Krylov methods 18

l

labeling, clusters 65
lack of replicability 298–301
Lagrangian particle tracking 166–169
Lagrangian time scale 171
Lagrangian time step
  –adaptive 170
  –droplet drying 179
large eddy simulations (LES) 155
large shrinkage values 36
latent heat, thermohydromechanical drying 135
lateral nondimensional displacement 187
layering concept, drying software 265
layering models, dryer models categorization 264
length scales
  –isothermal drying models 75–77
  –turbulent dispersion modeling 172
LES *see* large eddy simulations
less-common drying 29–30
  –radio-frequency heating 34
  –volume-averaging models 29–37
  –volumetric heating 29
level of complexity, dryer models categorization 264
light concrete, high-temperature convective drying 19–25
limited market 298–301
line-search strategy 16–17
linearized system iterations 17
liquid cluster, isothermal drying models 72
liquid distribution, thermohydromechanical drying 139
liquid flow, drying phases 12
liquid mass flow rates, isothermal drying models 73
liquid mobility tensor 14
liquid permeability tensor 6
liquid phase
  –continuity equation 3
  –isothermal drying models 64–65
  –volume-averaging models 1
liquid pressure 120–122
  –isothermal drying models 64

liquid redistribution, isothermal drying models 74
liquid velocity, length scales 75
liquid viscosity, pore network models 71–72
liquid volume fraction 111
liquid water fluxes, stack model 46
liquid/air interface 62
liquid/gas interface 86
local fluxes, stack model 46
local mass balance 139
local thermal equilibrium 126
local thermodynamic equilibrium, absence 42
low-temperature convective drying 7–10
lumber drying 11

m

macrochannels 86
macromolecular gels, continuous thermomechanical drying 106
macroscopic equations, volume averaging 6–7, 42–43
macroscopic properties, homogenization 39
macrothroats 98
magnetic permeability, microwave 136
magnetic viscosity, microwave 136
Magnus force 168
mass balance
  –equations 129, 277
  –pore network models 63
mass change rate 126
mass concentration, thermohydromechanical drying 126
mass conservation, numerical resolution technique 114
mass-conservation equations 109–112
mass content, thermohydromechanical drying 126
mass flow, milk drying 177
mass flow rate
  –droplet drying 180
  –isothermal drying models 73
mass transfer
  –drying phases 10
  –fluid phase modeling 164
  –Gibbs-Duhem equation 140
  –menisci 140
  –milk drying 176
  –thermohydromechanical drying 132–134, 139
mass-transfer coefficient 61
material models, drying software 265
matrix-type rule-based algorithms 289
Maxwell equations
  –less common drying configurations 30
  –microwave heat source 135
Maxwell model 142
mean particle volume, multidimensional population balances 250
mechanosorptive strain 142
  –drying quality 50
menisci
  –drying front 71–77
  –film flow 81
  –heat transfer 88
  –identification 74
  –isothermal drying models 63–66
  –mass transfer 140
MeshPore 41
method of lines (MOL) 233
microscopic foundations
  –homogenization 39
  –volume-averaging models 1–6
microwave drying 150–152
  –IDC diagrams 35
  –less common drying configurations 30
  –softwood 33
microwave heating 135–139
  –absorbing properties 136
  –burning period 31
  –convective drying 148
  –drying phases 11, 30
microwave power 138
microwave propagation 150
milk drying, diffusion 176, 178
milk products, drying models 176–178
minimum shell diameter, droplet drying 178
mixing length, fluid phase modeling 166
mixtures, thermohydromechanical drying 125–154
mobility tensor 15
model description, isothermal drying 58–67
model extensions, pore network 87–92
modeling depth, spray-dryer simulation 192
moist air flow 19, 22–25
moisture, continuous fluidised bed dryer 211
moisture content 115, 140
  –population balances 213
moisture content curves, stack model 48
moisture content evolution, stack model 51
moisture content fields 44
  –drying behaviour 37

moisture content loss, softwood drying behavior 28
moisture content profiles 119
moisture content variation, pneumatic conveying dryer 280
moisture distribution, convective drying 147
moisture flux 126
moisture-time drying curve, scale up principles 273
moisture transport coefficient 140
moisture two-dimensional predicted fields 116–118
MOL *see* method of lines
molecular viscosity 166
Mollier chart 286
momentum-conservation equations
 –averaged internal 109–112
 –continuous thermomechanical drying 103
 –numerical resolution technique 114
momentum equations 156
 –volume-averaging models 6
MorphoPore 41
moving sectional methods 237
multicomponent liquid 92
multicomponent medium, entropy 128
multidimensional population balances 247–255
multimodal temperature, errors 185
multiphase drying processes 155
multiphase flow modeling 158–160
multiphase systems, continuous thermomechanical drying 103
multiphase transport, thermohydromechanical drying 125
multiple surfaces, drying conditions 45
multiscale approach, model calculation 42–51

n
narrow radius distribution 75
net flux, multidimensional population balances 251
network flux, balance equations 127
network geometry, isothermal drying models 59–60
network heat flux, rate equations 133
network size, isothermal drying model 75

new dryers, software 301
newborn particles 217, 228, 239
Newton step 16
nonlinear functions 15–18
normalization 275
Norway spruce 25
nucleation and aggregation 243–244
nucleation and growth 245–247
number density
 –multidimensional population balances 247
 –pure aggregation 228
numerical calculation procedures, dryers 263–268
numerical flux, finite-volume scheme 223
numerical implementation, droplet-drying models 178
numerical methods, pure aggregation 226
numerical resolution technique, continuous thermomechanical models 114

o
oak
 –model calculation 41
 –soaking wood samples 42
Ohnesorge number
 –STD droplets 187
 –VD droplets 188
one-equation models, turbulence 157
one-way coupling, Euler–Lagrange approach 159
outer iterations, fluid-phase modeling 16–17

p
parametric models 266–268
partial vapor pressure, milk drying 177
particle change rate 218
particle collisions 181–192
 –dry particles 189–190
 –spray-dryer simulation 161
particle displacement, balance equations 127
particle morphology 201–203
particle number, cell-average technique 219
particle number concentration, spray-dryer simulation 197
particle-phase properties, two way coupling procedure 173
particle properties distributions 209
particle relaxation time, particle collisions 182
particle-size distribution
 –computational fluid dynamics 158
 –multidimensional population balances 255

–product properties prediction 200–201
–spray-dryer simulation 199
particle source terms, spray-dryer simulation 194
particle stokes number, stochastic collisions 184
particle tracking 169–171
particle transport equation, incremental models 277
particle turbulent dispersion modeling 171–173
particles
   –age of 211
   –dry *see* dry particles
   –newborn *see* newborn particles
   –point 159
   –volume averages 217
particulate gels, continuous thermomechanical drying 106
PBE *see* population balance equation
PDE problems, homogenization 40
perfectly wetting liquid phase, isothermal drying models 59
performance optimization, software 301
periodic displacement fields 41
periodic homogenization 38
permeability
   –identity drying card 33
   –volume-averaging models 6
permittivity, microwave heat source 137–138
perturbation parameter, Jacobian matrix 17
phase boundary evolution 70
phase boundary stabilization 77
phase distributions
   –isothermal drying models 68
   –length scales 77–79
   –pore structure influence 97
phase potential, generalized Darcy's law 7
phase source terms, transport equation 164
phase transitions
   –distribution functions 141
   –efficiency 140
   –identity drying card 33
   –rate equations 134
phases, thermohydromechanical drying 125
phenomenological Fourier's law 112

Picea abies, softwood drying behavior 26
pilot-plant experimental data 279
pneumatic conveying dryer 278–279, 282
point particles 159
polarization, microwave heat source 137
population balance equation (PBE) 210
   –binary breakage 220–221
   –cell-average technique 216–222
   –pure aggregation 225
   –pure breakage 214–225
   –pure growth 233
population balances
   –combined aggregation and breakage 239–242
   –combined aggregation and nucleation 242–244
   –combined growth and aggregation 244–245
   –combined growth and nucleation 245–246
   –multidimensional 247–255
   –numerical methods 209–260
   –pure aggregation 225–232
   –pure breakage 214–225
   –pure growth 232–239
population density function 209
pore effect 2
pore level drying 57–102
pore network models 57–102
   –extensions 87–92
   –influence of pore structure 92–100
   –isothermal 58–87
pore pressure 143
   –constitutive equations 130
pore shapes 92–94
pore-size distributions, bimodal 95–100
pore structure influence 92–100
pores
   –connectivity 59, 94
   –saturation 62
porosity, thermohydromechanical drying 126
porous media
   –deformable 125
   –drying phases 8
   –high-temperature convective drying 19
   –volume-averaging models 1
potato, drying behaviour 36
powder, particle morphology 201
Prandtl number 88, 166
Prandtl's eddy-viscosity theory 166
pressure difference, length scales 75
pressure gradients
   –convective drying 148
   –Lagrangian particle tracking 168

pressure linked equations   165
pressure profile   122
   –identity drying card   33
probability density function, isothermal drying models   67
problem solving, dryers   294–295
process manual, knowledge bases   295–296
process performance, computational fluid dynamics   157
process simulators   287–289
process-systems simulation tools   261–305
   –ancillary calculations   283–287
   –commercialization of drying software   296–301
   –detailed design models   276–283
   –expert systems and decision-making tools   289–295
   –heat and mass balances   268
   –knowledge bases and qualitative information   295–296
   –numerical calculation procedures   263–268
   –scaling methods   272–276
   –scoping design methods   269–272
   –simulators   287–289
product properties prediction   200–203
production, theory of mixtures   126
profiles comparison, pressure and water   122
psychrometric charts   286–287
   –ancillary calculations   284–287
pure aggregation   225–233
pure breakage   214–225
pure growth   233–239

q

qualitative information
   –decision making tools   292
   –knowledge bases   295
quality, drying   50–51

r

radio-frequency drying   30
radio-frequency heating   34
RANS see Reynolds-averaged Navier–Stokes
Ranz–Marshall relationship, milk drying   177
rapid fluid migration   44

rate equations, thermohydromechanical drying   132–134
real viscoelatic behavior   115–120
replicability, lack of   298–301
representative elementary volume (REV)   6, 100
   –thermohydromechanical drying   125
   –volume-averaging models   2
reversible deformation, constitutive equations   130
Reynolds-averaged Navier-Stokes (RANS) equations   155, 163
Reynolds equations   156
Reynolds number
   –computational fluid dynamics   156
   –drag force   167
   –droplet drying   180
   –isothermal drying models   60
Reynolds stress
   –fluid phase modeling   166
   –turbulence modeling   157
rheological behavior   119

s

Saffman force   169
sapwood, drying behavior   25–29
saturation
   –isothermal drying models   62
   –pore structure influence   93–94
   –thermohydromechanical drying   126
sawing-pattern plane, stack model   49
scale up
   –basic principles   273–274
   –dryers   157
   –integral models   274–276
scaling methods   272–276
scoping design methods   269–272
SCV see subcontrol volume
second drying period   10
semi-implicit method for pressure linked equations (SIMPLE)   165
shear forces, VD collisions   191
shear stress, drag force   167
Sherwood number   61
shrinkage   8, 50
shrinkage strains   139, 142
SIMPLE see semi-implicit method for pressure linked equations
simulations
   –continuous thermomechanical models   114–120
   –process-systems   261–305
   –spray-dryer   155–208
simulators, process   287–289

SIP *see* strongly implicit procedure
size distribution   160
size enlargement, product properties prediction   200
skeletal frame of reference   130
slip-rotation lift *see* Magnus force
slip-shear lift *see* Saffman force
smoothness sensor   15
soaking wood samples   42
software
– dryer designs   301
– dryers   263
– drying   296–301
– drying modeling   261
softwood, drying behavior   25–29
solid mass fraction, spray-drying process   185
solid skeleton, thermohydromechanical drying   125–126
solid velocity   104
– averaged internal equations   112
solids modeling difficulties   297–298
solids processing challenge   262–263
solids residence time   270
solvent, mass flow rate   180
solvent content, particle collisions   189
solvent evaporation, particle morphology   202
source terms, spray-dryer simulation   198
spatial averaging theorem   3
spatial deviation variables   5
spatial discretization, spray dryer   192
spherical volumetric strain tensor   114
spray droplet   158
spray-dryer geometry   192–193
spray-dryer problems   282
spray-dryer simulation   155–208
– chamber design   160
– collisions of particles   181–192
– droplet-drying models   173–181
– Euler–Lagrange approach   162–173
– examples   192–200
– prediction of product properties   200–203
– state of the art   160–162
stack model   43–51
standalone calculation programs, dryers   263
state equations, averaged internal   108

STD droplets *see* surface-tension dominated droplets
STD-STD collisions   190
stochastic collision models   182–187
Stokes number   184
Stokesian drag force   191
strain   116–118
strain components, drying   141
strain deviator   142
strain rate tensor, balance equations   129
strain tensor
– averaged internal equations   110
– drying quality   50
– numerical resolution technique   114
– rate equations   132
streaming period, microwave heating   30
stress deviator, thermohydromechanical drying   126
stress profile inversion   115
stress tensor
– averaged internal equations   109
– balance equations   128
– constitutive equations   130
strongly implicit procedure (SIP)   165
structure influence, pore network models   57–102
subcontrol volume (SCV)   12
subgrid-scale models   156
superficial averages, volume-averaging models   3
superheated steam flow   25
– high-temperature convective drying   19–22
surface evaporation, heat transfer   91
surface tension   141
surface-tension dominated droplets (STD droplets)   161, 187–188, 190
swelling strain   142
symmetrical second-order stress tensor   109

t

temperature distribution, microwave drying   150
temperature evolution, convective drying   147
temperature gradient, rate equations   133
temperature patterns, computational fluid dynamics   283
temperature pressure sensors   26
temperature profile, identity drying card   33
temperature two-dimensional predicted fields   116–118
tensor evaluation
– Arnoldi process   17
– CV face   14–15

tensors
- deformation gradient   36
- diffusion   14
- diffusivity   6
- effective stress   130
- Eulerian correlation   172
- gas permeability   6
- liquid mobility   14
- liquid permeability   6
- mechanosorptive strain   50
- mobility   15
- permeability   6
- Reynolds-stress   166
- spherical volumetric strain   114
- strain   50, 110, 114, 132
- strain rate   129
- stress   109–17, 128, 130
- symmetrical second-order stress   109
- total strain   110
- viscoelastic strain   50, 110
- viscous drag   6
- viscous stress   110

theorem
- averaging   3, 5
- gauss divergence   13
- Gauss-Ostrogradsky   128
- general transport   3
- spatial averaging   3

theory of mixtures, thermohydromechanical drying   125–154

thermal conduction   24
- pore network models   88

thermal equilibrium, thermohydromechanical drying   126

thermal shocks, thermohydromechanical drying   125

thermodynamic equilibrium, constitutive equations   130

thermodynamic fluxes, rate equations   133

thermohydromechanical drying
- chemical potential   139
- constant drying rate period   140
- constituent velocity of moisture   126
- constitutive equations   130–132
- creep strain   134
- deformations   125
- dielectric properties   126

thermohydromechanical model
- continuous   125–154
- global balance equations   126–130
- heat and mass transfer   132–141
- kaolin cylinder   144–152
- skeletal frame of reference   130–132
- thermomechanical equations   141–144

thermomechanical models, continuous   103–124

thick films, pore networks   80
thin films, pore networks   80

throats
- isothermal drying models   59
- pore network models   57
- refilling   73
- saturation   62, 88
- size distribution   67, 78

time discretization
- heat transfer   89
- particle tracking   170

time scale
- fluid phase modeling   170–171
- turbulence   182

time step, isothermal drying models   65
tissues, model calculation   42
total gaseous pressure   10
total momentum   143

total momentum conservation   114
- averaged internal equations   110
- continuous thermomechanical drying   103
- numerical resolution technique   114

total strain tensor, averaged internal equations   110

transfer coefficients, milk drying   177
TransPore   17, 46

transport equations   1–5
- Euler–Lagrange approach   163
- film flow   82
- particle   277
- phase source terms   164

transport phenomena, volume-averaging models   1

transport properties, identity drying card   33

tree-search approach
- decision making tools   289, 292–294

troubleshooting
- dryers   294
- heat and mass balances   268

turbulence modeling, approaches   157
κ–ε turbulence models   157, 163
turbulence scales, computational limits   156
turbulence time scale, particle collisions   182
turbulent dispersion modeling   171

turbulent flows 155
turbulent kinetic energy 166
turbulent viscosity 157
two-dimensional predicted fields 116–118
two-equation models, turbulence 157
two-fluid approach, Euler-Euler 158
two-phase flows
 –Euler–Lagrange approach 159
 –stochastic collisions 182
two-phase system, continuous thermomechanical drying 107
two-way coupling 173
 –fluid phase modeling 164
 –stack model 47
two-zone process, drying phases 9

u

underrelaxation procedure 173
unit step function 218

v

vacuum drying 10–11
van der Waals forces 191
van Leer flux limiter 15
vapor
 –convective drying 145
 –enthalpy 113
 –rate equations 134
vapor concentration, milk drying 177
vapor condensation, heat transfer 89
vapor diffusion 24
vapor mass balance, stack model 46
vapor phase, diffusivity 5
vapor transfer 63–64
 –convective drying 145–146
 –isothermal drying models 58, 61
vapor/liquid interface
 –convective drying 146, 148
VD droplets see droplets dominated by viscous forces
velocity, solid phase 104
velocity, of the skeleton 126
velocity field, Euler–Lagrange approach 163
virtual mass force, Lagrangian particle tracking 168
viscoelastic case, continuous thermomechanical drying 104
viscoelastic material 142
viscoelastic strain tensor

 –averaged internal equations 110
 –drying quality 50
viscoelatic behavior 115–120
viscosity 141
 –magnetic 136
 –pore network models 71
viscosity change, VD collisions 190
viscous drag tensor 6
viscous droplets, particle collisions 161, 190–191
viscous effects, length scales 75
viscous forces 188
viscous stress tensor 110
volume-averaging models
 –comprehensive drying 1–52
 –continuous thermomechanical drying 103–124
 –convective drying 18–29
 –homogenization 37–42
 –less-common drying 29–37
 –macroscopic equations 6–7, 11–18
 –microscopic foundations 1–6
 –multiscale approach 42–51
 –physical phenomena 7–11
volume conservation, averaged internal equations 108
volume fraction, thermohydromechanical drying 126
volumetric heating 29
volumetric saturation 126
volumetric strain 134
von Mises stress 116–118

w

wall deposition rates, spray-dryer simulation 160
water, microwave heat source 138
water activity, milk drying 177
water conservation 13
 –volume-averaging models 6
water content profiles 122
water-vapor transport equations 1
wave equation, microwave heat source 136
wet porous medium 1
wet solid, rheological behavior 103
wettability effects 83–84
wind-tunnel dryer 19
wood variability 49–49

y

y-periodic functions 40
Ytong 19